Green Energy and Technology

For further volumes:
http://www.springer.com/series/8059

Silvio de Oliveira Jr.

Exergy

Production, Cost and Renewability

Springer

Silvio de Oliveira Jr.
Mechanical Engineering Department
Polytechnic School of the University
of São Paulo
São Paulo
Brazil

ISSN 1865-3529 ISSN 1865-3537 (electronic)
ISBN 978-1-4471-4164-8 ISBN 978-1-4471-4165-5 (eBook)
DOI 10.1007/978-1-4471-4165-5
Springer London Heidelberg New York Dordrecht

Library of Congress Control Number: 2012939479

Printed on acid-free paper

Springer is part of Springer Science+Business Media (www.springer.com)

Our main challenge is to live better, more and using less Exergy!

Preface

This book was conceived and designed to be a contribution to bridge the gap between Second Law of Thermodynamics derived concepts and their application to engineering practice by means of using the property exergy and the exergy balance as a tool for analyzing and improving the performance of energy conversion processes.

With the exergy analysis, it is possible to evaluate the performance of energy conversion processes not only on a thermodynamics basis but also by including economic and environmental aspects and impacts of the studied processes. This comprehensive approach of the energy resources utilization has, as one of the most important features, the identification of sustainable ways of energy resources utilization.

This idea has been deepened and improved during my graduate course on Exergy and Thermoeconomic Analysis, offered at the Mechanical Engineering Department of the Polytecnic School of the University of São Paulo since 1992, and some Efficiency Energy Use in Industrial Processes courses taught to engineers in the last 15 years.

Based on a detailed presentation of the fundamentals of the exergy concept, its calculation, graphical representations, and exergy balances evaluation, the book contains eight chapters dedicated to describe the application of detailed exergy and thermoeconomic analysis to power plants and polygeneration systems, petroleum production and refining plants, chemical plants, biofuel production routes, combined production of ethanol and electricity, aircraft systems design, environmental impact mitigation processes, and human body behavior. The presented case studies, developed by the author and his co-workers, aim at providing engineers with guidelines to the utilization of the exergy and thermoeconomic analysis to model, simulate, and optimize real processes and industrial plants.

I wish to express my gratitude to my co-workers and former students who have contributed to the writing and final manuscript of this book:

Prof. Dr. Marco van Hombeeck
Dr. Elisio F. Caetano
Dr. Celso Y. Nakashima
Dr. Luiz F. Pellegrini
Prof. Dr. Hector I. Velásquez Arredondo
Prof. Dr. Carlos Mora Bejarano
Dr. Ricardo Gandolfi
Prof. Dr. Juan Burbano Jaramillo
Dr. Marcela Loboguerrero Larrazábal
MSc. Julio Augusto Mendes da Silva
MSc. Samuel Borelli
MSc. Ralf Lattouf
MSc. Uirajara Vieira Junior
MSc. Flavio E. Cruz
MSc. Maurilo de Sousa Teixeira
MSc. Wilson Garagatti Arriola
Eng. Carlos Eduardo Keutenedjian Mady
Eng. Cesare Tona
Eng. Paolo Antonio Raviolo
Eng. João Francisco Peral Cespedes

Contents

Chapter 1
Introduction

The sustainable use of energy resources is one of the most important challenges and constraints that societies have to deal with. Although always emphasized this issue is not properly pursued.

The sustainable use of energy is intrinsically linked to the efficiency of the systems that convert raw materials into valuable goods. The technology of these energy converters has a direct influence in the quality and quantity of their energy consumption, products, and also the wastes discharged into the environment. In this sense, the utilization of energy sources, regardless of being fossil, or the so-called renewable ones, must be conducted by means of efficient equipment, not only during its end use, but in the whole chain of energy conversion processes, from the energy source production and processing up to the wastes deactivation and disposal.

The assessment of energy utilization cannot be restricted to the criteria of energy efficiency and production cost. The increasing environmental constraints impose that the energy resources utilization and the wastes deactivation and disposal be taken also into account in the evaluation of the energy use. These needs require a broader definition of efficiency and value associated to every energy quantity to allow a common basis for weighing the inputs, effects and outputs of energy conversion systems, regardless of the type of utilized energy resource, fossil, or the so-called renewable source. For instance, instead of discussing the overall conversion efficiency of the transportation by car, one intends to maintain this environmentally harmful way of displacement by using the so-called biofuels.

This requirement is particularly important in some developing countries that will increase the use of fossil as well as renewable energy sources in their energy matrix.

In this scenario, the concept of exergy, which is a thermodynamic standard of the quality of energy, is useful to identify and quantify the inefficiencies of the energy conversion processes that cause the reduction in the value of the energy and the increase of the wastes discharged into the environment. This is possible because the determination of the exergy implies the combined use of the First and the Second laws of Thermodynamics, and it is by means of the concept of entropy

S. de Oliveira Jr., *Exergy*, Green Energy and Technology,
DOI: 10.1007/978-1-4471-4165-5_1,

generation or exergy destruction that one can quantify the inefficiencies of a given energy conversion process and the reduction of value of the energy utilized in this process.

Traditionally, in undergraduate Thermodynamics courses, the time spent in teaching the fundamentals and applications of the Second Law of Thermodynamics and the entropy and entropy balances concepts is smaller than the time dedicated to present and utilize First Law and the energy balances. This fact, allied to the intrinsic difficulty found by students to understand and use the property entropy in the analysis and evaluation of energy systems can explain why, after more then a century after the formulation of the Gouy-Stodola theorem, most engineers are not yet familiar and convinced of the powerful utility of the Second Law of Thermodynamics and its derived concepts to study the quality of energy conversion process and the possibilities and limitations to have sustainable energy systems.

Probably, as used to say Prof. P. Le Goff, if the concept of exergy had been formulated previously to the concept of entropy, nowadays, the utilization of the exergy balances, or the balances of thermodynamic utility would be incorporated to engineering practice, as the energy balances.

This book presents the foundations and some applications of the exergy analysis of energy conversion processes to demonstrate the potential of application of this energy systems assessment framework to improve the rational use of energy resources, taking into account the efficiency of the conversion processes, the production costs, and the environmental impacts due to the wastes generated during the conversion processes.

The environmental exergy indicator defined and employed in this book is based on the concept of reversible process to properly quantify the renewability of the energy conversion process. This approach provides an objective way to quantify the environmental quality of a given process. With this approach, the concept of renewability is associated to the process instead of to the product or to a particular energy source. The concept of environmental quality is linked to the general quality concept that must consider the sustainable maintenance of human life in the Earth.

The foundations and applications of the exergy analysis are presented in Chaps. 2–10, as follows.

Chapter 2 introduces the foundations of the exergy, exergy production cost, and renewability analysis of energy conversion processes. Some graphical representations are shown in which it is possible to determine or represent exergy and exergy balances. The exergy efficiency is introduced based on a general definition of efficiency, and the balance of cost is presented as an additional balance equation to be used in the performance analysis of energy systems. A brief discussion on cost partition criteria is presented to aid the analysis of the cost formation processes of the products of energy conversion processes. Finally, the renewability of energy conversion processes is analyzed by means of a renewability exergy index that takes into account the type of inputs, renewable or fossil, the wastes, and the destroyed exergy of a given energy conversion process.

Chapter 3 is dedicated to the exergy and thermoeconomic analysis of components of power plants, refrigerating, and polygeneration systems to characterize the performance of such systems as well as to determine their products cost formation processes. Based on the general formulation of efficiency, presented in Chap. 2, the expressions of the exergy-based performance parameters of the components of these systems are derived. These concepts are applied to evaluate the electricity cost formation of a combined cycle power plant, and the comparative performance and production costs of steam and electricity of cogeneration plants configurations for chemical and dairy industries. Finally a comparative exergoeconomic study of trigeneration systems to produce electricity, steam, and chilled water is described and discussed.

In Chap. 4 it is presented the application of exergy analysis to primary and refining petroleum processes. It is described the exergy and thermoeconomic analysis of an offshore platform. The comparative performance of two artificial lift systems is presented and discussed in detail through a scenario of 25 years of operation. A utilities plant of a petroleum refinery is studied in order to characterize its exergy and cost interactions with the hydrocarbons derived production processes, such as the combined distillation, fluid catalytic cracking (FCC), delayed coking, hydro-treating (HDT) and hydrogen generation, and sulfur recovery. The hydrogen production in a petroleum refinery to purify diesel oil, based on the steam reforming of natural gas, is analyzed in detail to evaluate its thermodynamic performance as well as to describe the cost formation of the produced hydrogen.

The use of the exergy analysis to evaluate the performance of chemical process is described in Chap. 5 where it is assessed the acetaldehyde production by ethanol partial oxidation. With the conclusions of such assessment, an improved plant configuration is proposed in order to use more efficiently the required process thermal exergy.

Chapter 6 compares, in an exergy basis, current technological paths used in sugarcane mills, with new ones, which can lead to a more renewable use of energy to produce sugar, ethanol, and electricity. Such technologies include: more efficient steam cycles (such as high pressure and supercritical steam ones) and biomass gasification-based combined cycles.

In Chap. 7 an exergy-based comparative analysis of four biofuels production routes is described and discussed. The selected feed stocks are glucose and sugarcane syrups, the fruit and flower stalk of banana tree and palm oil. For each production route, the effect of process variables on the exergy efficiency and the renewability exergy index (presented in Chap. 2) are determined allowing the identification of possible ways to optimize the production of such biofuels.

As a tendency of the commercial aeronautical industry is to develop more efficient aircraft in terms of fuel consumption and direct operational costs, Chap. 8 describes applications and a method based on exergy analysis for conception and assessment of aircraft systems. The method can support the design of the complete vehicle as a system and all of its subsystems in a common framework.

Chapter 9 is concerned with the definition and application of exergy indices to assess the performance of environmental impact mitigation technologies, based on the cumulative consumption of exergy and the environmental exergy efficiency.

The first approach is applied to analyze three scenarios: air emissions treatment, remediation of a contaminated site, and management of solid wastes. The environmental exergy efficiency and the renewability exergy indicator are used to evaluate three wastewater treatment technologies: biological (aerobic and anaerobic) and physicochemical processes.

Chapter 10 describes two applications of the exergy analysis in order to evaluate the behavior of human body and its systems. In the first one, the exergy destruction rate and the exergy efficiency of the human body and its systems are determined. In the second study, it is described a model of the human respiratory and thermal system developed to perform the exergy analysis of the human body under physical activities. The analysis quantifies the rate at which oxygen is supplied to the lung and transported by the blood to the tissues, and the rate of carbon dioxide elimination.

Chapter 2
Exergy, Exergy Costing, and Renewability Analysis of Energy Conversion Processes

Symbols

$\mathbf{B}$	Exergy (kJ)
B	Exergy rate/flow rate (kW)
b	Specific exergy (kJ/kg)
$B_{emissions}$	Exergy rate of wastes that are not treated or deactivated (kW)
$B_{destroyed}$	Destroyed exergy rate (kW)
$B_{deactivation}$	Destroyed exergy rate of additional natural resources during waste de-activation (kW)
$B_{disposal}$	Exergy rate or flow rate related to waste disposal of the process (kW)
B_{fossil}	Non-renewable exergy rate consumed on production processes chain (kW)
$B_{nat,res}$	Exergy rate of the natural resources consumed by the processes (kW)
$B_{processing}$	Exergy rate or flow rate required for extraction and preparation of the natural resources (kW)
$B_{product}$	Exergy rate or flow rate associated to the products and byproducts/useful effect (kW)
B_{reject}	Exergy rate or flow rate of the rejects (kW)
$B_{utilities}$	Exergy rate or flow rate required by the utilities of the process (kW)
$\mathbf{C}$	Cost ($)
C	Cost rate ($/s)
c	Specific heat, J/(kg K), specific cost (kJ/kJ, $/kJ, $/kg)
c_p	Specific heat at constant pressure, J/(kgK)
C_{eq}, C_r	Equipment cost of a given capacity ($); Equipment cost of a reference capacity ($)
$\mathbf{E}$	Energy (kJ)
E	Energy rate/flow rate (kW)
$f_{O\&M}$	Annual operational and maintenance factor
f_a	Capital recovery factor
g	Gravitational acceleration (9.8 m/s^2); molar Gibbs free energy of formation (kJ/kmol)

S. de Oliveira Jr., *Exergy*, Green Energy and Technology,
DOI: 10.1007/978-1-4471-4165-5_2, © Springer-Verlag London 2013

ΔG_o	Gibbs free energy variation in the direction of a given chemical reaction (kJ/kmol)
H; h	Enthalpy flow rate (kW); specific enthalpy (kJ/kg)
HR	Enthalpy of reactants (kJ/kmol)
HP	Enthalpy of products (kJ/kmol)
I_e	Energy investment (kJ)
I_{VC}	Investment rate of equipment inside control volume ($/h, $/s)
i	Interest rate (%)
LHV	Lower heating value (kJ/kg)
m	Mass flow rate (kg/s)
N	Capital recovery period, operating time (year)
N_i	Number of moles of species i
Q; q	Heat rate (kW); heat rate per unit of mass flow rate (kJ/kg)
P	Pressure (kPa)
R, $\bar{R}$	Ideal gas constant (kJ/kg K), universal gas constant (kJ/kmol K)
S; s	Entropy rate/flow rate (kW/K); specific entropy (kJ/kg K)
S, S_r	Component size, component reference size (see Table 2.8)
S_{ger}; s_{ger}	Entropy generation rate (kW/K); entropy generation rate per unit of mass flow rate (kJ/kg K)
T	Temperature (°C, K)
U; u	Specific internal energy (kJ/kg), internal energy (kJ)
v	Specific volume (m^3/kg); value scale
V	Volume (m^3)
W; w	Power (kW); power per unit of mass flow rate (kJ/kg)
x	Mole or mass fraction
z	Elevation (m)

Greek symbols

α	Angle in Fig. 2.11, percent excess air, exponent of Eq. 2.84
γ_i	Activity coefficient of species i
Δ_{comb}	Ratio between destroyed exergy and reactants exergy
η	Efficiency
θ	Carnot factor
λ	Renewability exergy index
μ_i	Chemical potential of species i (J/mol)
ν	Stoichiometric coefficient
υ	Velocity (m/s)
φ	Ratio between chemical exergy and lower heating value

Superscripts

–	Molar
*	Restricted reference state
s	System

Subscripts

0	Dead state; reference state
00	partial pressure
a	Input
ab	Absorber
air	Inlet air
B, b	Exergy, reboiler
btt	Heat transformer
carnot	Related to Carnot cycle
C, c, cd	Condenser
ch	Chemical
coreactants	Coereactants
de	Desorber
dest	Destroyed
e	Outlet, exit, electricity, energy
ef	Effective, effluents
env	Environmental
eq, equipment	Equipment
eqt	Equipment total
ev	Evaporator
f	Fossil
flue gases	Related to flue gases
fuel	Fuel
i	inlet, input, component
j, k	Component, species
kin	Kinetic
H	enthalpy
hp	High pressure
lp	Low pressure
m	Average, mass basis
max	Maximum
mix	Mixer
mr	Reversible engine
net	Net
o	Operational, outlet
P, p	Product, pump, perfection, process
ph	Physical
pot	Potential
process	Process
products; prod	Products
Q, q	Heat
r	Reference, renewable
reactants, react	Reactants
sep	Separator
sg	Steam generator

t	Thermal, during lifetime, turbine
tt	Heat transformer
u	Useful
VC	Control volume
w	Waste
W	Work
water	Water
wp	Production waste
wu	Utilization waste

2.1 Exergy, Quality, and Efficiency

The concern with efficiency and quality of energy conversion processes is already present in the beginnings of the development of thermal engines. Biot in his Traité de Physique Expérimentale et Mathématique, Tome 4 Chap. VI [1, 10], draw attention to the following aspects:

> Le plus important est la détermination de la temperature à laquelle il est le plus convenable d'employer la vapeur. En effet, plus elle est chaude, plus sa force élastique est considérable, et par conséquent plus elle produit d'effort sur la surface du piston qu'elle presse, le vide étant toujours de l'autre côté. Mais aussi il faut consommer plus de charbon pour produire une vapeur plus chaude; en sorte que le profit ou le désavantage de la température est um element à determiner.

This paragraph highlights the dependence of the performance of a heat engine with the fuel consumption and with the temperature level at which steam is generated, in other words, with the thermodynamic quality of steam.

The concept of exergy has its origin in the formulation presented by Gouy [6] and Stodola [18] which shows that the maximum potential for a system to perform work is a function of its internal energy and the ambient conditions (pressure, temperature, and composition). This concept was improved throughout the twentieth century by several authors, as indicated by Rivero [16] and showed in Table 2.1, where some exergy and related concepts are presented.

The search for thermodynamic quality standards is one of the most interesting challenges of the Engineering Thermodynamics. In the last decades, this concern is no more restricted to the performance quantification of a given process or equipment, but it extends to all the boundaries of any energy conversion process, including its energy inputs and wastes. This implies that, for every process and equipment and a given output:

- a larger energy input utilization as smaller is the conversion efficiency, and,
- a larger quantity of wastes, and consequently a higher potential to generate environmental impacts, as smaller is the conversion efficiency.

Although fundamental in the quantification of the terms of the efficiency expression, the energy balance, or the First Law of Thermodynamics, as well as, the concept of energy, are not enough to identify and correctly quantify what is usually

Table 2.1 Exergy and related concepts [16]

Year	Author	Designation
1824	Carnot	Puissance motrice du feu
1872	Thomson (Kelvin)	Motivity
1873	Gibbs	Available energy of the body and medium
1881	Gouy	Energie utilisable
1898	Stodola	Freie technische energie
1925	Debaufre	Available energy
1935	Bonsjakovic	Technische arbeitsfähigkeit
1944	Thring	Virtue of energy
1953	Schmidt	Technische maximale arbeit
1955	Gibert	Energie non dégradée
1956	Grigull	Ekthalpie
1956	Rant	EXERGIE

called inefficiencies. What is called inefficiency is a consequence of the occurrence of an irreversible process, in which there is entropy generation, for instance, the head loss in an isolated duct, where enthalpy is conserved and entropy increases provoked by an irreversible process: conversion of the work flow into internal energy.

In this sense, the use of the entropy balance, or the Second Law of Thermodynamics as complementary information to the energy balance, is of fundamental importance to quantify and to well understand the inefficiencies of energy conversion processes, the cost formation process of the products and wastes.

As discussed by Moore [11], the efficiencies of some systems and equipment are limited by restrictions imposed by the Second Law of Thermodynamics. These limits exist, for example, in conversions of:

- heat into work, as it happens in a heat engine;
- internal energy/enthalpy variation into work (turbine);
- chemical energy into work (fuel cell);
- chemical energy into kinetic energy (rocket);

Or, as characterized by Kotas [8], the conversion of ordered energy (work, kinetic energy, and potential energy) into disordered energy (internal energy, heat, and chemical energy) can be accomplished completely, but the inversion of these processes are always limited by the Second Law of Thermodynamics and the properties of the environment (P_0, T_0, $\mu_{0,i}$) in which the system is placed.

2.2 Exergy and Exergy Balance

2.2.1 Reversible Work

The concept of exergy can be introduced by using the concept of reversible work. It can also be shown that the so-called exergy balance is merely a linear

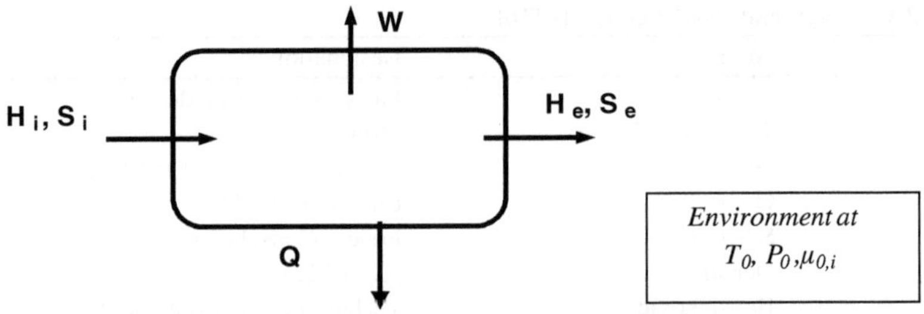

Fig. 2.1 Control volume with one inlet and one outlet

combination of the entropy and energy balances. These developments will be presented here for a simple case of processes that take place, in steady state, in a control volume with one inlet and one outlet, and neglecting the kinetic, potential, as well as electric, and magnetic effects, as represented in Fig. 2.1.

Equations 2.1 and 2.2 present the energy and entropy balances, per unit of mass flow rates that cross the boundaries at 'i' (inlet) and 'e' (outlet).

$$q = h_e - h_i + w \tag{2.1}$$

$$s_e - s_i = \int \frac{\delta q}{T} + s_{ger} \tag{2.2}$$

For the chosen control volume, one could ask the following question: 'What is the maximum work that can be performed giving the same thermodynamic states at the inlet and outlet sections of the control volume?' This maximum work is determined under the following conditions:

- all the processes within the control volume are reversible;
- the maximum work that can be generated using the heat rejected by the control volume is the one obtained from a Carnot engine, operating between temperatures T_i and T_e and the environment temperature where the control volume is placed (T_0).

The determination of the maximum work is done, by considering another control volume which contains the preceding control volume and the Carnot engine. This new control volume is crossed by the same mass flow rates, and q_0 is the rejected heat of the Carnot engine. The energy and entropy balances for this new control volume are given by Eqs. 2.3 and 2.4:

$$q_0 = h_e - h_i + w_{max} \tag{2.3}$$

$$s_e - s_i = q_0/T_0 \tag{2.4}$$

Substituting q_0 by $(s_e–s_i)T_0$, one obtains:

$$w_{max} = h_i - h_e - T_0(s_i - s_e) \tag{2.5}$$

Or, the term $[h_i - h_e - T_0(s_i - s_e)]$ characterizes a maximum potential to perform work giving the states 'i' and 'e'.

This development could be done for the original condition where there are irreversible processes inside the control volume as well as the original values of w and q. In this case, the maximum work could be obtained in a similar way: including a Carnot engine that operates by using q and rejecting q_0' to the environment.

For the reversible engine, it can be written:

$$\oint \frac{\delta Q}{T} = 0 = \int_{T_i}^{T_e} \left(\frac{\delta q}{T}\right) + \int \left(\frac{\delta q_0'}{T_0}\right) \tag{2.6}$$

And, by taking this expression into the entropy balance for the original control volume with the Carnot engine, one gets:

$$s_e - s_i = q_0'/T_0 + s_{ger} \tag{2.7}$$

Also, it is possible to write the energy balance for the region that encompasses the reversible engine and the original control volume:

$$q_0' = h_e - h_i + w + w_{mr} \tag{2.8}$$

In Eq. 2.8 the specific work of the Carnot engine, taking into account that heat has different signs for the heat engine control volume, is

$$w_{mr} = q - q_0' \tag{2.9}$$

By substituting q_0' and w_{mr}, one gets:

$$T_0(s_e - s_i) - T_0\, s_{ger} = h_e - h_i + w + q\eta_{mr} \tag{2.10}$$

where

$$w_{mr} = q\eta_{mr} \tag{2.11}$$

In this way, Eq. 2.10 can be written giving:

$$h_i\ h_e - T_0(s_i - s_e) = w + q\eta_{mr} + T_0\, s_{ger}$$

with:

$$\eta_{mr} = 1 - \frac{T_0}{q} \int_{T_i}^{T_e} \left(\frac{\delta q}{T}\right) \tag{2.12}$$

It is evident that, as shown by Eq. 2.5, the term $[h_i - h_e - T_0(s_i - s_e)]$ also characterizes a potential for doing work that in the occurrence of irreversible processes is reduced by the product $T_0 s_{ger}$, or the so-called irreversibility. Eventually, this potential is completed destroyed, or:

$$h_i - h_e - T_0(s_i - s_e) = T_0\, s_{ger} \tag{2.13}$$

This scenario happens in dissipative processes, for instance, during the flow through an expansion valve where: $h_i = h_e,\ w = 0,\ q = 0$ and $s_{ger} > 0$.

2.2.2 Exergy and Exergy Balance

Equation 2.10 can simply be obtained by multiplying the entropy balance by $-T_0$ and adding this equation to the energy balance. The result, a linear combination of the entropy and energy balances, gives:

$$[h_i - h_e - T_0(s_i - s_e)] = w + q\left[1 - \frac{T_0}{q}\int_{T_i}^{T_e}\left(\frac{\delta q}{T}\right)\right] + T_0\, s_{ger} \tag{2.14}$$

In the case where all processes inside the control volume are reversible, the control volume interacts only with the environment, and the mass flow rate at the outlet section is in thermodynamic equilibrium with the environment ($h = h_0$ and $s = s_0$), Eq. 2.14 becomes the expression of the maximum reversible work associated to the inlet flow rate state of the control volume:

$$[h_i - h_e - \mathrm{T}_0(s_i - s_e)] = w + q\left[1 - \frac{T_0}{q}\int_{T_i}^{T_e}\left(\frac{\delta q}{T}\right)\right] = w_{max} \tag{2.15}$$

This maximum reversible work is called specific exergy of the flow rate at the inlet section of the control volume:

$$b = [h_i - h_0 - T_0(s_i - s_0)] \tag{2.16}$$

It is worth to note that the specific exergy is a function of properties of two thermodynamic states: the state in which the mass flow rate is and the state where this mass flow rate would be in thermodynamic equilibrium with the environment components. In that sense, the specific exergy is a special thermodynamic property, function of two thermodynamic states, and its value gives the maximum potential for doing work.

For a general situation, as shown in Fig. 2.2, where there are several inlet and outlet sections, heat exchanges at temperatures T_j, unsteady state regime and deformable control volume, the linear combination of the energy and entropy balances, gives:

$$\begin{aligned}\frac{d}{dt}(\boldsymbol{E} + P_0V - T_0S) = \frac{d\boldsymbol{B}}{dt} &= \sum_j\left(1 - \frac{T_0}{T_j}\right)Q_j - W_{ef} + \sum_i m_i(h_i - T_0s_i)\\ &\quad - \sum_e m_e(h_e - T_0s_e) - T_0S_{ger}\end{aligned} \tag{2.17}$$

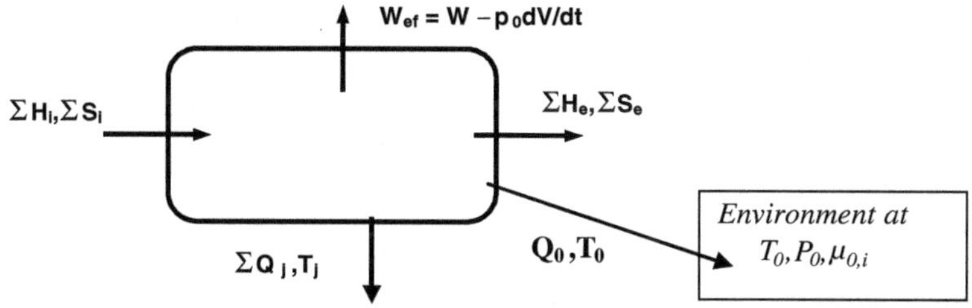

Fig. 2.2 Control volume with several inlets and outlets, heat transfer, and temperatures at T_i

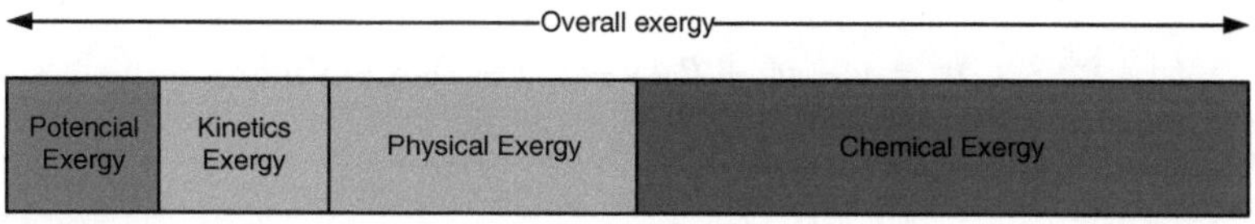

Fig. 2.3 Exergy components

This equation, called the exergy balance, is composed of the following terms:

$$\frac{\mathrm{d}}{\mathrm{d}t}(\boldsymbol{E} + P_0V - T_0S) = \frac{\mathrm{d}\boldsymbol{B}}{\mathrm{d}t} \text{ : control volume exergy variation rate}$$

$\sum_i H_i - \sum_e H_e - T_0(\sum_i S_i - \sum_e S_e)$: exergy flow rate variation between inlet and outlet control volume sections.

$B_{Q_j} = Q_j(1 - T_0/T_j)$: exergy rate associated to heat transfer rate (maximum work that could be performed by a Carnot engine operating between T_j and T_0, receiving Q_j at T_j, and rejecting heat to the environment at T_0).

W_{ef} = pure exergy rate: effective work rate performed or received by the control volume.

$B_{dest} = T_0S_{ger}$: destroyed exergy rate or irreversibility rate (rate of available work destroyed due to irreversible processes that take place inside the control volume).

Equation 2.17 can be considered as the law of energy degrading because it allows quantifying the reduction of the work capacity of a given energy conversion process (or processes) caused by the occurrence of irreversible processes.

As a consequence it is possible to conclude that the capacity for doing work (exergy) is not conservative, being always reduced every time that any irreversible process happens inside the analyzed control volume, such as non resisted expansion, head loss, mixture, heat transfer, and chemical reaction.

Neglecting nuclear, magnetic, electric, and surface tension effects, one can divide the exergy in four components: kinetic, potential, physical, and chemical, as shown in Fig. 2.3 and Eq. 2.18.

$$b = b_{kin} + b_{pot} + b_{ph} + b_{ch} \tag{2.18}$$

Kinetic and potential exergy correspond, respectively, to kinetic and potential energy, because both energy forms can be completly converted into work:

$$b_{\text{kin}} = \frac{\upsilon^2}{2} \tag{2.19}$$

$$b_{\text{pot}} = gz \tag{2.20}$$

Physical exergy is equivalent to the maximum work that can be performed when a system or a flow rate moves from a given thermodynamic state to the restricted reference state (indicated as *) where it is in thermal and mechanical equilibrium with the environment. For a system, the physical exergy is given by:

$$b^s = u - u^* + P_0(v - v^*) - T_0(s - s^*) \tag{2.21}$$

And for a mass flow rate:

$$b = h - h^* - T_0(s - s^*) \tag{2.22}$$

It is important to notice that the term $P_0(v - v^*)$ that appears in Eq. 2.21 is a non available work that the system performs, or receives, due to its interaction with atmospheric pressure.

For a perfect gas, with constant pressure specific heat, c_p, the expression of the physical exergy is:

$$b_{\text{ph}} = c_p\left[(T - T_0) - T_0 \ln\left(\frac{T}{T_0}\right)\right] + \text{R}T_0 \ln\left(\frac{P}{P_0}\right) \tag{2.23}$$

For solids and liquids, when the behavior of incompressible substance can be assumed, the physical exergy is done, by Eq. 2.24, where it is supposed constant specific heat, c and v_m is the average specific volume between P and P_0:

$$b_{\text{ph}} = c\left[(T - T_0) - T_0 \ln\left(\frac{T}{T_0}\right)\right] + v_{\text{m}}(P - P_0) \tag{2.24}$$

As indicated by Eqs. 2.22–2.24, the determination of the physical exergy requires only that the environment be defined by its ambient pressure and temperature, or reference pressure and temperature (P_0 and T_0).

The concept of chemical exergy means the maximum work that it is possible to be performed when a given system (or mass flow rate) is taken from the restricted reference state up to the dead state in which there is a complete thermodynamic equilibrium: mechanical, thermal, and chemical equilibrium. In that evolution, the system is submitted to reversible processes interacting only with the environment. The chemical exergy of a system is given by:

$$B^s_{\text{ch}} = U^* + P_0\, V^* - T_0 S^* - \sum_{i=1}^{n} \mu_{0,i}\, N_i = \sum_{i=1}^{n} (\mu_i^* - \mu_{0,i}) N_i \tag{2.25}$$

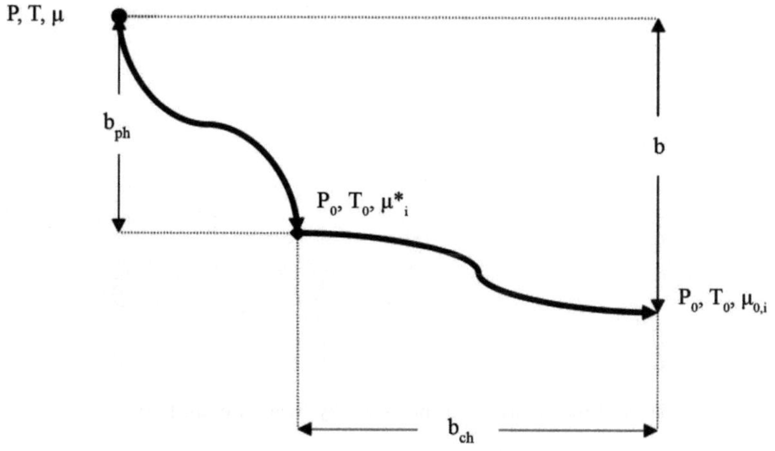

Fig. 2.4 Physical, chemical, and total exergy

And the chemical exergy of a mass flow rate is:

$$\overline{b}_{\mathrm{ch}} = \overline{h}^* - T_0\, \overline{s}^* - \sum_{i=1}^{n} \mu_{0,i}\, x_i \tag{2.26}$$

With the expressions of the physical and chemical exergies, the total system exergy is:

$$B^s = U + P_0 V - T_0 S - \sum_{i=1}^{n} \mu_{0,i} N_i \tag{2.27}$$

And for a mass flow rate:

$$\overline{b} = \overline{h} - T_0\, \overline{s} - \sum_{i=1}^{n} \mu_{0,i}\, x_i \tag{2.28}$$

Figure 2.4 summarizes the procedure to determine the total exergy, by calculating the physical and chemical components:

2.2.3 Chemical Exergy Calculation

The chemical exergy determination requires a model of the standard environment, as proposed by Szargut et al. [19]. This model is composed of the following reference substances, given an environment at $T_0 = 298.15$ K and $P_0 = 101.325$ kPa:

- Gaseous components of the atmosphere (O_2, N_2, CO_2, H_2O, D_2O, Ar, He, Ne, Kr, Xe);

Fig. 2.5 Components of the environment proposed by Szargut et al. [19]

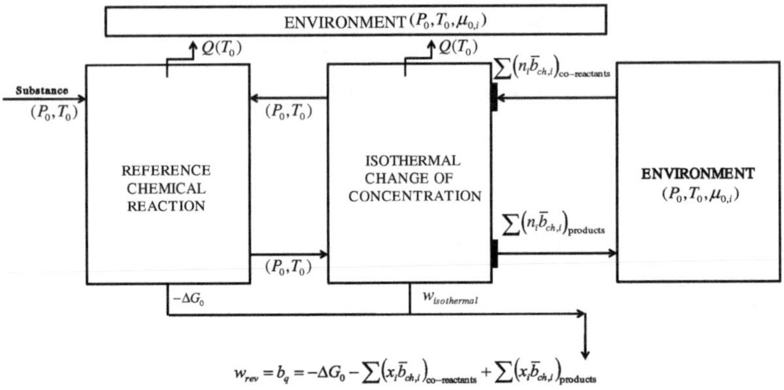

Fig. 2.6 Chemical exergy determination of substances that does not exist in the reference environment

- Solid reference substances of the crust of the Earth;
- Ionic reference substances of the sea;
- Molecular reference substances, nonionizable of the sea.

As stated above, the reference environment is characterized by substances of the atmosphere, earth crust, and seas, as shown in Fig. 2.5.

The chemical exergy of reference gases that compose the standard atmosphere is given by the isothermal expansion work that could be performed in a turbine that exchanges heat with the environment at T_0 and takes the gas from P_0 up to its partial pressure in the standard atmosphere (p_{00}), as presented by Eq. 2.29:

$$\bar{b}_{ch} = \bar{R}T_0 \ln\left(\frac{P_0}{P_{00}}\right) \tag{2.29}$$

For a pure substance that does not exist in the reference environment, the chemical exergy is determined considering two reversible processes. The first one is a reference chemical reaction that happens at P_0 and T_0 where the substance reacts with substances brought from the environment (coreactants) and generating substances that exist in the reference environment (products). The second process is a change in the concentrations of coreactants (from the state they are in equilibrium in the reference environment to the state they are pure at P_0 and T_0) and the concentration of the products (from P_0 and T_0 to the state they are in equilibrium in the reference environment) as shown in Fig. 2.6. The chemical exergy is then calculated as indicated in Eq. 2.30:

$$\bar{b}_{ch} = -\Delta G_0 - \left[\sum_i x_i\,\bar{b}_{ch,i}\right] \text{coreactants} + \left[\sum_j x_j\,\bar{b}_{ch,j}\right] \text{products} \tag{2.30}$$

with,

$$\Delta G_0 = \sum_j \nu_j\, g_j - \sum_k \nu_k\, g_k \tag{2.31}$$

In Eq. 2.31, ν_k, ν_j, g_k and g_j are, respectively, the stoichiometric coefficients and the free Gibbs energy in molar basis. Index j represents the products and index k represents the reactants of the reaction (substance under study and coreactants).

For a mixture, the chemical exergy can be calculated as:

$$\bar{b}_{ch,mist} = \sum_i x_i\,\bar{b}_{ch,i} + \mathrm{R}T_0 \sum_i x_i\, \ln \gamma_i x_i \tag{2.32}$$

where x_i is the molar fraction of component "i", $\bar{R}$ is the universal gas constant, and γ_i is the activity coefficient. For ideal mixtures, the activity coefficient is equal to one.

The chemical exergy of fuels can be evaluated as a function of the lower heating value (LHV), by means of expressions such as:

$$b_{ch} = \phi \mathrm{LHV} \tag{2.33}$$

where ϕ can be determined by using correlations based on the composition of the fuel, as presented by Szargut et al. [19] and Kotas [8]. As examples, ϕ is about 1.04 for natural gas [8] and 1.08 for petroleum [17].

Example 2.1 Determine the chemical exergy of methane. The reference reaction for the determination of the chemical exergy of methane is the combustion reaction of methane with oxigen:

$$CH_4 + 2O_2 \rightarrow 2H_2O + CO_2$$

The determination of the chemical exergy of methane will be done by using two procedures. The first one determines the chemical exergy by solving Eqs. 2.30 and 2.31:

$$\bar{b}_{ch\,CH_4} = -\Delta G_0 - 2b_{ch\,O_2} + 2b_{ch\,H_2O} + b_{ch\,CO_2}$$

Table 2.2 Properties of some ideal gases of the reference atmosphere

	h (T_0 = 25 °C) (kJ/kmol)	s (T_0 = 25 °C, P_0 = 101.325 kPa) (kJ/kmol K)	s (T = 25 °C, $P = P_{00}$) (kJ/kmol K)	b_{ch} (kJ/kmol)
CO_2	−393,486	213.7	280.3[a]	19,867
H_2O	−241,811	188.7	220.6[b]	9,494
N_2	0	191.5	193.9[c]	720
O_2	0	205.0	218.4[d]	3,974

[a] $P_{00\ CO_2}$ = 0.03 kPa
[b] $P_{00\ H_2O}$ = 2.2 kPa
[c] $P_{00\ N_2}$ = 75.78 kPa
[d] $P_{00\ O_2}$ = 20.39 kPa

And

$-\Delta G_0$ = Enthalpy (CH_4; T = 25 °C) + 2 Enthalpy (O_2; T = 25 °C) − Enthalpy (CO_2; T = 25 °C) − 2 Enthalpy (H_2O; T = 25 °C) − 298.15 [Entropy (CH_4; T = 25 °C; P = 101.325 kPa) + 2 Entropy (O_2; T = 25 °C; P = 101.325 kPa) − Entropy (CO_2; T = 25 °C; P = 101.325 kPa) − 2 Entropy (H_2O; T = 25 °C; P = 101.325 kPa)]

The values of enthalpies, entropies, and chemical exergies of the coreactants and products presented in the reference reaction and considered as ideal gases are shown in Table 2.2. This table also presents the entropy of the coreactants and products at the state they are in the reference environment, or dead state. In this table, the chemical exergy of these ideal gases is calculated according to Eq. 2.29:

These properties together with the enthalpy and entropy of methane at T_0 and P_0:

h_{CH_4} = −74,595 kJ/kmol
s_{CH_4} = 186.3 kJ/kmol K
allows the calculation of $\bar{b}_{ch\,CH_4}$:

$$\bar{b}_{ch\,CH_4} = 831{,}862\,\text{kJ/kmol CH}_4$$

The second procedure to determine $\bar{b}_{ch\,CH_4}$, that requires only the values of enthalpies and entropies of the substances involved in the reference reaction, consists in applying energy and entropy balances to the control volume indicated in Fig. 2.7 in which there are reversible processes and the heat transfer interactions occur with the environment at T_0. Methane (at P_0, T_0), and oxygen (at $P_{00\ O_2}$, T_0) enter the control volume, while water (at $P_{00\ H_2O}$, T_0) and carbon dioxide (at $P_{00\ CO_2}$, T_0) leave it. The reversible work that can be obtained from the control volume is the chemical exergy of methane, given by:

$\bar{b}_{ch\,CH_4}$ = Enthalpy (CH_4; T = 25 °C) + 2 Enthalpy (O_2; T = 25 °C) − Enthalpy (CO_2; T = 25 °C) − 2 Enthalpy (H_2O; T = 25 °C) − 298.15 [Entropy (CH_4; T = 25 °C; P = 101.325 kPa) + 2 Entropy (O_2; T = 25 °C; P = 20.43 kPa)

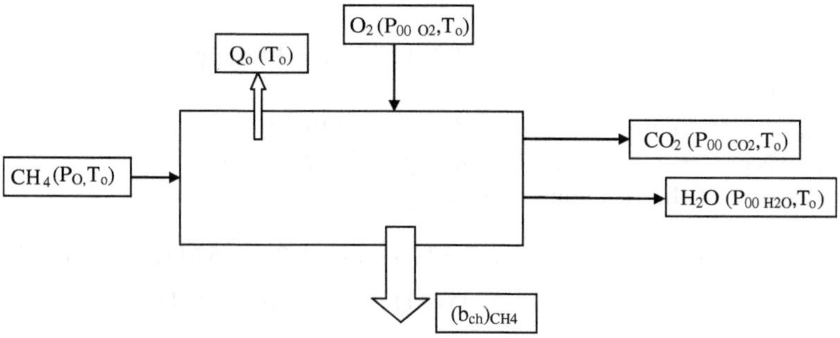

Fig. 2.7 Thermodynamic scheme for calculation of the chemical exergy of methane (maximum work capacity of methane at P_0 and T_0)

− Entropy (CO_2; $T = 25$ °C; $P = 0.03347$ kPa) − 2 Entropy (H_2O; $T = 25$ °C; $P = 2.2$ kPa)]

$$\bar{b}_{\mathrm{Ch\,CH_4}} = 831{,}874\,\mathrm{kJ/kmol\,CH_4}$$

The small difference between the values of $\bar{b}_{\mathrm{ch\,CH_4}}$ is due to some approximations made in the calculation of the thermodynamic properties.

Note that this second procedure for the determination of the chemical exergy of the methane is similar to calculation of the chemical exergy by using the following equation:

$$\bar{b}_{\mathrm{ch\,CH_4}} = h_{\mathrm{CH_4}} - h_{\mathrm{CH_40}} - T_0(s_{\mathrm{CH_4}} - s_{\mathrm{CH_40}})$$

where:

$h_{\mathrm{CH_40}}$ = Enthalpy (CO_2; $T = 25$ °C) + 2 Enthalpy (H_2O; $T = 25$ °C) − 2 Enthalpy (O_2; $T = 25$ °C)

$s_{\mathrm{CH_40}}$ = Entropy (CO_2; $T = 25$ °C; $P = 0.03347$ kPa) + 2 Entropy (H_2O; $T = 25$ °C; $P = 2.2$ kPa) − 2 Entropy (O_2; $T = 25$ °C; $P = 20.43$ kPa)

Example 2.2 Methane is burnt with air. The mixture air methane enters the reactor at $T_i = T_0 = 25$ °C and $P_i = 101.325$ kPa. The combustion products leave the reactor, considered adiabatic, at $P_e = 101.325$ kPa and at the adiabatic flame temperature (T_e). Determine the exergy destruction per kmol of methane, considering that reactants and products can be modeled as ideal gases and with percent excess air ranging from 0 up to 180 % ($1 \leq \alpha \leq 2.8$).

The combustion reaction of methane with air is given by:

$$CH_4 + 2\alpha(O_2 + 3.762N_2) \rightarrow 2H_2O + CO_2 + 7.524\alpha N_2 + 2(\alpha - 1)O_2$$

As the reactor is supposed to be adiabatic, the temperature of the combustion gases, the adiabatic flame temperature, T_e, is determined by solving the reactor energy balance in which the enthalpy of the reactants is equal to the enthalpy of products:

$$H_{\mathrm{reactants}} = H_{\mathrm{products}}$$

With:

$H_{reactants}$ = Enthalpy (CH_4; $T = T_0$) + 7.524 α Enthalpy (N_2; $T = T_0$) + 2 α Enthalpy (O_2; $T = T_0$)

$H_{products}$ = Enthalpy (CO_2; $T = T_e$) + 2 Enthalpy (H_2O; $T = T_e$) + 7.524 α Enthalpy (N_2; $T = T_e$) + 2(α−1) Enthalpy (O_2; $T = T_e$)

The exergy of the reactants is done by:

$$B_{reactants} = \bar{b}_{ch\,CH_4} + 2\,\alpha\,\bar{b}_{ch\,O_2} + 7.524\,\alpha\,\bar{b}_{ch\,N_2} + 298.15\,\mathrm{M}\,\bar{\mathrm{R}}\,[1/\mathrm{M}\ln(1/\mathrm{M}) + 7.524\,\alpha/\mathrm{M}\ln(7.524\alpha/\mathrm{M}) + 2\,\alpha/\mathrm{M}\ln(2\,\alpha/\mathrm{M})]$$

With:

$$M = 1 + 9.524\,\alpha\ (\text{number of mols of reactants})$$

$$\bar{R} = 8.314\,\mathrm{kJ/kmolK}$$

The exergy of the reactants can also be determined, as shown in Example 2.1, by:

$$B_{reactants} = H_{reactants} - H_{reactants0} - 298.15\,(S_{reactants} - S_{reactants0})$$

In this equation, $S_{reactants}$, $H_{reactants0}$, and $S_{reactants0}$ are, respectively, the entropy of reactants, enthalpy, and entropy of reference of the reactants at the dead state, given by:

$$S_{reactants} = \text{Entropy}\,(CH_4; T = T_0; P = P_{CH_4}) + 7.524\,\alpha\,\text{Entropy}\,(N_2; T = T_0; P = P_{N_2}) + 2\,\alpha\,\text{Entropy}\,(O_2; T = T_0; P = P_{O_2})$$

$$H_{reactants0} = 2(\alpha - 1)\,\text{Enthalpy}\,(O_2; T = T_0) + \text{Enthalpy}\,(CO_2; T = T_0) + 2\,\text{Enthalpy}\,(H_2O; T = T_0) + 7.524\,\alpha\,\text{Enthalpy}\,(N_2; T = T_0)$$

$$S_{reactants0} = 2(\alpha - 1)\,\text{Entropy}\,(O_2; T = T_0; P = P_{00\ O2}) + \text{Entropy}\,(CO_2; T = T_0; P = \mathrm{P}_{00\ CO_2}) + 2\,\text{Entropy}(H_2O; T = T_0; P = P_{00\ H_2O}) + 7.524\,\alpha\,\text{Entropy}\,(N_2; T = T_0; P = P_{00\ N_2})$$

With:

$$P_{CH_4} = P_i/M$$
$$P_{N_2} = 7.524\,\alpha\,P_i/M$$
$$P_{O_2} = 2\,\alpha\,P_i/M$$

Analogously, the exergy of products is given by:

$$B_{product} = H_{product} - H_{products0} - 298.15\,(S_{product} - S_{products0})$$

Table 2.3 Characteristic parameters of the combustion of methane with air

Excess air (%)	T_e (°C)	$B_{reactants}$ (kJ/kmol CH_4)	$B_{products}$ (kJ/kmol CH_4)	B_{dest} (kJ/kmol CH_4)	Δ_{comb} (%)
0	2,051	824,903	597,285	227,618	27.6
20	1,795	824,718	574,361	250,357	30.3
40	1,598	824,597	554,466	270,131	32.8
60	1,441	824,524	536,471	288,053	34.9
80	1,314	824,487	519,987	304,500	36.9
100	1,208	824,480	504,766	319,714	38.8
120	1,118	824,496	490,628	333,868	40.5
140	1,042	824,532	477,435	347,097	42.1
160	975	824,584	465,077	359,507	43.6
180	917	824,650	453,463	371,187	45.0

With:

$$H_{products} = \text{Enthalpy}\,(CO_2; T = T_e) + 2\,\text{Enthalpy}\,(H_2O; T = T_e) + 7.524\,\alpha\,\text{Enthalpy}\,(N_2; T = T_e) + 2(\alpha - 1)\,\text{Enthalpy}\,(O_2; T = T_e)$$

$$S_{products} = \text{Entropy}\,(CO_2; T = T_e; P = P_{CO_2}) + 2\,\text{Entropy}\,(H_2O; T = T_e; P = P_{H_2O}) + 7.524\,\alpha\,\text{Entropy}\,(N_2; T = T_e; P = P_{N_2}) + 2(\alpha - 1)\,\text{Entropy}\,(O_2; T = T_e; P = P_{O_2})$$

Where:

$$P_{CO_2} = P_e/M$$
$$P_{H_2O} = 2P_e/M$$
$$P_{N_2} = 7.524\,\alpha\,P_e/M$$
$$P_{O_2} = 2(\alpha - 1)P_e/M$$

Taking into account the products of the combustion reaction:

$$H_{reactants0} = H_{products0}$$

$$S_{reactants0} = S_{products0}$$

As the reactor is adiabatic, $H_{reactants} = H_{products}$, then:

$$B_{dest} = B_{reactants} - B_{products} = 298.15\,(S_{products} - S_{reactants}) = T_0 S_{ger}$$

The relative exergy destroyed during the combustion reaction can be calculated as:

$$\Delta_{comb} = B_{dest}/B_{reactants}$$

Table 2.3 presents the values of T_e, $B_{reactants}$, $B_{products}$, B_{dest} and Δ_{comb} for excess air ranging from 0 up to 180 % ($1 \leq \alpha \leq 2.8$)

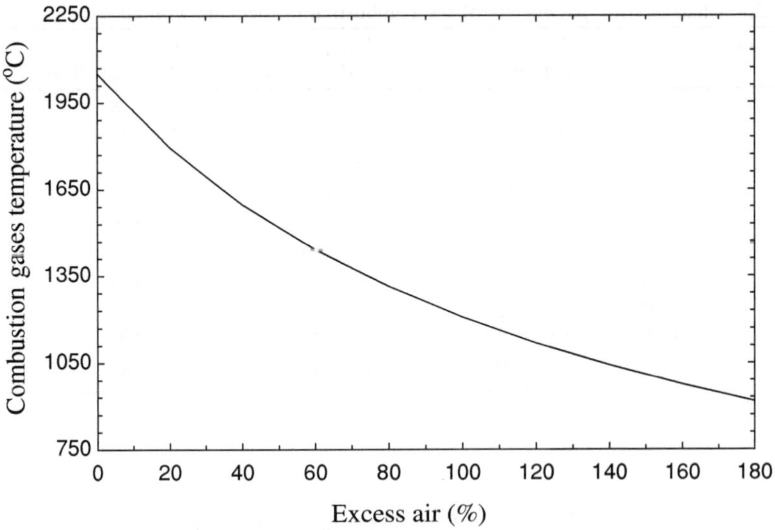

Fig. 2.8 Temperature of combustion gases as a function of percentage of excess air

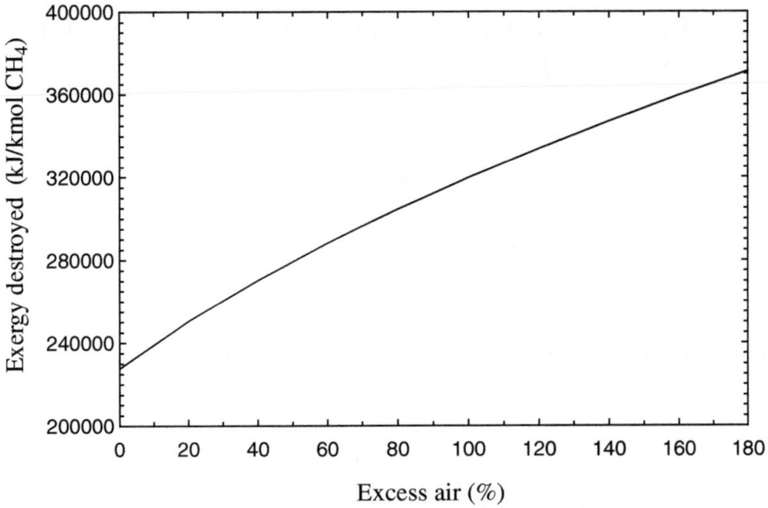

Fig. 2.9 Exergy destroyed in the reactor as a function of percentage of excess air

Figures 2.8 and 2.9 present the behavior of T_e and B_{dest} with the percent excess air, showing that the higher the percent excess air, the greater the exergy destroyed because the excess air mass dilutes the combustion gases, lowering the thermal exergy of the mixture of gases, and thus, reducing its capacity to perform work. This conclusion is particularly useful when studying gas turbines combustors, because the exergy destroyed in the combustor is a direct measure of the reduction capacity of the turbine to generate mechanical or electrical power. This result also evidences that the quality of combustion is proportional to its adiabatic flame temperature [21].

2.3 Exergy: Graphical Representations and Exergy Diagrams

2.3.1 Introduction

The use of thermodynamic diagrams to represent processes, properties, as well as to determine work and heat quantities is a common practice in Engineering Thermodynamics. In the case of the Exergy Analysis of energy conversion processes, diagrams such as entropy–enthalpy and exergy–enthalpy are very useful to help the understanding and the evaluation of these processes.

In the next sections, the most important diagrams are presented for the determination of physical exergy, the exergy transferred and destroyed, the exergy of separation, and mixing of a binary mixture.

2.3.2 The Physical Exergy in the Enthalpy–Entropy Diagram

The physical exergy of a pure substance is easily determined in a Mollier diagram, as shown in Fig. 2.10. Giving the states represented by points *m* (*P*, *T*) and *O* (P_0 and T_0) and drawing the line from point *O* and with angle α, T_0 can be determined by:

$$(\delta h/\delta s)_{P_0} = T_0 = \mathrm{tg}\alpha \tag{2.34}$$

This segment is called reference line and it is tangent to the isobaric that passes by P_0. Then it is possible to write Eq. 2.35:

$$(s_m - s_o)\mathrm{tg}\alpha = (s_m - s_o)T_0 \tag{2.35}$$

And introducing the expression of the physical exergy, one gets:

$$b_{ph} = h_m - h_o - (s_m - s_o)\mathrm{tg}\alpha \tag{2.36}$$

2.3.3 Diagram Carnot Factor-Enthalpy

The first reference to the use of the diagram Carnot factor-enthalpy is attributed to Thring [21] for the study of combustion processes, where he introduced the concept of virtue of energy. This diagram is particularly useful for the study of heat transfer processes. In the abscissae axis it is represented the heat transfer rate

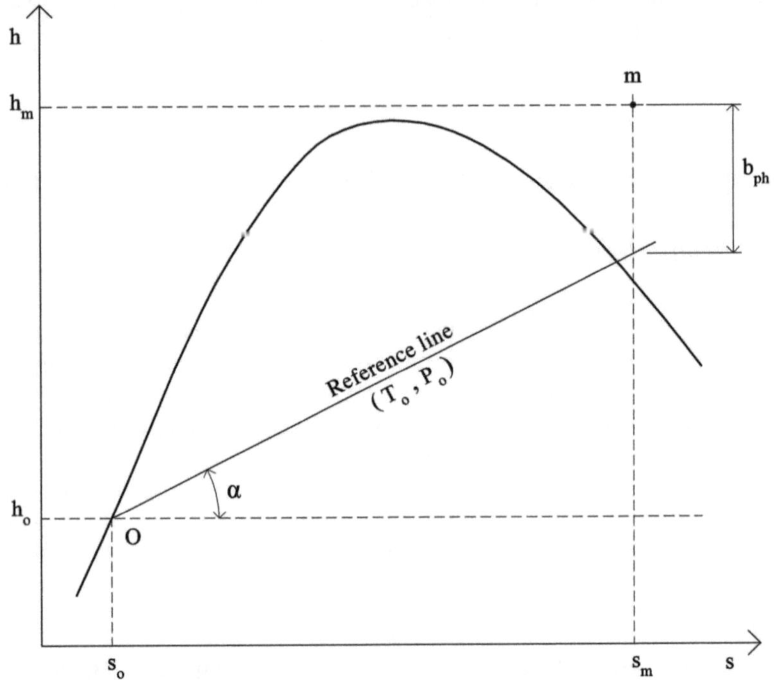

Fig. 2.10 Physical exergy determination in a Mollier diagram

from one mass flow, determined by the variation of the enthalpy flow rate of this mass flow. The Carnot factor ($\theta = T_0/T$) is represented in the ordenate axis in order to characterise the temperature level in which the heat trasfer takes place, as shown in Fig. 2.11 where it is represented the heat transfer process in a sugarcane juice heater of a sugar and alchool mill. In this heat exchanger steam is used to heat the juice extracted from sugarcane.

By using the expression of the physical exergy (Eq. 2.22) and calculating the partial derivative, as indicating in Eq. 2.37, one can show that the Carnot factor indicates the change of the physical exergy with respect to enthalpy in a isobaric process.

$$(\partial b/\partial h)_p = 1 - T_0/T = \theta \tag{2.37}$$

It can be seen in this diagram that the area under the upper curve corresponds to the exergy transferred from the steam (heating agent) and the area under the lower curve is the exergy received by the sugarcane juice. The difference between these two values is the destroyed exergy rate in the heat transfer process.

This type of diagram was utilized by Le Goff et al. [9] for the study of the distillation of a binary mixture, as shown in Fig. 2.12. The corresponding diagram $\theta - \Delta H$ is presented in Fig. 2.13.

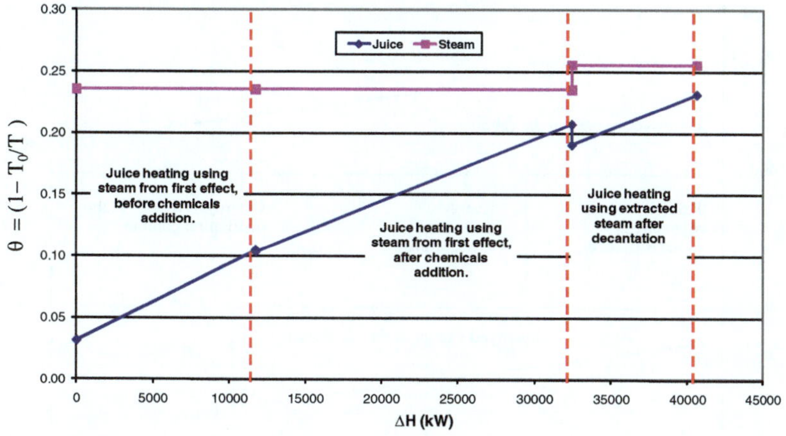

Fig. 2.11 Diagram Carnot factor (θ)–ΔH for a heating juice system of a sugar and alcohol mill [14]

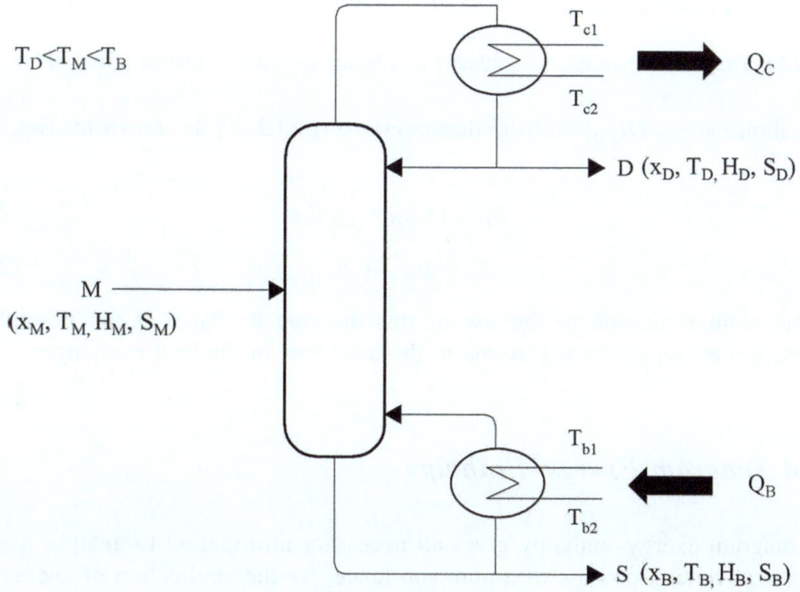

Fig. 2.12 Typical scheme of a binary distillation set [9]

Considering that in the reboiler the heating agent passes from temperature T_{b1} to T_{b2}, in order to vaporize the liquid at T_B, and the condenser is cooled by a fluid in which temperature increases from T_{c1} to T_{c2}, to condensate the vapor at T_D, it is possible to represent in the diagram θ – H the heat rate quantities Q_B *e* Q_C (see Fig. 2.13). The area between the line at θ_B and the line at θ_D is the net thermal exergy rate supplied to the separation process (B_t) and it is composed of the

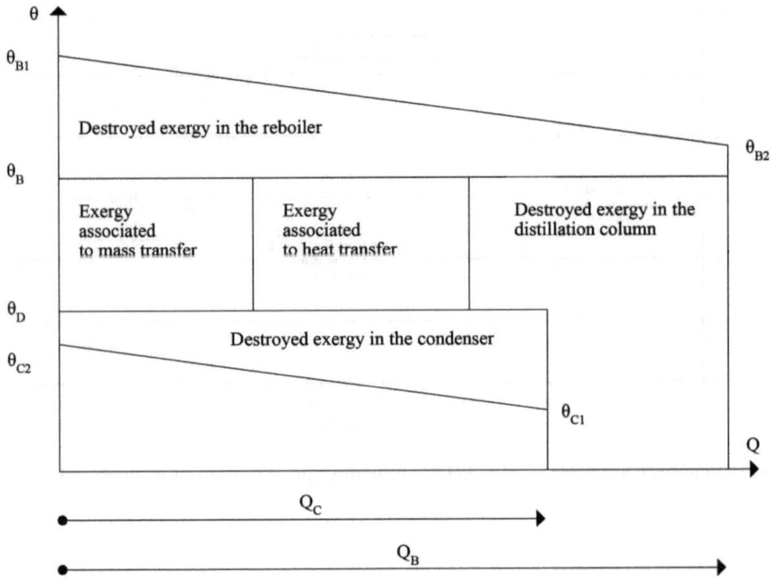

Fig. 2.13 Diagram Carnot factor—enthalpy of a binary mixture distillation process [9]

separation exergy (B_{sep}) and the destroyed exergy (B_{dest}) as shown by Eqs. 2.38 and 2.39.

$$B_t = Q_B\theta_B - Q_C\theta_D \tag{2.38}$$

$$B_t = B_{\text{dest}} + B_{\text{sep}} \tag{2.39}$$

The main limitation to the use of this diagram is that it is not possible to represent the exergy destroyed due to the head loss in the heat exchanger.

2.3.4 Diagram Exergy–Enthalpy

The diagram exergy–enthalpy gives all necessary information to analyze a given energy conversion process of a pure substance, for the application of energy and exergy balances. Figure 2.14 presents a typical exergy–enthalpy diagram. As pointed out in Sect. 2.3.3, based on the physical exergy definition, one can obtain the following expression:

$$(\partial b/\partial h)s = 1 \tag{2.40}$$

That is, in this diagram, the isentropic lines are inclined at 45° with respect to the enthalpy axis when both variables are represented with the same scale.

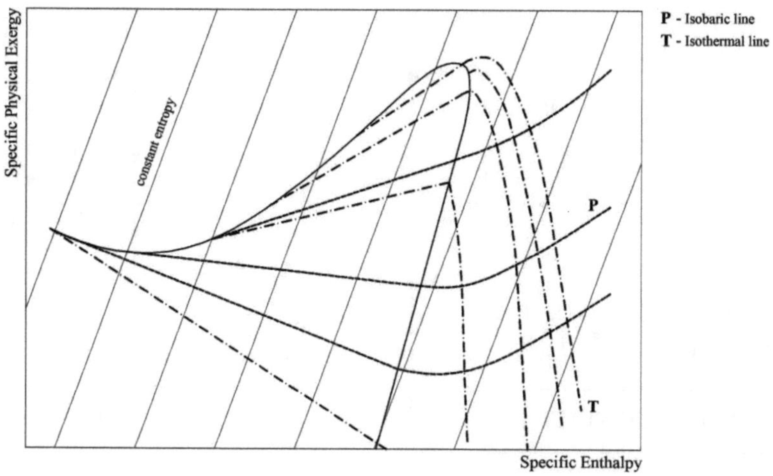

Fig. 2.14 Diagram specific physical exergy—specific entalphy

2.3.5 Diagrams Exergy-Composition and Exergy–Enthalpy for Binary Mixtures

The diagram exergy-composition for binary mixtures is analogous to the enthalpy-concentration diagram, being particularly useful in the study and optimization of absorption refrigerating and heat pump systems that work with mixtures H_2O—LiBr and NH_3—H_2O. Oliveira [13] proposed a simple procedure to calculate the exergy of a solution and to build these diagrams based on the definition of equilibrium solutions at P_0, T_0, the reference environment pressure and temperature. This procedure is adequate for systems that operate in closed circuit, such as absorption heat pumps and refrigerating systems, avoiding the necessity to use the reference conditions proposed by Szargut et al. [19].

First it is considered the calculation of the exergy of a binary mixture in which its components exist pure at P_0, T_0. Then it is considered a process where the studied mixture changes its thermodynamic state from state M (T_M, p_M, X_M) up to a state where $P = P_0$ and $T = T_0$, maintaining the same concentration. Then the mixture is completely separated in a isothermal process ($x = 0$ and $x = 1$). Finally, every component will change its pressure up to P_0, as shown in Fig. 2.15. This sequence of processes can be associated with those that take place in a separation process presented in Fig. 2.16. The separation exergy of the mixture M is in fact the exergy of the state M, given by Eq. 2.41:

$$b_{\mathrm{m}} = [h_{\mathrm{M}} - (1 - x_{\mathrm{M}})h_{A0} + x_{\mathrm{M}}h_{\mathrm{B0}}] - T_0[s_{\mathrm{M}} - (1 - x_{\mathrm{M}})s_{A0} + x_{\mathrm{M}}s_{\mathrm{B0}}] \quad (2.41)$$

In this case, the reference enthalpy and entropy of the mixture M are given by Eqs. 2.42 and 2.43:

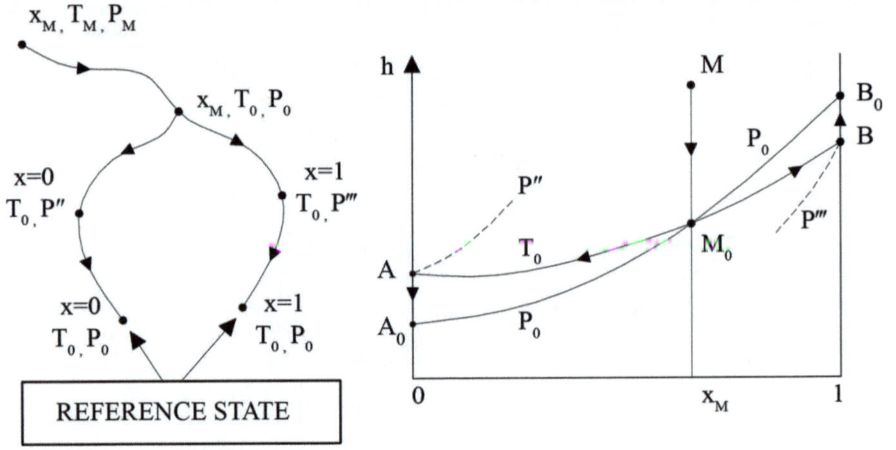

Fig. 2.15 Separation process [13]

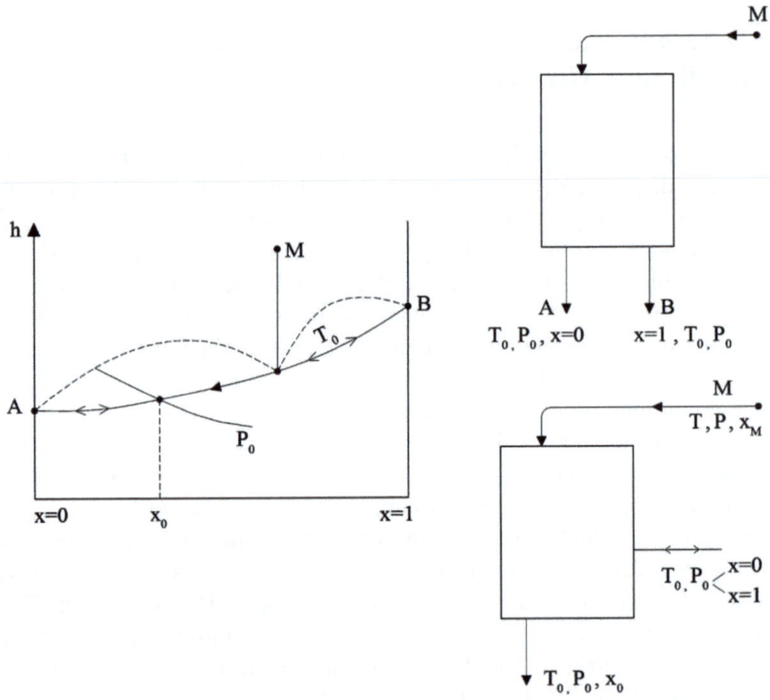

Fig. 2.16 Determination of the exergy of a mixture at state *M* [13]

$$h_{M0} = (1 - x_M)h_{A0} + x_M h_{B0} \tag{2.42}$$

$$s_{M0} = (1 - x_M)s_{A0} + x_M s_{B0} \tag{2.43}$$

and in the particular case of the exergy determination of pure substances ($x_M = 0$ or $x_M = 1$), one has:

$$h_{M0} = h_{A0} \text{ or } h_{M0} = h_{B0}$$

$$s_{M0} = s_{A0} \text{ or } s_{M0} = s_{B0}$$

When there is a mixture at T_0 and P_0, as is the case for mixtures NH_3—H_2O, at the exit of the separator there will be a mixture at x_0 and one pure substance ($x = 0$ or $x = 1$), as indicated in Fig. 2.16. Now there are two possible reference conditions:

- $T_0,\ P_0,\ x_0$ and $T_0,\ P_0,\ x = 0$
- $T_0,\ P_0,\ x_0$ and $T_0,\ P_0,\ x = 1$

For the first case, the reference enthalpy and entropy are given by Eqs. 2.44 and 2.45:

$$h_{M0} = \left(\frac{x_m}{x_0}\right) h_0 + \left(1 - \frac{x_m}{x_o}\right) h_{A0} \tag{2.44}$$

$$s_{M0} = \left(\frac{x_m}{x_0}\right) s_0 + \left(1 - \frac{x_m}{x_o}\right) s_{A0} \tag{2.45}$$

For the second case, one has:

$$h_{M0} = \left(\frac{1 - x_m}{1 - x_0}\right) h_0 - \left(\frac{x_0 - x_m}{1 - x_o}\right) h_{A0} \tag{2.46}$$

$$s_{M0} = \left(\frac{1 - x_m}{1 - x_0}\right) s_0 - \left(\frac{x_0 - x_m}{1 - x_o}\right) s_{A0} \tag{2.47}$$

The described procedure was utilised in the development of diagrams exergy–enthalpy of binary mixtures H_2O—LiBr and NH_3—H_2O shown in Figs. 2.17 and 2.18. In these diagrams, the partial derivative $(\partial b/\partial h)p$ gives:

$$(\partial b/\partial h)_p = 1 - T_0(\partial s/\partial h)_p \tag{2.48}$$

This term can be considered a modified Carnot factor for a mixture/solution, and the analysis of its behavior is useful in the determination of the optimal operating conditions of a separator or a mixer [13].

2.3.6 Grassmann Diagram

The Grassmann diagram [7] is a graphical representation similar to the well-known Sankey diagram employed in the energy analysis of processes, where the width of the bands is associated to the exergy rate or exergy flow rate of a given energy

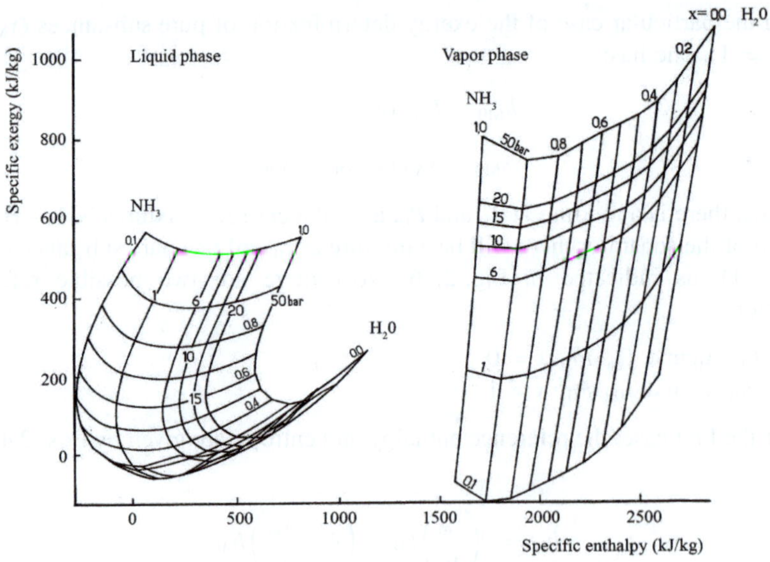

Fig. 2.17 Diagram exergy–enthalpy for NH_3—H_2O solutions ($T_0 = 20$ °C, $P_0 = 1$bar, $x_{0A} = 0$ and $x_0 = 34$ % [13]

conversion process. Considering a cogeneration plant shown in Fig. 2.19, the Sankey and Grassmann diagrams are presented, respectively, in Figs. 2.20 and 2.21.

It is interesting to notice that in both diagrams the difference of the values of energy and exergy bands. The Grassmann diagram evidences that the component of the plant that is the main responsible for the exergy destruction in the processes is the boiler, because of the highly irreversible processes that take place inside it: the combustion reaction and the heat transfer between the combustion products and the water. It is due to this exergy destruction that the overall efficiency conversion: fuel chemical exergy into mechanical/electrical work is about 30 % in conventional thermal power plants that operates based on Rankine cycles.

2.4 Exergy Efficiency

2.4.1 Balance of the Energy Value

Given an energy conversion system, shown in Fig. 2.22, that receives an energy rate amount E_i, supplies a useful energy rate E_u, and rejects E_w, the energy balance is done by Eq. 2.49:

$$E_i = E_p + E_w \tag{2.49}$$

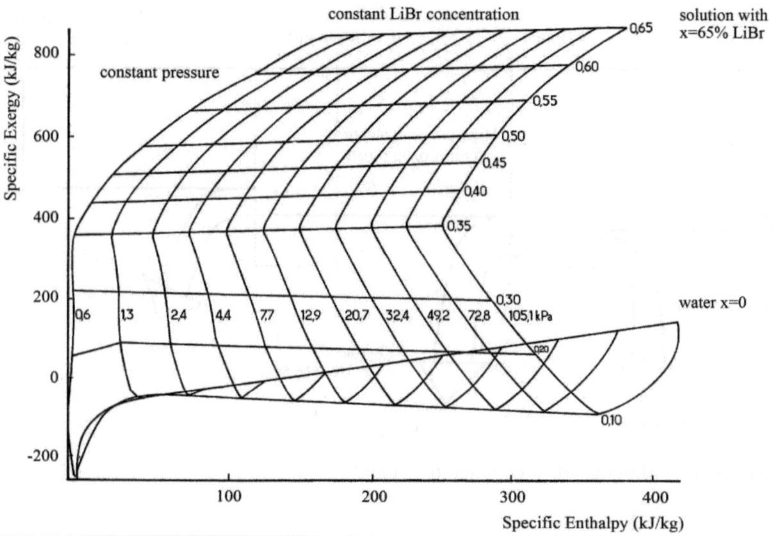

Fig. 2.18 Diagram exergy–enthalpy for water–lithium–bromide solutions (T_0 = 25 °C, $P_0 = 1$bar, $x_{0A} = 0$ and $x_0 = 20$ %) [13]

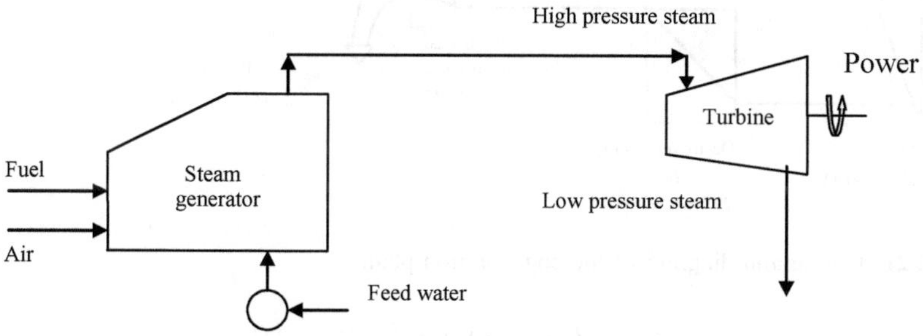

Fig. 2.19 Simplified cogeneration plant

As emphasized by Le Goff et al. [9], an economist would see this process in a different way. It consists of a system that converts goods of small value into goods with added value by means of consuming expensive goods (such as raw materials, energy, information, capital, etc.) and generating wastes with reduced value that are rejected into the environment, as represented in Fig. 2.22 in the production and utilization phases.

The balance of value of this process can be written as:

$$E_i v_i + C_o = E_p v_p + E_w v_w \tag{2.50}$$

In this expression the terms v are the 'values' that are given to every energy rate or flow rate and C_o is the operational energy cost rate (see Eq. 2.51)

Combining Eqs. 2.49 and 2.50 one obtains the expression of the operational cost rate:

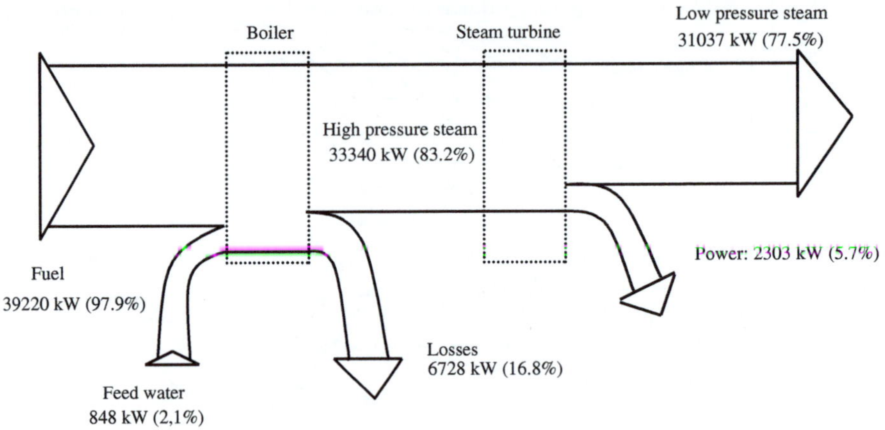

Fig. 2.20 Sankey diagram of a cogeneration plant

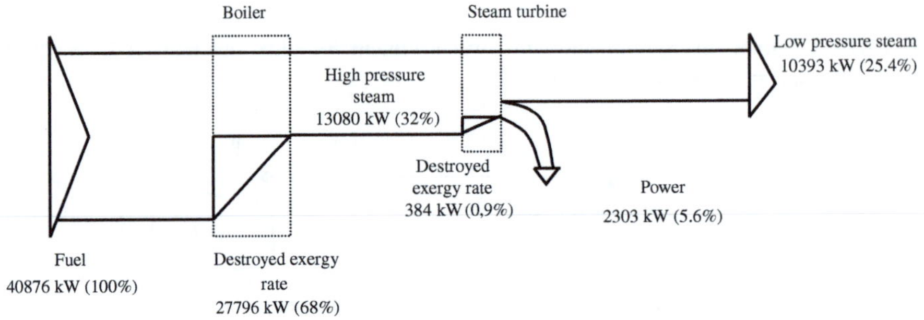

Fig. 2.21 Grassmann diagram of the cogeneration plant

$$C_o = E_p(v_p - v_i) + E_w(v_w - v_i) \tag{2.51}$$

Dividing Eq. 2.51 by E_u, one obtains the specific operational cost:

$$c_o = \frac{C_o}{E_u} = \underbrace{(v_p - v_i)}_{\text{value gain of the products}} + \overbrace{\frac{E_W}{E_u}(v_w - v_i)}^{\text{value loss of the wastes}} \tag{2.52}$$

The first term in the right side of Eq. 2.52 is the value gain of the products and the second term is the loss of value of the wastes.

2.4.2 General Definition of Efficiency

The performance of a given process or equipment can be defined by the following expression of efficiency:

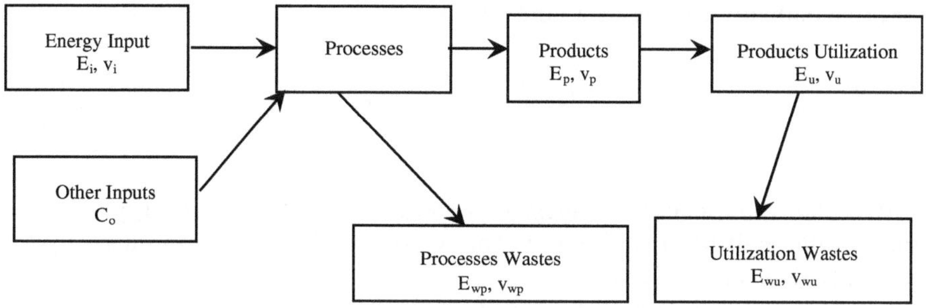

Fig. 2.22 Energy and value balance

$$\eta = \sum(\text{value of products}) / \sum(\text{value of inputs}) \tag{2.53}$$

For an energy system whose objective is to perform a given energy conversion, its efficiency can be given as:

$$\eta = \frac{\sum(E_u v_u)}{\sum(E_i v_i)} \tag{2.54}$$

It is important to note that this definition does not imply, necessarily, that $\eta < 1$ as will be further discussed.

According to the utilized energy value scale, Eq. 2.54 can provide different information.

The first scale postulates that every energy form has the same value.

$$v_a = v_u = v_w = v_i = v_f \tag{2.55}$$

In this case, the value balance is equal to the energy balance, or the First Law of Thermodynamics, and the efficiency expression is the energy efficiency. The energy efficiency of thermal engines is the thermal efficiency, which is always lower than one, and the energy efficiency of heat pumps and refrigerating systems is the coefficient of performance, that can be higher or lower than one.

A second value scale consists in associating to a given energy quantity its corresponding value of exergy, for instance:

– For work:

$$v_W = 1 \tag{2.56}$$

– For heat:

$$v_Q = \left(1 - \frac{T_0}{T}\right) \tag{2.57}$$

– For an enthalpy change of a pure substance:

$$\nu_H = 1 - T_0\left(\frac{s_1 - s_2}{h_1 - h_2}\right) \tag{2.58}$$

With this value scale, the efficiency expression becomes the exergy efficiency and its value will be always lower than one (it will be one for an internally and externally reversible process).

It is interesting to note that when the energy value scale is used, the specific operational cost will be zero, as a consequence of the First Law of Thermodynamics. Also, when the exergy value scale is employed for a heat engine, its specific operational cost will be zero for a Carnot engine and will be negative, for a real heat engine, due to the occurrence of irreversible processes (the output exergy is lower than the input exergy).

2.4.3 Exergy Efficiency

In the exergy literature there are many expressions to quantify the exergy performance of energy conversion processes ([16, 19, 22]). Three definitions will be presented next. They are commonly applied to analyze the performance of chemical processes, thermal engines, and dissipative processes.

The first expression (Eq. 2.59), called degree of perfection by Szargut et al. [19] is useful to evaluate chemical processes:

$$\eta_P = \frac{\text{Exergy of useful products}}{\text{Feeding exergy}} \tag{2.59}$$

The second expression is similar to Eq. 2.59 and is indicated in the analysis of thermal processes:

$$\eta_{b1} = \frac{\text{Useful exergy effect}}{\text{Driving exergy}} \tag{2.60}$$

For thermal engines and refrigerating system that interact with the environment, Eq. 2.60 can be rewritten as:

$$\eta_{b2} = \frac{\eta_e}{\eta_{carnot}} = \frac{\eta_e}{\eta_{e\,max}} \tag{2.61}$$

The third expression is recommended to be used to quantify the performance of processes and equipment where the only effect is the exergy destruction, such as the flow of steam trough an expansion valve, or the heat rejection that takes place in a condenser of a thermal power plant, or in the condenser of a refrigerating system. These processes are usually called pure dissipative processes.

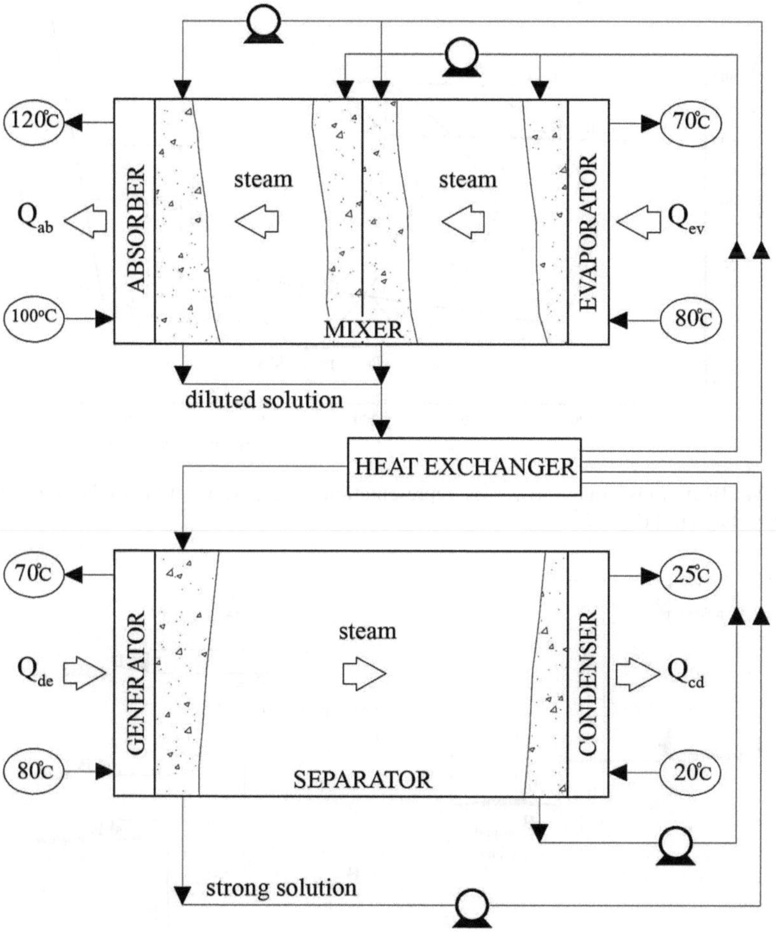

Fig. 2.23 Double effect heat transformer [13]

$$\eta_{b2} = \frac{\sum(\text{exit exergy})}{\sum(\text{inlet exergy})} \tag{2.62}$$

In order to apply the concept of the exergy efficiency, consider the determination of the exergy efficiency of a two stage heat transformer that has to heat up a particular mass flow rate from 100 to 120 °C, by using a thermal waste available at 85 °C. This heat transformer operates according to an absorption cycle with water-lithium bromide as the working pair, as represented in Fig. 2.23.

The mixer (absorber–evaporator set) is composed of two evaporation/absorption sets in thermal series and in a parallel arrangement for the solution and water flow rates that enters the mixer (see Fig. 2.23). The aqueous solutions of lithium bromide enter and leave the absorbers with salt concentrations of, respectively, 60 and 50 %. The water saturation pressures in both effects are 35.3 and 97.6 kPa,

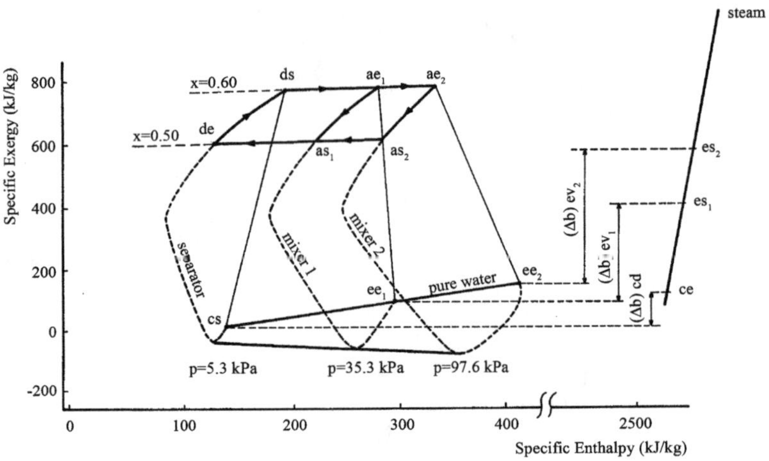

Fig. 2.24 Heat transformer processes representation in the water lithium bromide exergy–enthalpy diagram [13]

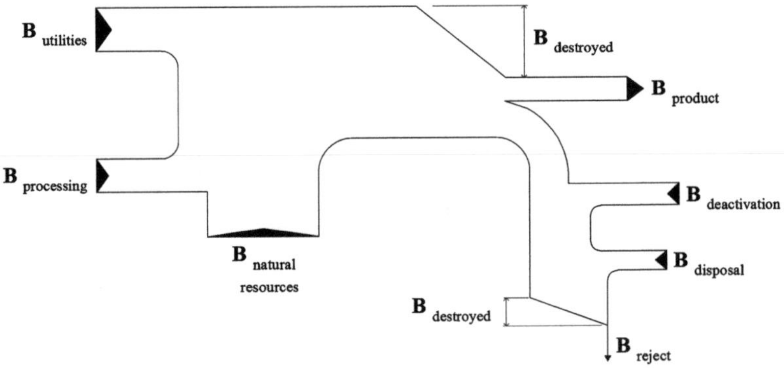

Fig. 2.25 Exergy balance of energy conversion processes

respectively. The separator (generator–condenser set) operates under a pressure of 5.3 kPa.

The processes that take place in the mixer and in the separator are represented in a diagram exergy–enthalpy, shown in Fig. 2.24 ($T_0 = 25$ °C, $P_0 = 1$bar, $x_{0A} = 0$ and $x_0 = 20$ %). Based on the data of states indicated in Fig. 2.25, it is possible to develop the energy and exergy balances for every component of the heat transformer, and calculate the parameters presented in Table 2.4. Considering additionally the inlet and outlet temperatures of the mass flow rates indicated in Fig. 2.24, a minimum temperature difference of 3 °C in the heat and mass exchangers and the required power (W_p) in the solution and water pumps, one can determine the exergy efficiencies of the separator (η_{sep}), mixer (η_{mix}), and the heat transformer (η_{btt}).

Table 2.4 Results of the energy and exergy balances

Energy balances (kJ/kg water)	Exergy balances (kJ/kg water)
$Q_{cd} = 5069$	$\Delta b_{cd} = 112.8$
$Q_{de} = 5809$	$\Delta b_{sep} = 627.6$
$Q_{ev1} = 2329$	$\Delta b_{ev1} = 317.9$
$Q_{ev2} = 2471$	$\Delta b_{ev2} = 436.0$
$Q_{ab} = 2628$	$\Delta b_{mix} = 607.9$ (for both stages)

$$\eta_{sep} = \frac{\Delta b_{sep}}{Q_{de}\theta_{de} - Q_{cd}\theta_{cd}} = 0.504$$

$$\eta_{mix} = \frac{Q_{ab}\theta_{ab} - Q_{ev}\theta_{ev}}{\Delta b_{mix}} = 0.353$$

$$\eta_{btt} = \frac{Q_{ab}\theta_{ab}}{W_p + Q_{de}\theta_{de} + Q_{ev}\theta_{ev}} = 0.462$$

The obtained results allow determining the exergy performance of the proposed heat transformer, as well as the performance of its components. With this information it is possible to identify which are the most important components in the overall performance of the system.

2.4.4 Environmental Exergy Efficiency ($\eta_{b,env}$)

The environmental impact of the energy conversion processes can be reduced by the increase of the exergy efficiency of these processes. An increase in the exergy efficiency would have as a consequence, a decrease in the consumption of resources and thus a reduction of the wastes and the emissions to the environment. This implies an improvement in the environmental performance of these processes.

The environmental exergy efficiency is defined as the ratio of the final product exergy (or useful effect of a process) to the total exergy of natural and human resources consumed, including all the exergy inputs in a given energy conversion process. This ratio is also an indication of the theoretical potential of future improvements for a process. The environmental exergy efficiency is calculated in agreement with Eq. 2.63:

$$\eta_{b,env} = \frac{B_{product}}{B_{nat,res} + B_{processing} + B_{utilities} + B_{deactivation} + B_{disposal}} \tag{2.63}$$

where:

$B_{product}$ exergy rate of the useful effect of a process

$B_{nat,\ res}$ exergy rate of the natural resources consumed by the processes

$B_{processing}$ exergy rate or flow rate required for extraction and preparation of the natural resources

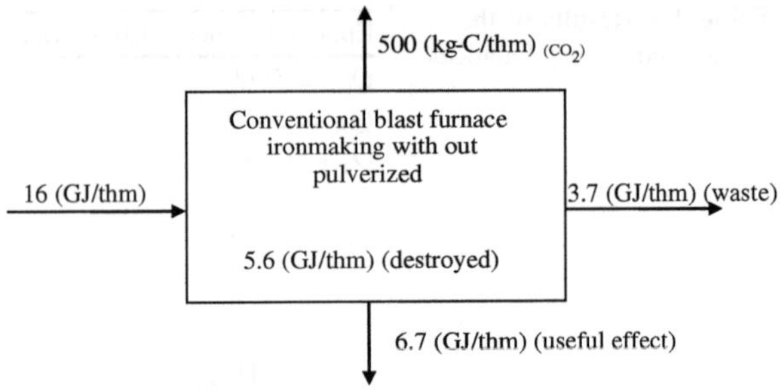

Fig. 2.26 Exergy balance and CO_2 emission of the conventional blast furnace ironmaking without pulverized coal injection (thm = ton of hot metal) [12]

$B_{utilities}$ exergy rate or flow rate required by the utilities of the process

$B_{disposal}$ exergy rate or flow rate related to waste disposal of the process

$B_{deactivation}$ destroyed exergy rate of additional natural resources during waste deactivation

Figure 2.25 illustrates the Grassmann diagram of a given energy conversion process including the terms of Eq. 2.63 and the exergy flow rate of the reject sent to the environment B_{reject}.

The differences found in the literature, between the conventional definitions of the exergy efficiency and the environmental exergy efficiency, are based on the choice of different control volumes for each one of them. This fact determines the inclusion or exclusion of some terms of the total exergy consumed by human and natural resources, with the consequence that the values of the indexes are influenced by the definition of the boundaries of the considered system.

Mora and Oliveira [12] tried to encompass the exergy and environmental evaluation of iron making processes by using the environmental exergy efficiency. This study was based on data presented by Yagi and Akiyama [27]. They applied the exergy analysis to a conventional blast furnace iron making process without and with injection of 110 kg of pulverized coal, and examined the effect of high pulverized coal injection on the total exergy loss and on the emission of greenhouse effect gases. They demonstrated that the total exergy loss decreased when an amount of coke was replaced equally by pulverized coal, improving combustion with oxygen enrichment. With respect to the emission of the greenhouse effect gases, it was found that the injection of pulverized coal decreased by 5 % the total emission of these gases, or 132 kg-C/thm (ton of hot metal). All greenhouse effect gases emitted (CO, CO_2, CH_4, etc.) were determined and converted into equivalent mass of CO_2, for the process without injection of pulverized coal, on the basis of the carbon mass (kg-C). They presented the data of the input, destroyed and lost exergy in GJ/thm (ton of hot metal) for both cases.

The exergy output for both processes was calculated by applying the exergy balance for these two processes. These data together with those obtained from the work of Yagi and Akiyama [27], the destroyed and lost (consumed) exergy and the emission of CO_2, are presented in Fig. 2.26.

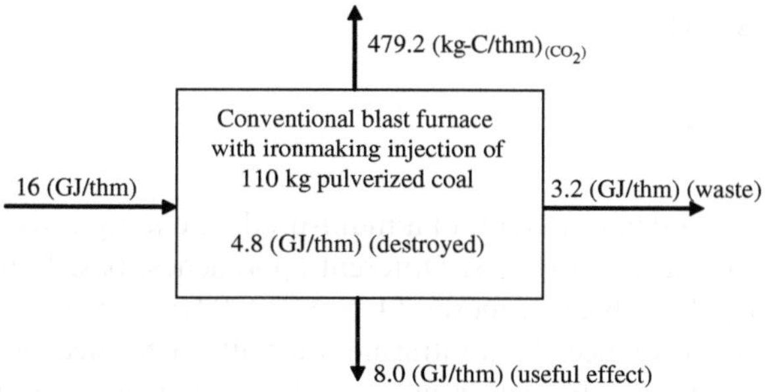

Fig. 2.27 Exergy balance and CO_2 emission of the conventional blast furnace ironmaking with injection of 110 kg pulverized coal (thm = ton of hot metal) [12]

Table 2.5 Values of the environmental exergy efficiency, destroyed and lost exergy, and CO_2 emissions for two ironmaking processes

Process	$\eta_{b,env}$	Destroyed and lost exergy (GJ/thm)	CO_2 (kg-C/thm)
Conventional blast furnace ironmaking without pulverized coal injection	0.42	9.3	500.0
Conventional blast furnace ironmaking with injection of 110 kg pulverized coal	0.50	8.0	479.2

To calculate the emission of CO_2 of the conventional blast furnace ironmaking process with injection of 110 kg of pulverized coal, it was taken as the reference the data of the decreasing rate of emission of 5 % for 132 kg-C/thm of high pulverized coal injection presented by Yagi and Akiyama [27]. In this case, with 110 kg-C/thm of injected coal, it was obtained a decreasing emission of 4.16 %, given a CO_2 emission value of 479.2 kg-C/thm. These results are shown in Fig. 2.27.

Observing the results of Figs. 2.26 and 2.27, it can be concluded that the process that causes the smallest impact in the environment is the conventional blast furnace ironmaking process with injection of 110 kg pulverized coal, due to the fact that it has the smallest destroyed and lost exergy (8.0 GJ/thm), and the smallest emission of CO_2 (479.2 kg-C/thm). These conclusions are coincident with that based on the values of the environmental exergy efficiency, as summarized in Table 2.5.

In this comparison, the values of the environmental exergy efficiency quantifies properly the quality of environmental solutions (pulverized coal injection) that aim at the decreasing the emissions of the greenhouse effect gases in the conventional blast furnace ironmaking process.

2.5 Exergy Costing

2.5.1 Introduction

The production cost determination in a multi-product energy conversion plant has several methodological challenges. Different approaches, based on the use of the concept of exergy have been proposed [4, 5, 8, 22, 24] regarding the cost partition methods in order to reduce the arbitrariness usually presented in the procedures that characterize the cost formation process of the products of a given plant.

Exergy-based cost analysis aims at determining the costs of products and irreversibilities (exergy destroyed) generated in energy conversion processes, by applying cost partition criteria which are function of the exergy content of every energy flow that takes place in the studied process. In this approach one studies the cost formation processes by valuing the products according to its exergy content and the destroyed exergy during the energy conversion processes. This combination of exergy analysis with economic concepts is called thermoeconomic analysis when monetary costs are used and exergoeconomic analysis when exergy costs are employed.

From the point of view of the modeling, simulation and optimization procedures of energy systems, the thermoeconomic or the exergoeconomic analysis aggregates, basically, two sets of equations: the cost balances (in fact, cost rate balances) for components/equipment or processes and the exergy-based cost partition criteria. As already mentioned by Tsatsaronis [22], the steps of thermoeconomic analysis comprise:

- detailed exergy and economic analysis of the components and the overall energy system;
- exergy costing;
- exergoeconomic evaluation of every component; and of the overall system.

It is important to emphasize that the more detailed the cost balances are, the better the results provided by the thermoeconomic or the exergoeconomic analysis will be.

The information generated by the thermoeconomic analysis are quite unlike from those provided by the traditional methods used for assessing the economic viability of energy-saving projects, such as those that determine, the net present value (NPV), the internal rate of return (IRR), and the payback period [3]. The thermoeconomic analysis allows the determination of production costs based on the quality of the energy conversion processes by using a rational criterion of costs distribution along the processes that is the thermodynamic value of each product, or its exergy [5, 22].

2.5.2 Cost Balance

The cost balance for a given component or equipment, operating in steady state, is done by Eq. 2.64:

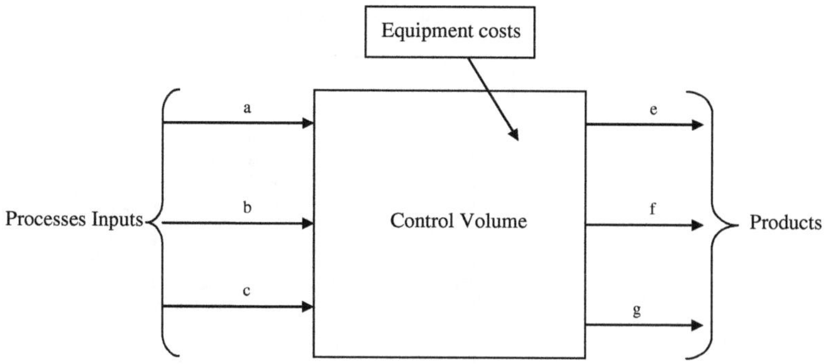

Fig. 2.28 Control volume for cost balance application

$$\Sigma(\text{Rate of expenses}) = \Sigma(\text{Cost rate of the products}) \tag{2.64}$$

The expenses are due to capital costs, operational, and maintenance costs.

For the control volume that embodies equipment and processes, shown in Fig. 2.28, with inputs a, b, and c (such as fuel, electricity, water, air, etc.) and products e, f, and g (such as steam, chilled water, electricity, mechanical power, etc.), the cost balance is done by Eq. 2.65.

$$C_a + C_b + C_c + C_{equipment} = C_e + C_f + C_g \tag{2.65}$$

Every term of Eq. 2.65 has dimension of cost per unit of time ($/s), in a similar form used for mass, energy, and exergy balances in steady state conditions.

The average unitary cost, c_i, is defined by Eq. 2.66 as:

$$c_i = {}^{C_i}/_{N_i} \tag{2.66}$$

In Eq. 2.66, N_i can be a quantity such as mass flow rate, volumetric flow rate, energy rate/energy flow rate, exergy rate/exergy flow rate. Considering the control volume shown in Fig. 2.28, one can rewrite Eq. 2.65 as:

$$c_a N_a + c_b N_b + c_c N_c + C_{equipment} = c_e N_e + c_f N_f + c_g N_g \tag{2.67}$$

When there are more than one product, as shown in Fig. 2.28, the values of c_e, c_f, and c_g cannot be determined only by the cost balance equation, being necessary additional equations based on a cost partition criterion such as:

- the products are for exclusive use of the analyzed plant;
- the products will be sold in the market;
- one product is considered a 'primary product' of the process;
- one product is considered to be a 'by-product' of the process, etc.

In the case there is only one product in the process, the cost balance determines the value of the average unitary cost:

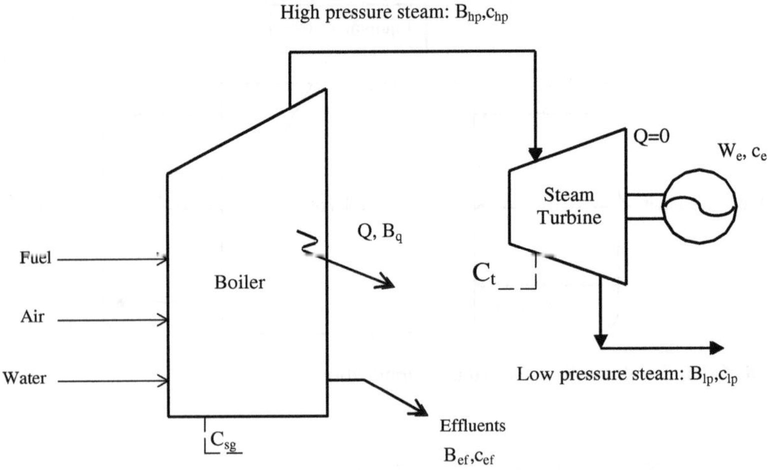

Fig. 2.29 Cogeneration system

$$c_i = \frac{\text{Overall production cost}}{N_i \text{ product units}} \tag{2.68}$$

2.5.3 Exergy-Based Cost Partition Criteria

When applying a thermoeconomic analysis one intends to use exergy-based cost partition criteria for the determination of the production costs in a multi-products equipment, components, or processes. A simple and effective approach to this problem consists in utilizing the equality method or the extraction method.

These two methods will be described by taking a cogeneration plant operating with a backpressure steam turbine, as shown in Fig. 2.29. In this plant there are two products: electricity and low pressure steam for heating purposes.

The cost balances for the steam generator and the steam turbine are given, respectively, by Eqs. 2.69 and 2.70.

$$c_{\text{hp}}B_{\text{hp}} + c_{\text{ef}}B_{\text{ef}} + c_{\text{q}}B_{\text{q}} = c_{\text{fuel}}B_{\text{fuel}} + c_{\text{air}}B_{\text{air}} + c_{\text{water}}B_{\text{water}} + C_{\text{sg}} \tag{2.69}$$

$$c_{\text{e}}W_{\text{e}} + c_{\text{lp}}B_{\text{lp}} = c_{\text{hp}}B_{\text{hp}} + C_t \tag{2.70}$$

In Eqs. 2.69 and 2.70, the exergy rates (B_{q} and W_{e}) and exergy flow rates (B_{hp}, B_{ef}, B_{fuel}, B_{air}, B_{water}, B_{lp}) have been previously determined by the application of the energy and exergy balances to the steam generator and turbine.

The terms C_{sg} and C_t are, respectively, the cost rates of the steam generator and steam turbine, including capital, operational and maintenance costs of these components, obtained by an economic analysis. The costs of the inputs, fuel, water, and air, are considered known in the forthcoming analysis.

For the sake of simplicity, it will be taken into account that B_q and B_{ef} are wastes, or they are not products of the cogeneration plant. Then:

$$c_{ef} = c_q = 0 \tag{2.71}$$

The determination of the exergy costs of the high pressure steam (c_{hp}), low pressure steam (c_{lp}), and electricity (c_e) requires three independent equations. As there are only two cost balances, one for each component, the third equation will be provided by a cost partition criterion.

Equality method

In this method the analyzed equipment/component must supply all exergy demands and consequently all products have the same exergy average cost. Then all the costs are divided among the products as a function of their exergy content. In the case of the steam turbine of Fig. 2.29, both products, electricity, and low pressure steam, must supply the exergy needs of the plant. In this way, the steam turbine capital, operational, and maintenance costs are shared to the two products, giving the additional equation:

$$c_e = c_{lp} \tag{2.72}$$

And the average cost of the products is:

$$c_e = c_{lp} = c_{hp}\left(\frac{B_{hp}}{B_{lp} + W_e}\right) + \frac{C_t}{(B_{lp} + W_e)} \tag{2.73}$$

In this case, one can define an exergy efficiency of the steam turbine as given by Eq. 2.74:

$$\eta_{\text{b turbine}} = \frac{W_e + B_{lp}}{B_{hp}} \tag{2.74}$$

Extraction method

In this method, the studied equipment or component has only one function and the product of this function is charged with its capital, operational, and maintenance costs. In this way, the user of this product will pay the exergy rate spent as well as the capital, maintenance, and other operational costs. Then, the product of the steam turbine of the cogeneration plant is electricity, and the exergy cost of the low pressure steam is not affected by the existence of the turbine. This consideration provides the additional equation:

$$c_{hp} = c_{lp} \tag{2.75}$$

It must be emphasized that although having the same exergy costs, the high pressure, and low pressure steam have different mass-based costs, because:

$$(c_{hp})_m = c_{hp} b_{hp} \tag{2.76}$$

and,

$$(c_{\mathrm{lp}})_{\mathrm{m}} = c_{\mathrm{lp}} b_{\mathrm{lp}} \tag{2.77}$$

Then,

$$(c_{\mathrm{hp}})_{\mathrm{m}} > (c_{\mathrm{lp}})_{\mathrm{m}}$$

This result characterizes the higher thermodynamic value of the high pressure steam, and consequently, its higher mass-based cost.

In the case, the studied steam turbine is a condensation-extraction one, this criterion would give:

$$c_{\mathrm{lp}} = c_{\mathrm{cd}} \tag{2.78}$$

and, as shown previously:

$$(c_{\mathrm{lp}})_{\mathrm{m}} > (c_{\mathrm{cd}})_{\mathrm{m}}$$

Taking Eq. 2.75 in the cost balance of the steam turbine, it is determined as the electricity cost:

$$c_{\mathrm{e}} = c_{\mathrm{hp}} \left(\frac{B_{\mathrm{hp}} - B_{\mathrm{lp}}}{W_{\mathrm{e}}} \right) + \frac{C_{\mathrm{t}}}{W_{\mathrm{e}}} \tag{2.79}$$

It is worthy of note that when applying the extraction method, the only product of the steam turbine is electricity (or mechanical work) and then its exergy efficiency is now defined as:

$$\eta_{\mathrm{b\,turbine}} = \frac{W_{\mathrm{e}}}{B_{\mathrm{hp}} - B_{\mathrm{lp}}} \tag{2.80}$$

2.5.4 Application of the Thermoeconomic Analysis

The presented study describes how an exergy and thermoeconomic analysis can be applied to a cogeneration plant in order to determine the production cost of steam and electricity. Figure 2.30 represents this plant that supplies electricity and steam to an industrial process as described by Pellegrini et al. [15]

The plant operational data are the following:

- Steam generation (section 1): 490 °C at 100 bar;
- Percent excess air in the boiler: 100 %
- Methane flow rate: 1 kg/s;
- Methane cost: US$ 10.00/MWh (US$ 144.00/t)
- Methane lower heating value: 50,146 kJ/kg;
- Stack gases temperature: 240 °C;

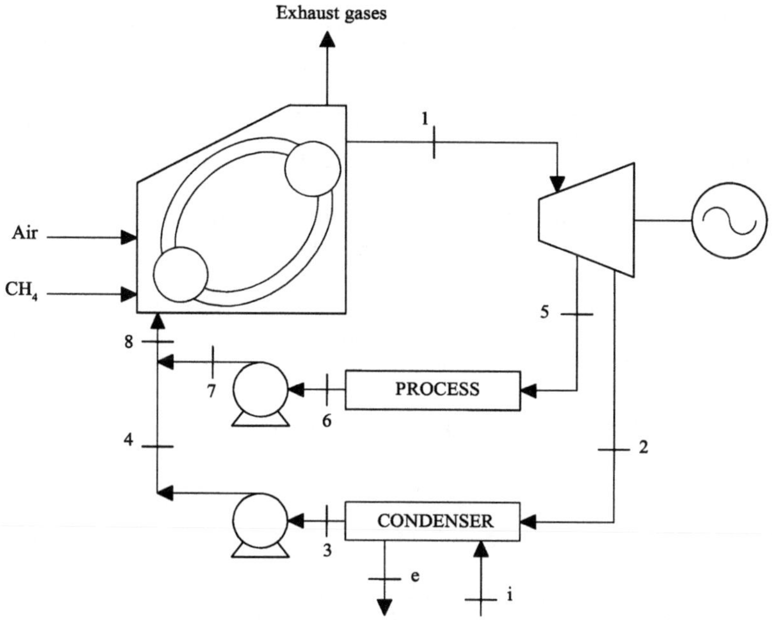

Fig. 2.30 Studied cogeneration plant [15]

- Air temperature: 25 °C;
- Water temperature at the condenser inlet: 25 °C;
- Water temperature at the condenser outlet: 45 °C;
- Extraction mass ratio: $m_5/m_1 = 0.5$;
- Steam state at turbine section 5: 240 °C@2 bar;
- Condensation pressure: 0.07 bar;
- Quality at turbine section 2: 0.97;
- Pumps Isentropic efficiency: 80 %.
- Environment conditions: 25 °C and 1 bar

Based on these operational data and Eqs. 2.81 and 2.82, applied to each component of the plant and to the overall plant, it is possible to determine the thermodynamic properties of water (the working fluid) at every section indicated in Fig. 2.30, as shown in Table 2.6:

Mass balance:

$$\sum_{\text{inlet}} m_{\text{i}} = \sum_{\text{outlet}} m_{\text{e}} \tag{2.81}$$

Energy balance:

$$\sum_{\text{inlet}} m_{\text{i}} h_{\text{i}} + Q_{\text{VC}} = W_{\text{VC}} + \sum_{\text{outlet}} m_{\text{e}} h_{\text{e}} \tag{2.82}$$

Table 2.6 Cogeneration plant data [15]

Section	Mass flow rate (kg/s)	Pressure (bar)	Temperature (K)	Specific enthalpy (kJ/kg)	Specific entropy (kJ/kg K)	Specific exergy (kJ/kg)
1	13.94	100.00	763.20	3348.00	6.56	1396.00
2	6.97	0.07	312.20	2499.00	8.04	106.10
3	6.97	0.07	312.20	163.40	0.56	1.24
4	6.97	100.00	313.00	175.90	0.57	11.41
5	6.97	2.00	513.20	2950.00	7.67	668.50
6	6.97	2.00	393.40	504.70	1.53	53.06
7	6.97	100.00	394.80	517.70	1.54	64.06
8	13.94	100.00	354.10	346.80	1.08	29.36

Then Eqs. 2.83–2.85 can be solved to determine the energy and exergy performance of the plant.

Exergy balance:

$$\sum_{\text{inlet}} m_{\mathrm{i}}\, b_{\mathrm{i}} + Q_{\mathrm{VC}}\left(1 - \frac{T_0}{T}\right) = W_{\mathrm{VC}} + \sum_{\text{inlet}} m_e\, b_e + I_{\mathrm{VC}} \tag{2.83}$$

Energy efficiency:

$$\eta_{\mathrm{e}} = \frac{W_{\mathrm{net}} + Q_{\mathrm{process}}}{m_{\mathrm{CH_4}}\,\mathrm{LHV}} \tag{2.84}$$

Exergy efficiency:

$$\eta_{\mathrm{b}} = \frac{W_{\mathrm{net}} + B_{\mathrm{process}}}{m_{\mathrm{CH_4}}\, b_{\mathrm{CH_4}}} \tag{2.85}$$

Table 2.7 presents the performance parameters of the cogeneration plant.

The components capital cost can by estimated by using, for every component, the so-called exponential form as described by Bohem [2] and presented by the cost curve given in Eq. 2.86, where C_{r} is the cost of reference size component, C_{eq} is the cost of a size of interest, α is the exponent on cost size relationship, S_{r} is the component reference size, and S is the component size of interest (see Table 2.8).

$$C_{\mathrm{eq}} = C_{\mathrm{r}}\left(\frac{S}{S_{\mathrm{r}}}\right)^{\alpha} \tag{2.86}$$

The following hypotheses are considered in the determination of the components cost rates:

- Capital recovery period (N): 20 years;
- Interest rate (i): 15 % per year;
- Annual operating hours: 8000;
- Annual operational and maintenance factor ($f_{\mathrm{O\&M}}$): 1.06

Table 2.7 Performance parameters of the cogeneration plant [15]

Performance Parameter
Turbine power: 8,684 kW
Pumping power: 87 and 90 kW
Net power plant: 8,507 kW
Process heat rate: 17,041 kW
Process exergy rate: 4,288 kW
Plant energy efficiency: 51 %
Plant exergy efficiency: 25 %

Table 2.8 Cost components parameters [15]

Component	α	Cr (U$1000)	Sr	S	
Steam generator	0.59	320	10_5	m	(lb/h)
Steam turbine	0.68	25	10_3	W	(kW)
Condenser	0.55	3	10	W	(kW)
Pumps	0.58	7.5	100	W	(kW)

Table 2.9 Components costs [15]

Equipment	C_{eq} (US$)	C_{eqt} (US$/s)
Steam generator	339603	0.001997
Steam turbine	108713	0.000639
Condenser	124001	0.000729
Pump 1	6941	0.000040
Pump 2	7070	0.000041

With these parameters the cost rates are obtained by Eq. 2.87 and 2.88 and the components values are presented in Table 2.9:

$$C = \frac{C_{eq} f_a f_{O\&M}}{3600\,8000} \tag{2.87}$$

$$f_a = \frac{i}{1 - (1+i)^{-N}} \tag{2.88}$$

With this information the cost balances can now be written for every component:

$$\sum_{\text{inlet}} c_i B_i + C = \sum_{\text{outlet}} c_e B_e \tag{2.89}$$

In the cost balance equation of the steam turbine there are three unknowns: c_2, c_5, and c_e, being necessary two additional equations to establish the relationships among c_1 and the three unknown variables. In this study both criteria described before will be used: the equality and the extraction methods, aiming at showing the differences they cause in the exergy-based cost values distribution.

The equality method gives the following equations:

$$c_2 = c_e \tag{2.90}$$

$$c_2 = c_5 \tag{2.91}$$

The extraction method considers that the product of the turbine is the electricity, then:

$$c_1 = c_2 \tag{2.92}$$

and also:

$$c_1 = c_5 \tag{2.93}$$

The process cost balance also requires a relationship between c_5 and c_6, because it is considered in the analysis as a black box (there is not available information about the use of the thermal exergy of the condensing steam when it changes its thermodynamic state from state 5 to state 6). Nevertheless, as the user of this transferred exergy from the steam is the process, it is charged with this 'consumed exergy', and the extraction method can be applied here. Then the additional equation is:

$$c_5 = c_6 \tag{2.94}$$

The cost of the expended exergy in the process, c_p, is determined by the process cost balance:

$$c_5 B_5 = c_p B_p + c_6 B_6 \tag{2.95}$$

Table 2.10 presents the results given by the thermoeconomic analysis with both cost partition criteria.

As expected, the extraction method causes a higher value of the electricity generated in the steam turbine because this criterion charges this product with the turbine costs as well as the cost of the destroyed exergy inside the turbine. On the other hand, the exergy-based costs given by the equality method for electricity and exergy transferred to the process are the same, as a consequence of giving the same importance to both products.

The annual overall cost rate of the cogeneration plant is done by Eq. 2.96.

$$C_{\text{overall}} = \sum_{\text{component}} C + c_{CH_4} B_{CH_4} = \text{US\$4, 258,013/year} \tag{2.96}$$

This cost rate must be distributed between both products of the plant: electricity and heat transferred to the process, according to the chosen cost partition criteria:

$$C_{\text{total}} = C_{\text{electricity}} + C_{\text{heat}} \tag{2.97}$$

Table 2.10 Results of the thermoeconomic analysis [15]

Cost	Extraction method	Equality method
1 (US$/t)	11.38	11.63
2 (US$/t)	0.86	1.23
3 (US$/t)	0.97	1.33
4 (US$/t)	1.14	1.48
5 (US$/t)	5.45	7.72
6 (US$/t)	0.43	0.61
7 (US$/t)	0.61	0.77
8 (US$/t)	0.88	1.13
Electricity (US$/MWh)	47.77	41.60
Process heat (US$/MWh)	29.35	41.60

By the extraction method:

$$C_{\mathrm{total}} = 47.77 \cdot 8.5 \cdot 8{,}000 + 29.35 \cdot 4.3 \cdot 8{,}000 = 4{,}258{,}000/\mathrm{year}$$

By the equality method:

$$C_{\mathrm{total}} = 41.60 \cdot 8.5 \cdot 8{,}000 + 41.60 \cdot 4.3 \cdot 8{,}000 = 4{,}259{,}840/\mathrm{year}$$

The small difference of overall values is due to some approximations done during the calculations.

This study of a simple cogeneration plant evidences the importance of the thermoeconomic analysis in the determination of the utilities (process steam/heat and electricity) production costs, by means of using the concept of exergy as the basis to valuate each product of a given energy conversion plant.

Nevertheless, the useful information thermoeconomics can provide to a process analyst, it must be pointed out that it is simply one method of cost evaluation and not the panacea of the production cost methods. Even with the use of thermoeconomics, one is subjected to a given sort of arbitrariness due to the particular cost partition criteria chosen in the analysis.

2.6 Exergy and Renewability Analysis

In the last years, several discussions have been conducted about sustainable development, greenhouse gas emissions, environmental impact, and renewability of energy sources. However, until now, the concept of renewability has been associated to mass and energy balances, not taking into account the reduction of the quality of the energy (exergy destruction) related to energy conversion processes. The traditional definition of sustainability, that calls for policies and strategies that meet society's present needs without compromising the ability of future generations to meet their own needs [23], does not provide a rational way to quantify this ability.

As stated by different authors [20, 26], exergy, which originates from the contrast between sun and space, drives flows of energy and matter on the surface of the Earth. This exergy input is destroyed in order to keep the natural cycles responsible for recycling materials in the surface, and a small part is stored as fossil fuels and mineral ores. Recycling takes time and exergy to be accomplished, but total recycling is not possible due to the second Law of Thermodynamics.

Currently, human development is based on the use of fossil fuels at a greater rate than that at which the deposit of fossil fuels have been generated. Also, since total recycling is not possible, it is imperative to seek for technologies that make better use of exergy available from all sources, including the so-called "renewable sources". Thus, two aspects should be taken into account whenever discussing renewability of any product:

- Origin of its energy source;
- Efficiency of the energy conversion processes.

The use of the concept of reversible processes, the one that having occurred can be reversed to the initial system and surroundings states, in the analysis of the renewability of energy conversion processes can contribute to develop such analysis in a rational basis by using thermodynamic parameters.

Based on these aspects, one can define a renewability exergy index [25] that takes into account the exergy associated to the useful products of a given energy conversion process (or a set of processes), the destroyed exergy, the exergy associated to the fossil fuels required, the needed exergy to deactivate the wastes, and the exergy of by-products and not treated wastes, according to Eq. 2.98:

$$\lambda = \frac{\sum B_{\text{product}}}{B_{\text{fossil}} + B_{\text{destroyed}} + B_{\text{deactivation}} + B_{\text{disposal}} + \sum B_{\text{emissions}}} \tag{2.98}$$

where:

- B_{product} represents the net exergy associated to the products and by-products.
- B_{fossil} is the non-renewable exergy consumed on production processes chain.
- $B_{\text{destroyed}}$ is the exergy destroyed inside the system, punishing the process for its inefficiencies.
- $B_{\text{deactivation}}$ is the deactivation exergy for treating wastes, when they are carried to equilibrium conditions with the environment. It accounts for exergy required for passing the streams leaving the system, considered as wastes, to no harmful environmental conditions.
- B_{disposal} is exergy rate or flow rate related to waste disposal of the process.
- $B_{\text{emissions}}$ is the exergy of wastes that are not treated or deactivated.

Depending on the value of the renewability exergy index, it indicates that:

- Processes with $0 \leq \lambda < 1$ are environmentally unfavorable.
- For internal and externally reversible processes with non-renewable inputs, $\lambda = 1$.

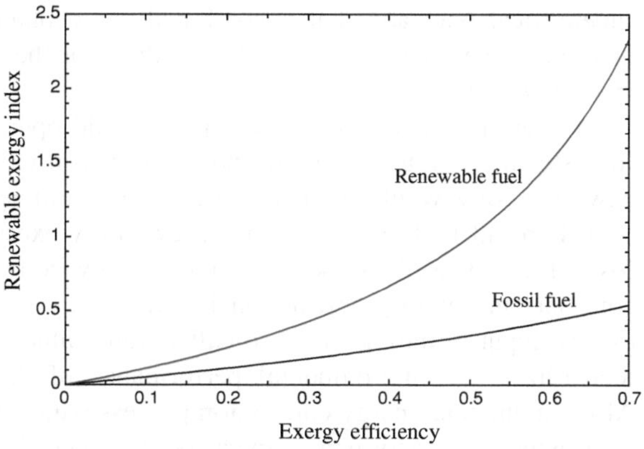

Fig. 2.31 Dependence of λ with η_b for thermal power plants using fossil and renewable fuels

- If $\lambda > 1$, the process is environmentally favorable, and additionally, increasing λ implies that the process is more environmentally friendly.
- When $\lambda \to \infty$, it means that the process is reversible with renewable inputs and no wastes are generated.

When λ is higher than 1, the exergy of the products could be used to restore the environment to its conditions before the process and yet have a net output of exergy. In this sense, the natural cycles would only be responsible for the recycling of the renewable sources, considered as those able to be recycled by natural cycles without causing immediate harm to the environment in human lifetime basis.

A first application of the renewability exergy index is its determination for thermal power plants. For these systems, λ is obtained as a function of the exergy efficiency of the power plant, taking into account that $B_{\text{deactivation}}$ and B_{disposal} are zero, as shown by Eqs. 2.99, 2.100 (when a fossil fuel is utilized), 2.101 (when a renewable fuel is utilized) and Fig. 2.31, where it is shown the behavior of λ with η_b for both fuels.

$$\eta_{\text{b}} = \frac{W_{\text{net}}}{B_{\text{fuel}}} \tag{2.99}$$

$$\lambda_f = \frac{W_{\text{net}}}{B_{\text{fuel}} + B_{\text{dest}} + B_{\text{fluegases}}} = \frac{\eta_{\text{b}}}{(2 - \eta_{\text{b}})} \tag{2.100}$$

$$\lambda_r = \frac{W_{\text{net}}}{B_{\text{dest}} + B_{\text{fluegases}}} = \frac{\eta_{\text{b}}}{(1 - \eta_{\text{b}})} \tag{2.101}$$

Note that, one can verify that, as stated before:

- $\lambda = 1$ when $\eta_{\text{b}} = 1$, for a reversible power plant using fossil fuel, and
- $\lambda \to \infty$ when $\eta_{\text{b}} = 1$, for a reversible power plant using renewable fuel

Considering that a conventional power plant using coal as fuel has thermal efficiency of about 35 % for a Rankine subcritical plant and up to 50 % for a supercritical one, and that a combined cycle plant using natural gas can have thermal efficiency up to 60 %, the λ values for these types of power plants range from 0.18 to 0.43.

In case the supercritical power plant could operate with sugarcane bagasse as fuel (steam generate @ 600 °C and 294 bar, as it will be discussed in Chap. 6), the new value of λ would be 0.38, for an exergy efficiency of 0.28 [14].

This result highlights that the renewability exergy index penalizes the use of fossil fuels as well as the inefficient energy conversion processes. Or, since λ considers the exergy destruction in the energy conversion process, even if the exergy input comes from a so-called renewable source, an inefficient process jeopardizes the environmental performance of the energy conversion process. Also, an efficient energy conversion process with fossil exergy input may have an environmental performance even better than those with renewable sources, depending on the comparative values of exergy destroyed of the processes.

Finally, it is important to note that this index evaluates the renewability of a given energy conversion process and not a product. In this sense it is questionable the concept of renewable product usually employed to characterize energy sources and fuels.

References

1. Biot JB (1816) Traité de physique expérimentale et mathématique, tome 4, p 739, Paris, Deterville
2. Boehm RF (1987) Design analysis of thermal system. Wiley, New York
3. Eastop TD, Croft DR (1996) Energy efficiency for engineers and technologists. Longman, London
4. Erlach B, Serra L, Valero A (1999) Structural theory as standard for thermoeconomics. Energ Convers Manag 40:1627–1649
5. Gaggioli RA (1983) Efficiency and costing. ACS Symposium Series 235
6. Gouy G (1889) Sur lénergie utilisable. J Phys Theor Appl 8:501–518
7. Grassmann P (1959) The exergy and the flow diagram of the technically available power. Allg Wärmetech 9:79–86 (in German)
8. Kotas TJ (1995) The exergy method of thermal plant analysis. Krieger Publishing Company, Florida
9. Le Goff P, Rivero RR, Oliveira S Jr et al. (1990) Application of the enthalpy—carnot factor to the exergy analysis of distillation processes. In: Fundamentals of thermodynamics and exergy analysis, AES 19:21–28, New York, ISBN: 0-7918-0580-8
10. Locqueneux R (2009) Histoire de la thermodynamic classique : de Sadi Carnot à Gibbs. Belin, Paris
11. Moore DF (1981) Thermodynamic principles of energy degrading. The Macmillan Press, London
12. Mora BCH, Oliveira S Jr (2004) Exergy efficiency as a measure of the environmental impact of energy conversion processes. In: Proceedings of the 17th international conference on efficiency, costs, optimization, simulation and environmental impact of energy systems, Guanajuato
13. Oliveira Jr S (1991) Upgrading industrial thermal effluents—exergetic, entropic and economic analysis. Ph.D. Thesis. Polytechnic National Institute of Lorraine, Nancy (In French)

14. Pellegrini LF (2009) Analysis and thermo-economic and environmental optimization applied to the combined production of sugar, alcohol and electricity. Ph.D. Dissertation—Polytechnic School of the University of São Paulo, São Paulo, Brazil (in Portuguese)
15. Pellegrini LF, Costa RP, Oliveira Jr S (2005) Allocation of costs in energy systems: thermoeconomy as a basis. In: Proceedings of the 25th National Meeting of Production Engineering, Porto Alegre (in Portuguese)
16. Rivero RR (1993) L'analyse d'exergie: application à la distillation diabatique et aux pompes à chaleur à absorption. Ph.D. Thesis. Polytechnic National Institute of Lorraine, Nancy
17. Rivero RR, Rendón C, Monroy L (1999) The exergy of crude oil mixtures and petroleum frac- tions: calculation and application. Int J Thermodyn 2:115–123 (In French)
18. Stodola A (1898) Die Kreisprozesse der Gasmaschinen. Zeistschrift der VDI 23:1086–1091
19. Szargut J, David RM, Steward F (1988) Exergy analysis of thermal, chemical, and metallurgical processes. Hemisphere Publishing, New York
20. Szargut J (2002) Anthropogenic and natural exergy losses (exergy balance of the Earth's surface and atmosphere). Energy 28:1047–1054
21. Thring MW (1944) The virtue of energy, its meaning and practical significance. J Inst Fuel 17:116–123
22. Tsatsaronis G (1993) Thermoeconomic analysis and optimization of energy systems. Prog Energ Combust 19:227–257
23. United States environmental production agency. http://www.epa.gov/sustainability/basicinfo.htm#sustainability, Cited in 16 May 2011
24. Valero A, Lozano MA, Muñoz M (1986) A general theory of exergy saving: part I on the exergetic cost; Part II on the thermoeconomic cost; Part III exergy saving and thermoeconomics. In: Gaggioli RA (ed) Computer-aided engineering of energy systems—sec- ond law analysis and modelling. ASME, NY
25. Velásquez HI, Pellegrini LF, Oliveira S Jr (2008) Ethanol and sugar production process from sugarcane: renewability evaluation. In Proceedings of the 12th Brazilian congress of thermal science and engineering, Belo Horizonte
26. Wall G (1977) Exergy—a useful concept within resource accounting. Report no. 77-42, Institute of theoretical physics, Götenborg
27. Yagi J, Akiyama T (2001) Exergy analysis of iron making processes for reduction of carbon dioxide emission. In: Proceedings Japan-brazil symposium on dust processing-energy-environment in metallurgical industries, São Paulo

Chapter 3
Exergy and Thermoeconomic Analysis of Power Plants, Refrigeration and Polygeneration Systems

Symbols

b	Specific exergy (kJ/kg)
B	Exergy rate (kW)
c	Specific cost (US$/kWh, US$/kJ or US$/t)
C	Cost rate ($/s)
C_{oi}	Cost of equipment i (US$)
$C_{equip,i}$	Equipment i cost rate (US$/s)
COP	Coefficient of performance
C_{turb}	Steam turbine cost rate (US$/s)
E	Energy rate (kW)
f	The fraction of the rejected heat of the heat engine that is sent to the refrigeration system
f_i	Ratio of the exergy supplied to component i to the exergy consumed by the whole plant
f_l	Load factor
f_{om}	Annual operational and maintenance factor
f_t	Time factor
I	Investment cost rate (US$/h)
LHV	Lower heating value (kJ/kg)
m	Mass flow rate (kg/s)
n	Annual interest rate
N_h	8760 h/year
P	pressure (bar)
P_o	Reference pressure (bar)
Q	Heat rate (kW)
r	Capital recovery period (year); parameter defined by Eq. 3.24
T_o	Reference temperature (K)
W	Power (kW)

S. de Oliveira Jr., *Exergy*, Green Energy and Technology,
DOI: 10.1007/978-1-4471-4165-5_3, © Springer-Verlag London 2013

Greek symbols

α	Relation between chemical exergy and lower heating value
β	Relation between heat rate and power
Δ	Variation
η_b	Exergy efficiency
η_e	Energy efficiency
$\theta, \overline{\theta}$	Carnot factor, average Carnot factor

Subscripts

abs	Absorption refrigerating system
air	Combustion air
b	Exergy
c	Compressor
cc	Combined cycle; combustion chamber
cd	Condenser
chilled water	Related to chilled water
cp	Compressor
cpi	Compressor inlet
cpo	Compressor outlet
crs	Compression refrigerating system
e	Electricity, Energy
ev	Evaporator
excess	Excess electricity
eg	Exhaust gas
equip	Equipment
fuel	Related to fuel
fuelcc	Fuel consumption in the gas turbine combustion chamber
fuelhrsg	Fuel consumption in the heat recovery steam generator
G	Related to the whole plant
gas	Natural gas
gases	Combustion gases
ge	Generator of the absorption chiller
gt	Gas turbine
hrsg	Heat recovery steam generator
i	Inlet, component *i*
o	Outlet
overall	Related to the whole plant
proc	Process
p	Pump; process
pump	Pump
pump i	Pump inlet
pump o	Pump outlet
plant	Related to plant
process	Related to process

products	Combustion products
$p1$	Steam demanded by process 1
$p2$	Steam demanded by process 2
q, Q	Heat/chilled water
sb	supplementary burning
sc	Steam cycle
st	Steam turbine
steam	Steam
t	Turbine
ti	Turbine inlet
to	Turbine outlet

Abbreviations

ABS	Absorption chiller
CC	Combustion chamber
CHP	combined heat and power unit
COND	Condenser
CONDP	Condensate pump
CIRCP	Circulating pump
CP	Air compressor
CT	Combustion turbine
CT	Cooling tower
D	Duct, Dimension
DB	Supplementary firing module
DEAR	Deaerator
ECON	Economizer
EVAP	Evaporator
FH	Fuel heater
GEE_q	Gas engine with equality method
GT	Gas turbine; turbine of the gas turbine
GTE_q	Gas turbine with equality method
GTE_x	Gas turbine with extraction method
HP	High pressure
HPCON	High pressure economizer
HPECO2	High pressure economizer 2
HPEVAP	High pressure evaporator
HPPUMP	High pressure feed pump
HPSHR	High pressure superheater
HPSHT1	High pressure superheater 1
HPST	High pressure section
HRSG	Heat recovery steam generator
IP	Intermediate pressure
IPCON	Intermediate pressure economizer
IPPUMP	Intermediate pressure feed pump

IPST	Intermediate pressure section
IPSHT	Intermediate pressure superheater
IPSTH2	Intermediate pressure superheater 2
IPVAP	Intermediate pressure evaporator
LP	Low pressure
LPEVAP	Low pressure evaporator
LPSHT	Low pressure superheater
LPST	Low pressure section
MMBtu	10^6 Btu
OOC	Original operating condition
P	Pump
RH	Reheater
SHT	Superheater
ST	Steam Turbine
STE_q	Steam turbine with equality method
STE_x	Steam turbine with extraction method
TCR	Total cost rate (US$/h)
TR	Ton of refrigeration (3.5 kW)
WTHT	Water heater

3.1 Introduction

The increasing interest in combined heat and power systems as well as combined cycle plants points out the need for identifying the best components configurations in order to maximize their exergy efficiency.

The exergy analysis of thermomechanical conversion plants aims to characterize how the fuel exergy is used and destroyed in the energy conversion processes that take place in these plants.

3.2 Exergy Analysis of Cogeneration and Combined Cycle Plants

3.2.1 Exergy-Based Performance Parameters

Cogeneration systems are based on power generation plants that are modified to allow the utilization of the exergy associated to their thermal wastes.

The exergy performance of cogeneration systems can be derived from a general performance definition as discussed in Chap. 2, and presented in Eq. 3.1:

$$\eta = (\text{useful effect})/(\text{driving energy, exergy, etc.}) \quad (3.1)$$

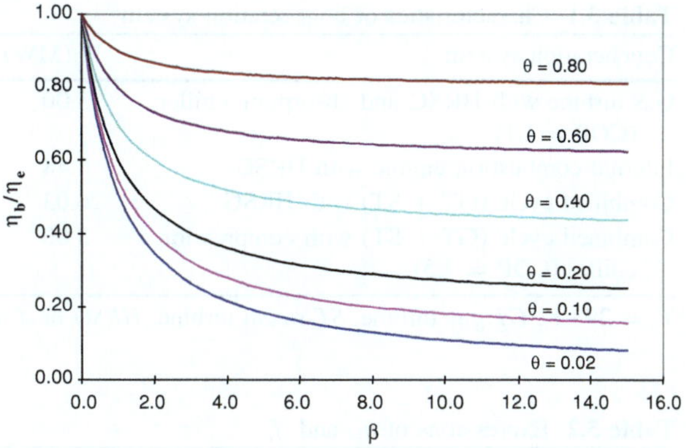

Fig. 3.1 Behavior of η_b/η_e as a function of β for several values of θ [13]

Thus, the expressions of the energy (η_e) and exergy (η_b) performances are:

$$\eta_e = (W + Q) / E_{fuel} \tag{3.2}$$

$$\eta_b = (W + B_Q)/B_{fuel} \tag{3.3}$$

where B_Q is the exergy transferred to a process for heating or cooling (chilled water "production") purposes.

Combining Eqs. 3.2 and 3.3, introducing the heat to work ratio, $\beta = Q/W$, the Carnot factor of the process, $\theta_p = 1 - T_0/T_p$, considering $B_Q = \theta_p Q$ and $\alpha = B_{fuel}/E_{fuel}$, one obtains:

$$\frac{\eta_b}{\eta_e} = \left(\frac{1 + \theta\beta}{1 + \beta}\right)\frac{1}{\alpha} \tag{3.4}$$

This expression allows to obtain easily the relation between the two performances parameters, given β, T_o, T_p, and knowing α.

Figure 3.1 shows the behavior of η_b/η_e as a function of β having θ as parameter, for a fuel with $\alpha = 1$. It is observed that when $\beta \to 0$ the relation between the performances tends to one because the useful effect of the system is the power generation (pure exergy). When $\beta \to \infty$ the relation between performances tends to θ, because the useful effect is essentially a heat transfer (heating/cooling). It can also be observed that for decreasing θ (in module), i.e., heat transfer processes with temperatures near to T_o (which is the case of an air conditioning systems), the relation between the performances drops for a same β, due to the lower exergy associated to the heat transfer.

Data presented in Fig. 3.1 are easily corrected for a fuel with $\alpha \neq 1$ by simply dividing the value of η_b/η_e of the graphic by the particular value of α (that can be obtained, for example, in the correlations presented by Szargut [1] and [2].

Table 3.1 shows the values of W, Q, $\bar{\theta}$ (average Carnot factor), η_e, and η_b for four types of cogeneration systems.

Table 3.1 Characteristics of cogeneration system

Cogeneration system	W (MW)	Q (MW)	$\lvert\bar{\theta}\rvert$	β	η_e	η_b
Gas turbine with HRSG and absorption chiller (COP = 1.1)	3.00	5.39	0.057	1.80	0.84	0.33
Internal combustion engine with HRSG	7.48	6.70	0.215	0.90	0.75	0.47
Combined cycle (GT + ST) with HRSG	26.03	32.21	0.273	1.24	0.77	0.46
Combined cycle (GT + ST) with compression chiller (COP = 4.5)	3.00	6.30	0.057	2.10	0.93	0.34

$T_o = 298$ K, *GT* gas turbine, *ST steam* turbine, *HRSG* heat recovery steam generator

Table 3.2 Expressions of η_{bi} and f_i

Component / parameter	η_{bi}	f_i
Compressor	$[\eta_b]_{cp} = \frac{B_{cpo} - B_{cpi}}{W_{cp}}$	$f_{cp} = \frac{W_{cp}}{B_{fuel}}$
Pump	$[\eta_b]_{pump} = \frac{B_{pump\,o} - B_{pump\,i}}{W_{pump}}$	$f_{pump} = \frac{W_{pump}}{B_{fuel}}$
Turbine	$[\eta_b]_t = \frac{W_t}{B_{ti} - B_{to}}$	$f_t = \frac{B_{ti} - B_{to}}{B_{fuel}}$
Combustion chamber	$[\eta_b]_{cc} = \frac{B_{products}}{B_{fuelcc} + B_{air}}$	$f_{cc} = \frac{B_{fuelcc} + B_{air}}{B_{fuel}}$
Heat recovery steam generator	$[\eta_b]_{hrsg} = \frac{\Delta B_{steam}}{B_{fuelhrsg} + \Delta B_{gases}}$	$f_{hrsg} = \frac{\Delta B_{gases} + B_{fuelhrsg}}{B_{fuel}}$
Compression refrigerating system	$[\eta_b]_{crs} = \frac{Q_{ev}\theta_{ev}}{W_{cp}}$	$f_{crs} = \frac{W_{cp}}{B_{fuel}}$
Absorption refrigerating system	$[\eta_b]_{abs} = \frac{Q_{ev}\theta_{ev}}{Q_{ge}\theta_{ge}}$	$f_{abs} = \frac{Q_{ge}\theta_{ge}}{B_{fuel}}$

In Table 3.1 it is interesting to notice the higher values of η_b for combined cycle and internal combustion engine systems with HRSG, due to the relative importance of the produced power.

Based on Eq. 3.1 it is possible to define the exergy efficiencies of the main components of a cogeneration and a power plant (compressor, turbine, pump, combustion chamber, steam generator, heat recovery steam generator, compression refrigerating system, and absorption refrigerating system). The expressions of these efficiencies are shown in Table 3.2.

To characterize the importance of each component in the exergy consumption of the plant it is introduced the factor f_i ([3]; see also [4]) defined as the relation between the exergy consumed by each component and the exergy consumed by the plant. The expressions of f_i for the main components of a cogeneration and power plant are summarized in Table 3.2.

With the definitions of η_{bi} and f_i it is possible to obtain an expression that relates the overall exergy efficiency of the plant, $\eta_{overall}$, with η_{bi}, and f_i.

For a cogeneration plant composed of a Rankine cycle with a back-pressure steam turbine the expression of $\eta_{b\ overall}$ is:

$$\eta_{b\,overall} = \frac{W_{st} - W_{pump} + B_q}{B_{fuel}} \tag{3.5}$$

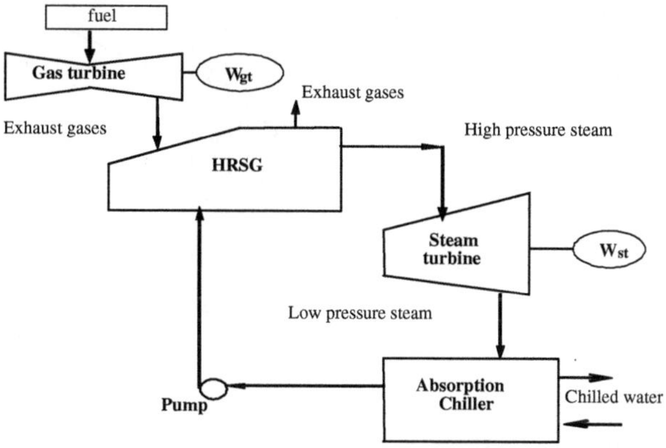

Fig. 3.2 Scheme of a cogeneration plant

This equation can be rewritten with the use of η_i and f_i, as shown by Eq. 3.6:

$$\eta_{\text{b overall}} = [\eta_b]_{st} f_{st} - f_{pump} + [\eta_b]_q f_q \tag{3.6}$$

In the case of a cogeneration plant composed of a gas turbine (gt is the turbine of the gas turbine) and a heat recovery steam generator the expression of $[\eta_b]_{\text{overall}}$ is:

$$\eta_{\text{b overall}} = [\eta_b]_{gt} f_{gt} - f_{cp} + [\eta_b]_q f_q \tag{3.7}$$

Equation 3.8 presents the expression of $\eta_{\text{b overall}}$ of a cogeneration plant, shown in Fig. 3.2 composed of a gas turbine, a heat recovery steam generator (HRSG), a back-pressure steam turbine and an absorption chiller that generates electricity, and chilled water for air conditioning purposes:

$$\eta_{\text{b overall}} = \frac{W_{gt} - W_{cp} + W_{st} - W_{pump} + B_q}{B_{fuel}} \tag{3.8}$$

The determination of $\eta_{\text{b overall}}$ can also be done with the use of η_i of each component of the plant.

$$\eta_{\text{b overall}} = \left([\eta_b]_{gt}\frac{\Delta B_{gt}}{B_{fuel}} - \frac{W_{cp}}{B_{fuel}}\right) + \left([\eta_b]_{st}\frac{\Delta B_{st}}{B_{fuel}} - \frac{W_{pump}}{B_{fuel}}\right) + [\eta_b]_q \frac{B_q}{B_{fuel}} \tag{3.9}$$

Writing $\eta_{\text{b overall}}$ in terms of η_i and f_i of each component of the plant gives Eq. 3.10

$$\eta_{\text{b overall}} = [\eta_b]_{gt} f_{gt} - f_{cp} + [\eta_b]_{st} f_{st} - f_{pump} + [\eta_b]_{abs} f_{abs} \tag{3.10}$$

For absorption refrigerating system η_{abs} can be given as:

$$[\eta_b]_{abs} = \text{COP}\left|\frac{\theta_{ev}}{\theta_{ge}}\right| \tag{3.11}$$

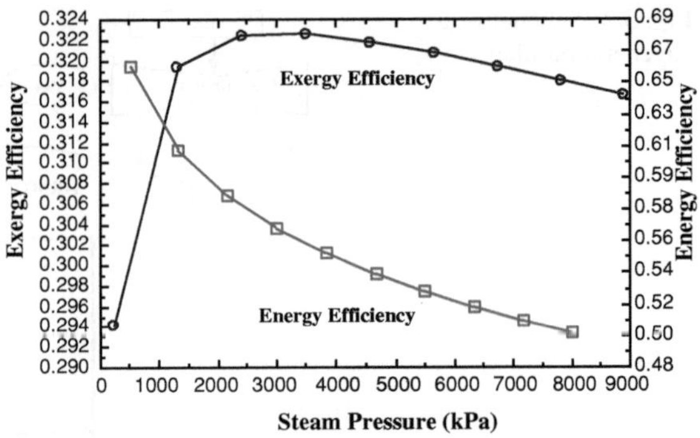

Fig. 3.3 Overall exergy and *energy efficiency* as a function of the *steam pressure*

Therefore the evaluation of the influence of each set of equipment in the overall exergy efficiency of the plant, $\eta_{b\ overall}$, can be done, indicating possible ways of increasing the overall efficiency of the plant.

3.2.2 Exergy Evaluation of a Cogeneration Plant

The methodology described before is applied to the cogeneration plant based on a combined cycle, shown in Fig. 3.2. The basic operational data of this plant are:

- fuel: gas with chemical exergy of 39,150 kJ/kg;
- gas turbine pressure ratio: 14:1;
- compressor isentropic efficiency: 0.90;
- gas turbine isentropic efficiency: 0.90;
- turbine combustor outlet temperature: 1,293 K;
- gas turbine air ratio: 270 %;
- gas turbine outlet temperature: 766 K;
- steam turbine outlet pressure: 100 kPa;
- steam turbine isentropic efficiency: 0.85;
- pump isentropic efficiency: 0.70;
- transmission, generator, and mechanic efficiency: 0.95;
- substances of the absorption chiller: H_2O-LiBr;
- evaporation temperature of the absorption chiller: 3 °C;
- condensation temperature of the absorption chiller: 40 °C;
- weak solution concentration of the absorption chiller: 55 %;
- strong solution concentration of the absorption chiller: 60 %;
- inlet chilled water temperature: 12 °C;
- outlet chilled water temperature: 6 °C.

Figure 3.3 presents the evolution of overall exergy efficiency of the cogeneration plant as a function of the steam pressure in the heat recovery steam generator (in these calculations the transmission, generator, and mechanic efficiencies were

Table 3.3 Values of η_{bi} and f_i for the components of the cogeneration plant

Component/parameter	CC	GT	CP	HRSG	ST	P	ABS
η_i	0.779	0.890	0.906	0.513	0.791	0.599	0.235
f_i	–	0.846	0.482	0.272	0.083	0.0006	0.057

CC combustion chamber, *GT* turbine of the gas turbine, *CP* air compressor, *HRSG* heat recovery steam generator, *P* pump, *ABS* absorption chiller

considered the same for both steam and gas turbines). The performance simulation of the plant indicates a maximum value of $\eta_{b\ overall}$ for a steam pressure of 3,000 kPa.

Table 3.3 summarizes the values of η_i and f_i for the components of the cogeneration plant in the condition of maximum $\eta_{b\ overall}$.

In the condition of maximum overall exergy efficiency, the gas turbine converts 24.4 % of the chemical exergy of the natural gas into electricity and the steam cycle converts 6.5 %. The absorption chiller transfers 1.3 % of the chemical exergy of the natural gas to the chilled water. Therefore, the overall exergy efficiency of the plant is 32.2 %. In this operating condition the energy efficiency is 56.5 % and the 'heat to electricity ratio' is 0.763.

3.3 Exergy Method for Determining the Electricity Cost Formation in Combined Cycle Power Plants

3.3.1 Introduction

The needs to evaluate the cost production processes in a combined cycle power plant or cogeneration plant can be rationally conducted if the exergy of the products of the plant: electricity generated in the gas turbine, in the steam turbine, and process steam, is taken as the value basis.

This is an interesting application of thermoeconomics concepts to evaluate and allocate the cost of exergy throughout the power plant energy conversion processes, considering costs related to exergy inputs and investment in equipment. Although the concept may be applied to any combined cycle or cogeneration plant, this section describes the mathematical modeling for three-pressure heat recovery steam generator configurations and total condensation of the produced steam. It is possible to study any $n \times 1$ plant configuration (n sets of gas turbine and heat recovery steam generators associated to one steam turbine generator and condenser) with the developed model, as shown in Fig. 3.4 and Table 3.4, assuming that every train operates identically and in steady state. The presented model was conceived from a complex configuration of a real power plant, over which variations may be applied in order to adapt it to a defined configuration under study [5] such as the use of reheat, supplementary firing and partial load operation. It is also possible to undertake sensibility analysis on geometrical equipment parameters.

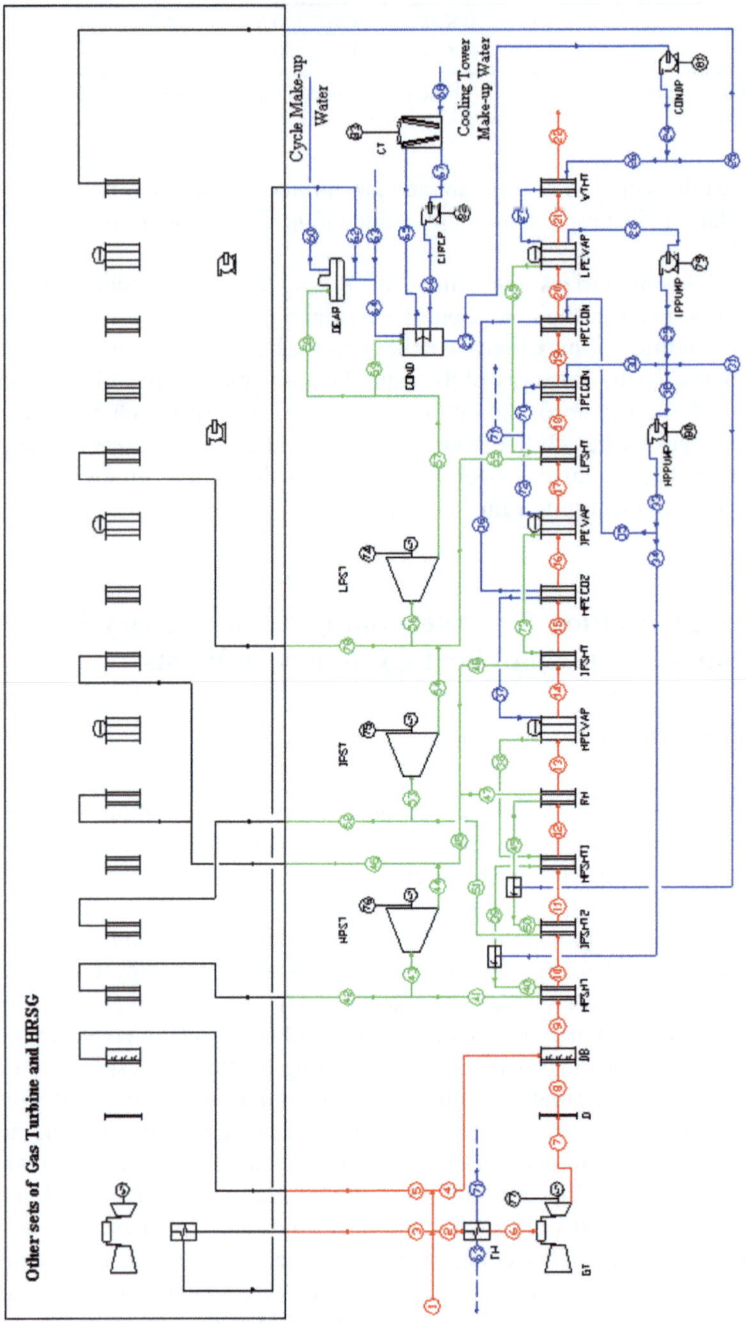

Fig. 3.4 Power plant model [5]

Table 3.4 Components nomenclature of the combined cycle shown in Fig. 3.4

Abbreviature	Component
GT	Gas turbine
D	Duct
HPPUMP	High pressure feed pump
IPPUMP	Intermediate pressure feed pump
CONDP	Condensate pump
CIRCP	Circulating pump
COND	Condenser
DEAR	Deaerator
CT	Cooling tower
Heat recovery steam generator	
DB	Supplementary firing module
HPSHT	High pressure superheater
IPSTH2	Intermediate pressure superheater II
HPSHT1	High pressure superheater I
RH	Reheating module
HPEVAP	High pressure evaporator
IPSHT	Intermediate pressure superheater
HPECO2	High pressure economizer II
IPEVAP	Intermediate pressure evaporator
LPSHT	Low pressure superheater
IPECON	Intermediate pressure economizer
HPECON	High pressure economizer
LPEVAP	Low pressure evaporator
WTHT	Water heater
FH	Natural gas heater
Steam turbine	
HPST	High pressure section
IPST	Intermediate pressure section
LPST	Low pressure section

3.3.2 Method Description

The first step is to collect information on the power plant configuration under study, detailing equipment and its capacity, transport properties of each process stream, and the consumed and generated power of the plant. The data that define each stream (i.e. mass flow rate, temperature, pressure, enthalpy, entropy and, consequently, exergy) may be collected using many different sources. One usual source is a computerized process simulator.

Nevertheless, data from the digital control system of operating plants and information from engineered energy balances also provide the necessary information.

The next step comprises using the equipment capacity to estimate costs, using a parametric method, which calculates an estimated cost based on a reference biome of cost and capacity [6, 7].

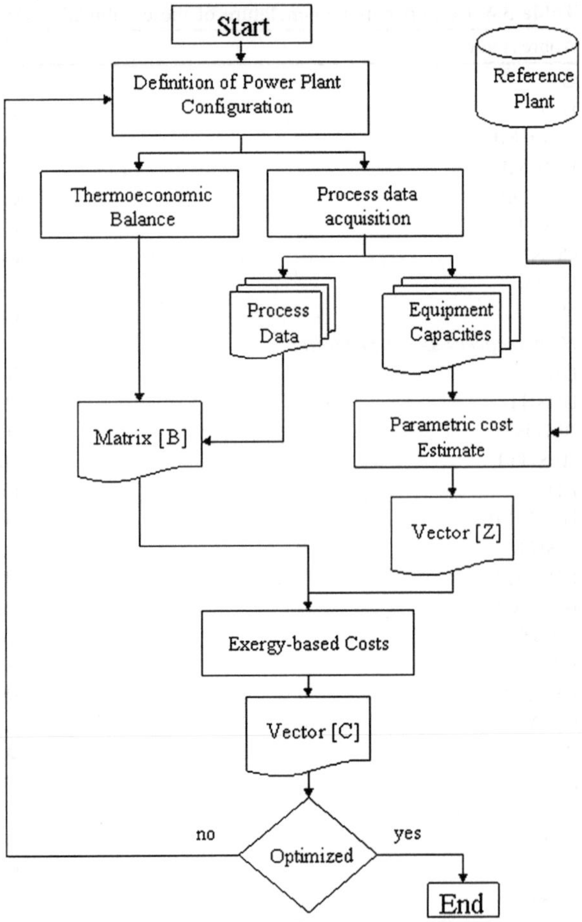

Fig. 3.5 Flowchart depicting the proposed method [5]

Finally, the system of linear equations (cost balances for every component of the plant and the required cost partition criteria, for instance, for gas and steam turbines) is obtained, and from the solution of Eq. 3.12, results a vector containing exergy-based costs for each stream. From this point on, several plant characteristics may be altered, which will ultimately recalculate matrix B and vector Z of Eq. 3.12, and therefore, resulting in a new exergy-based cost vector. Improving the power plant performance depends on defining which values from this vector are to be optimized and thus seeking alterations that will produce the desired effect. Figure 3.5 illustrates the method.

$$[B]_{nxn}[c] = [Z]_n \quad (3.12)$$

3.3.3 Cost Allocation Criteria for the Heat Recovery Steam Generator

Exhaust gases from the combustion turbine have monetary value because the steam that turns other turbines is generated from them. However, in the exhaust stack, the same gases are no longer used as far as the power plant is concerned, and for that reason cannot be assigned a value. As there is use for the exhaust gases from the combustion turbines and supplementary firing modules, these streams are assigned a cost (and the boiler stack exhaust must contain a null monetary value). Hence it is necessary to propose a criterion to distribute costs through the several modules of the heat recovery steam generator, taking into consideration the exergy of each stream and its variation.

For studying this problem, take a HRSG with n components (for instance, economizer, superheater, vaporizer, reheater), in which the reduction of the exergy flow rate of the combustion gases when flowing through a module i is ΔB_i. The overall exergy reduction in the HRSG is ΔB. If it is considered that the cost reduction of the combustion gases is a linear function of the exergy flow rate reduction in every component of the HRSG, then one can write for every module:

$$c_i = a_i c_{i-1} + b \tag{3.13}$$

where c_i is the specific exergy cost of the combustion gases at the exit of module i and c_{i-1} is the specific exergy cost of the combustion gases at the inlet of module i. The proposed relation does not imply that the combustion gases cost will have a linear reduction through the heat recovery steam generator, but only inside each module.

In order to assure that the variation of the cost in each component be proportional to the flow exergy variation, a_i is defined as [5]:

$$a_\mathrm{i} = \left(1 - \frac{\Delta B_i}{\Delta B}\right) \tag{3.14}$$

Moreover, knowing that at the outlet of the last module the value (cost) of the gases is zero:

$$b = \left(\frac{\Delta B_n}{\Delta B} - 1\right) c_{n-1} \tag{3.15}$$

Thus, the proposed variation law is shown in Eq. (3.16).

$$c_i = \left(1 - \frac{\Delta B_\mathrm{i}}{\Delta B}\right) c_{i-1} + \left(\frac{\Delta B_\mathrm{n}}{\Delta B} - 1\right) c_{n-1} \tag{3.16}$$

This relation provides a procedure to determine the variation of the cost of combustion gases (exergy transferred and destroyed) in each module of the HRSG. This procedure is shown in Fig. 3.6.

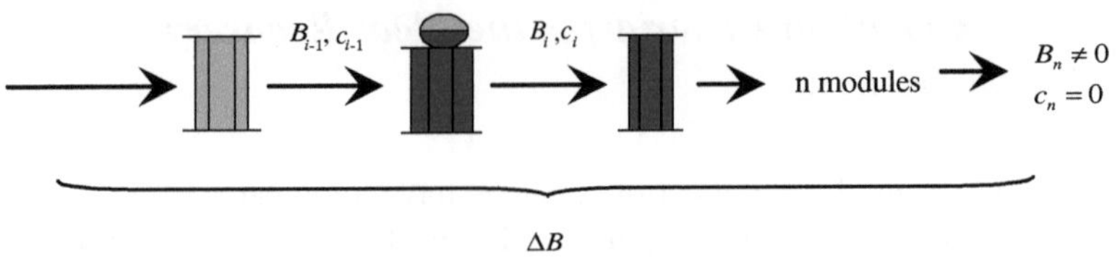

$$c_i = \left(1 - \frac{\Delta B_i}{\Delta B}\right) c_{i-1} + \left(\frac{\Delta B_n}{\Delta B} - 1\right) c_{n-1}$$

Fig. 3.6 Steam cost evaluation in the heat recovery steam generator [5]

Table 3.5 Combined cycle configurations with several gas turbines [30]

Gas turbine model	SWPC 501G	ABB GT24	SWPC 501FD	GE PG7421FA	SWPC 501DA	GE PG9171E
Gas turbine power (MW)	212.73	167.61	168.07	157.52	109.85	112.84
Power plant power (MW)	311.11	247.21	249.47	239.63	166.49	172.52
Gas turbine exergy efficiency	0.36	0.35	0.35	0.34	0.32	0.31
Power plant exergy efficiency	0.53	0.52	0.52	0.51	0.49	0.48
High pressure steam turbine power (MW)	29.02	23.94	23.91	24.25	14.82	15.82
Intermediate pressure steam turbine power (MW)	27.69	22.54	22.97	23.14	16.27	17.12
Low pressure steam turbine power (MW)	41.66	33.11	34.52	34.72	25.55	26.75
Instalation cost[a] (US$/kW)	324.67	314.05	336.39	348.70	385.16	382.63
Gas turbine electricity cost (US$/MWh)	17.60	17.18	17.72	17.59	18.23	18.51
High pressure steam turbine electricity cost (US$/MWh)	41.86	42.26	42.68	42.58	47.61	47.07
Intermediate pressure steam turbine electricity cost (US$/MWh)	45.17	45.42	45.88	45.82	50.84	50.23
Low pressure steam turbine electricity cost (US$/MWh)	61.47	61.50	62.52	62.48	66.45	66.01
Average electricity cost (US$/MWh)	28.20	28.12	28.90	29.35	31.43	31.64

[a] Installation costs presented are referred to power island only. Power island comprises gas turbine, steam turbine, steam generator, condenser, condensing, and steam generator feeding pumps [31]

3.3.4 Results

Table 3.5 shows results of comparisons made using 1 × 1 power plants using several combustion turbines technology, as a first application of the method. The configuration of the heat recovery boiler is equal for every simulation, and comprising three pressures and three modules in each pressure level, and no reheating.

Table 3.6 $N \times 1$ configurations study [30]

Configuration	1 × 1	2 × 1	3 × 1
Gas turbine power (MW)	167.61	167.61	167.61
Power plant capacity (MW)	247.21	490.12	567.96
Gas turbine exergy efficiency	0.35	0.35	0.35
Power plant exergy efficiency	0.52	0.52	0.52
High pressure steam turbine power (MW)	23.94	49.28	74.63
Intermediate pressure steam turbine power (MW)	22.54	44.98	67.42
Low pressure steam turbine power (MW)	33.11	66.05	98.99
Installation cost (power island) (US$/kW)	314.05	257.17	221.92
Electricity cost—Gas turbine (US$/MWh)	17.18	17.17	17.17
Electricity cost—High pressure steam turbine (US$/MWh)	42.26	38.84	37.51
Electricity cost—Intermediate pressure steam turbine (US$/MWh)	45.42	42.69	41.65
Electricity cost—Low pressure steam turbine	61.50	47.22	43.11
Average electricity cost (US$/MWh)	28.12	25.65	24.88

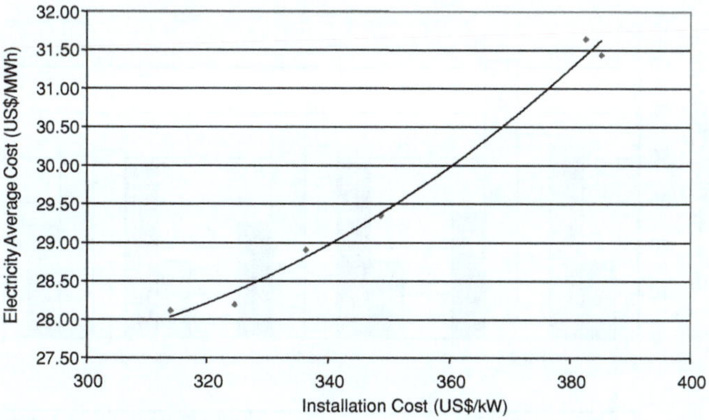

Fig. 3.7 Installation cost study [5]

It was allowed to the simulator [8] to adjust the modules within a range of geometrical characteristics to obtain best results for each configuration.

The electricity production costs were determined considering interest rate of 17 % per year, 20 years as the capital recovery period and a load factor of 92 %.

By analyzing Table 3.5 and Fig. 3.7, one can notice the relation between exergy efficiency and the average cost of electricity. The comparison between a plant based on ABB GT24 machine and another based on a SWPC 501G shows that the installation cost may sometimes compensate a lower efficiency.

A deeper analysis of Table 3.5 allows the study of the composition of the average cost as a function of the costs calculated in each generator. The lower value is related to the combustion turbine generator for all cases. Hence, the addition of more combustion turbines in 2 × 1 or 3 × 1 configurations results in lower average costs. Table 3.6 and Fig. 3.8 show results obtained by varying the number of ABB GT24 machines in the power plant.

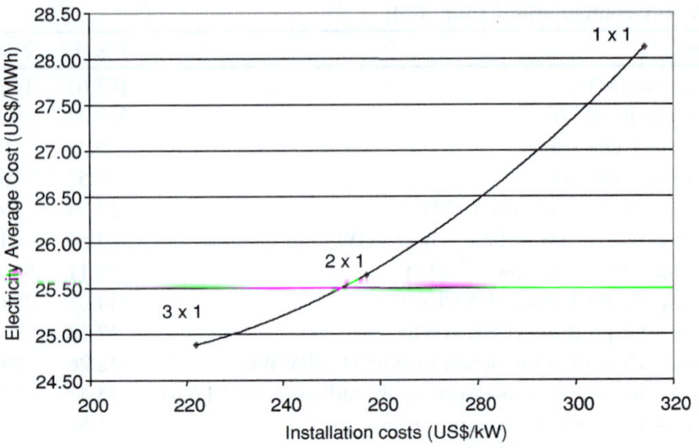

Fig. 3.8 '$n \times 1$' configuration study [5]

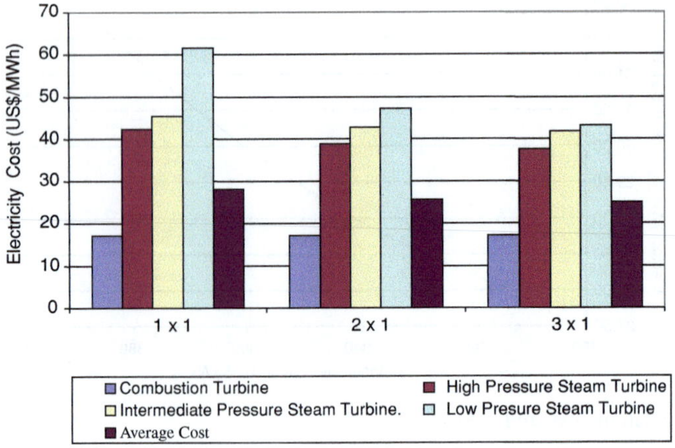

Fig. 3.9 Electricity cost composition [5]

By analyzing Table 3.6 and Fig. 3.8, it can be observed that adding more combustion turbines reduces the average cost. The installation costs also decreased, indicating a scale gain. Figure 3.9 depicts that the average costs are much influenced by the low pressure steam turbine in the 1×1 configuration. The other configurations show more equalized costs; however, the electricity generation in this section is the highest in considering all steam turbine sections. To solve this problem it is necessary either to decrease its production or to increase production in other sections, in such a way that the average cost will be reduced. Table 3.7 presents results obtained with three optimization studies of a 2×1 ABB GT24 power plant.

The first study shows results for the same configuration studied before. In Study 2, reheat modules were added, along with other high and intermediate pressure modules. The efficiency increase can be observed, but the addition of a reheat section

Table 3.7 Cycle optimization [30]

Study	1	2	3
Gas turbine power (MW)	167.61	167.48	167.48
Power plant capacity (MW)	490.12	505.81	512.01
Gas turbine exergy efficiency	0.35	0.36	0.36
Power plant efficiency	0.52	0.54	0.55
High pressure steam turbine power (MW)	49.28	41.15	41.15
Intermediate pressure steam turbine power (MW)	44.98	60.91	60.95
Low pressure steam turbine power (MW)	66.05	73.79	79.90
Installation cost (power island) (US$/kW)	257.17	264.18	260.98
Electricity cost—Gas turbine (US$/MWh)	17.17	17.28	17.28
Electricity cost—High pressure steam turbine (US$/MWh)	38.84	45.37	40.69
Electricity cost—Intermediate pressure steam turbine (US$/MWh)	42.69	30.97	28.45
Electricity cost—Low pressure steam turbine	47.22	45.73	36.01
Average electricity cost (US$/MWh)	25.65	25.21	23.41

rearranged the steam and power generation inside, producing new results. The average cost was reduced; however, the low pressure section continues to push its value upwards.

Study 3 was obtained taking the condenser pressure from 0.10 to 0.06 bar. Consequently, the destruction of exergy in this component was reduced, which reduced also the exergy-based costs of the condensate and the boiler feed water. Hence, the steam was generated at a lower cost, which pulled the overall costs downwards, especially in the low pressure section, which had its power production increased. In this sense, the combined result is a lower average electricity cost. Table 3.8 shows pinch points of evaporators and approach temperatures (temperature difference between gas inlet and steam or liquid outlet) of other heat exchangers of the heat recovery steam generators.

Table 3.9 summarizes the data of every section of the optimized 2 × 1 ABB GT24 power plant (shown in Fig. 3.4). Data of Table 3.9 indicate how is the electricity cost formation process, as well as the composition of the average electricity cost.

Figure 3.10 depicts the effect of the proposed law of variation over the exergy-based cost of the exhaust gases (see also Fig. 3.4). It can be observed that the stack exhaust carried no monetary value, and the cost is distributed to each module according to the exergy it captures. Hence the high pressure sections, which use more of the exergy of the gas exhaust stream, received larger portion of costs.

Aiming to demonstrate the applicability of the method to sensitivity analysis, Figs. 3.11 and 3.12 are presented. Both based on a 1 × 1 SWPC 501FD power plant.

Figure 3.11 shows the response of the electricity exergy-based cost and efficiency to a variation in the condenser pressure. As a lower condenser pressure implies in larger heat exchanger areas, pumps and cooling towers, the method allows the capital costs to vary to capture these changes, and produce results according to the heat balance modifications. Figure 3.12 depicts the behavior of the exergy-based cost according to the load condition of the power plant. Partial

Table 3.8 Heat recovery steam generators pinch points and approach temperatures [30]

	Study 1 (°C)	Study 2 and 3 (°C)
HPSHT	88	88
IPSTH2	[a]	32
HPSHT1	[a]	130
RH	[a]	31
HPEVAP	10	10
IPSHT	30	20
HPECO2	50	15
IPEVAP	10	10
LPSHT	10	16
IPECON	35	28
HPECON	[a]	13
LPEVAP	10	10
WTHT	40	29

[a] Nonoperating

load conditions produce naturally less efficient results, a consequence which can be captured by the method when allocating cost according to exergy destruction, thus resulting in higher costs.

The last set of results comprises the study on supplementary firing. The base power plant model for this study is a 2 × 1 SPWC 501F, which generates 590 MW using its full supplementary firing capacity, which corresponds to its design condition.

The fuel burnt in the heat recovery boilers is decreased until it is completely turned off, in which case the power plant produces approximately 500 MW. The installation cost is kept constant throughout the study in order to more accurately reflect the load variations and investment in excess capacity. Figures 3.13 and 3.14 present the variation of the exergy-based cost of electricity and according both to power generation and to plant efficiency.

It can be observed that the efficiency decreases with the increase of power generation, which is to be expected since the supplementary firing causes this effect on the overall cycle efficiency. However, the exergy-based cost of electricity generated finds a local minimum value, which can be interpreted as the balance between the loss in terms of efficiency and its compensation in capital cost utilization, since operating the supplementary firing in less than 100 % capacity implies using less capacity than the installed.

3.3.5 Comments on the Method Application

With the described method is possible to analyse the composition of the electricity costs in a variety of combined cycle power plants configurations. The parametric cost estimation allows the method to vary equipment costs accordingly to the capacity of the equipment used, providing parameters for a cost–benefit analysis.

Table 3.9 Characteristics of the 2 × 1 optimized configuration [30]

	m (kg/s)	*P* (bar)	*T* (°C)	*b* (kJ/kg)	*B* (kW)	c_b (US$/MWh)	c_m (US$/t)
1	17.98	34.47	25.00	52324.49	940590.72	10.08	146.51
2	8.99	34.47	25.00	52324.49	470295.33	10.08	146.51
3	8.99	34.47	25.00	52324.49	470295.33	10.08	146.51
4	–	34.47	25.00	52324.49	–	10.08	146.51
5	–	34.47	25.00	52324.49	–	10.08	146.51
6	8.99	34.47	141.12	52379.69	470791.47	10.44	151.90
7	373.29	1.03	654.59	675.00	251971.85	10.44	1.96
8	373.29	1.01	654.59	673.09	251258.76	10.44	1.95
9	373.29	1.01	654.59	673.09	251258.76	10.44	1.95
10	373.29	1.01	628.10	634.06	236691.52	9.36	1.65
11	373.29	1.01	611.21	609.44	227500.14	8.64	1.46
12	373.29	1.01	538.78	506.34	189013.94	7.20	1.01
13	373.29	1.01	489.93	439.33	163998.05	6.12	0.75
14	373.29	1.01	341.04	251.13	93745.47	3.96	0.28
15	373.29	1.01	337.16	246.62	92061.00	3.60	0.25
16	373.29	1.01	271.92	174.45	65119.93	2.88	0.14
17	373.29	1.01	242.30	144.28	53860.50	2.16	0.09
18	373.29	1.01	240.40	142.42	53162.94	1.80	0.07
19	373.29	1.01	233.39	135.57	50607.66	1.44	0.05
20	373.29	1.01	193.21	98.59	36803.79	1.08	0.03
21	373.29	1.01	156.63	68.62	25614.39	0.36	0.01
22	373.29	1.01	91.00	25.83	9642.43	–	–
23	132.25	0.06	33.96	8.66	1144.66	2180.77	5.24
24	132.25	4.55	34.07	9.15	1210.20	2067.70	5.25
25	66.13	4.55	34.07	9.15	605.10	2067.70	5.25
26	66.13	4.55	34.07	9.15	605.10	2067.70	5.25
27	66.13	4.35	127.75	97.50	6447.59	203.10	5.50
28	61.74	4.35	146.63	124.51	7687.61	119.55	4.13
29	61.74	29.38	147.22	127.71	7885.26	119.91	4.25
30	9.91	29.38	147.22	127.71	1265.05	119.91	4.25
31	–	29.38	147.22	127.71	–	119.91	4.25
32	51.84	29.38	147.22	127.71	6620.21	119.91	4.25
33	51.84	130.55	149.16	140.00	7257.54	114.51	4.45
34	–	130.55	149.16	140.00	–	114.51	4.45
35	51.84	130.55	149.16	140.00	7257.54	114.51	4.45
36	51.84	130.35	220.52	266.18	13798.32	66.26	4.90
37	51.84	130.35	322.30	528.39	27390.56	40.69	5.97
38	51.32	130.35	331.04	1694.05	86937.83	20.89	9.83
39	51.32	130.35	482.00	1547.03	79393.01	31.33	13.46
40	51.32	130.35	482.00	1547.03	79393.01	31.33	13.46
41	51.32	125.14	566.80	1692.80	86873.66	33.13	15.57
42	51.32	125.14	566.80	1692.80	86873.66	33.13	15.57
43	102.64	125.14	566.80	1692.80	173747.32	33.13	15.57
44	102.64	28.01	347.66	1267.41	130086.36	33.13	11.66
45	51.32	28.01	347.66	1267.41	65043.18	33.13	11.66

(continued)

Table 3.9 (continued)

	m (kg/s)	P (bar)	T (°C)	b (kJ/kg)	B (kW)	c_b (US$/MWh)	c_m (US$/t)
46	51.32	28.01	347.66	1267.41	65043.18	33.13	11.66
47	57.69	28.01	344.68	1263.53	72898.13	33.13	11.62
48	6.37	28.01	321.18	1232.78	7858.11	34.57	11.83
49	57.69	26.89	507.74	1480.81	85433.89	32.77	13.48
50	57.69	26.89	507.74	1480.81	85433.89	32.77	13.48
51	57.69	26.89	566.00	1566.85	90397.86	33.85	14.73
52	57.69	26.89	566.00	1566.85	90397.86	33.85	14.73
53	115.39	26.89	566.00	1566.85	180795.72	33.85	14.73
54	115.39	4.17	305.83	1200.13	138480.23	33.85	11.28
55	4.38	4.17	226.54	931.88	4084.17	117.39	30.38
56	124.15	4.17	300.21	1006.58	124969.78	41.05	11.47
57	124.15	0.06	36.18	512.62	63644.10	41.05	5.84
58	0.04	0.06	36.18	276.73	12.23	41.05	3.15
59	124.11	0.06	36.18	276.73	34344.85	41.05	3.15
60	1.17	1.03	15.56	2.42	2.82	3075.26	2.06
61	1.21	0.06	36.18	9.71	11.75	792.58	2.14
62	3.47	29.18	44.21	16.89	58.56	76.34	0.36
63	3.47	29.18	44.21	16.89	58.56	76.34	0.36
64	8.14	0.06	36.18	39.89	324.85	56.18	0.62
65	6788.52	2.20	34.83	9.27	62939.88	2180.77	5.61
66	6788.52	2.20	24.83	5.18	35139.54	3932.66	5.65
67	6788.52	1.01	24.83	5.06	34318.10	4025.93	5.65
68	133.10	2.00	15.00	2.39	318.40	1875.05	1.25
69	4.38	4.35	146.63	866.04	3795.64	119.55	28.75
70	9.91	29.18	212.30	241.50	2392.19	76.34	5.12
71	3.47	29.18	212.30	241.50	837.27	76.34	5.12
72	6.44	29.18	212.30	241.50	1554.92	76.34	5.12
73	6.37	29.18	232.30	1110.13	7076.34	31.69	9.77
74					79902.97	36.01	–
75					60948.65	28.45	–
76					41147.11	40.69	–
77					167483.90	17.28	–
78	4.38	4.17	226.54	931.88	4084.17	117.39	30.38
79					252.80	23.41	–
80					763.72	23.41	–
81					113.63	23.41	–
82					954.25	23.41	–
83					1744.16	23.41	–

The application of cost balances showed its benefits in terms of process analysis, allowing a detailed study of each stream that comprises the power plant. The proposed law of variation on the exergy-based cost of the exhaust gases aided to identify the costs, which affect significantly the composition of the electricity average cost.

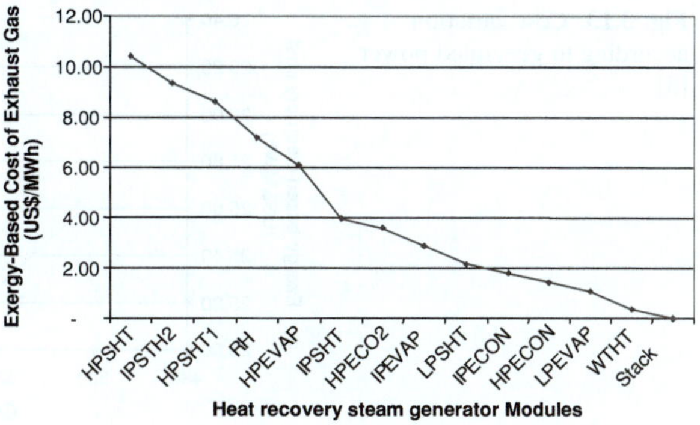

Fig. 3.10 Effect of the combustion gases cost law variation [5]

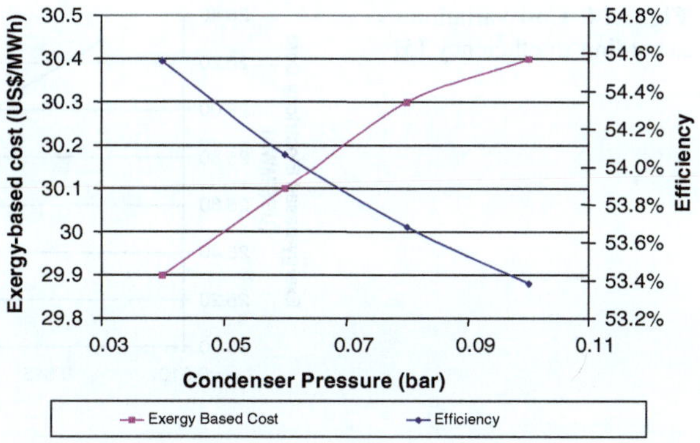

Fig. 3.11 Sensitivity analysis: condenser pressure [5]

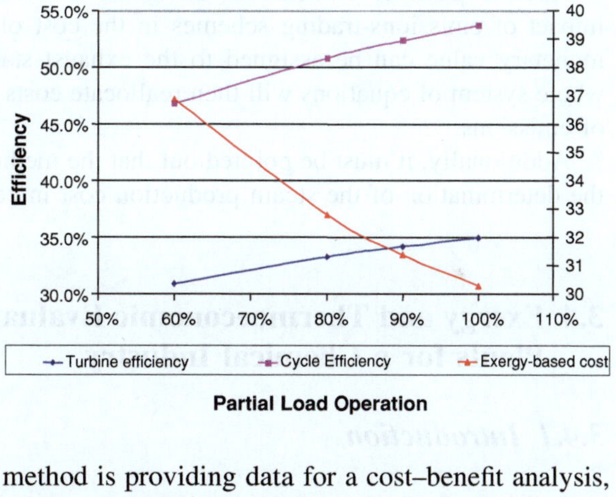

Fig. 3.12 Sensitivity analysis: partial load operation [5]

Another advantage of the method is providing data for a cost–benefit analysis, combining thermoeconomics, and parametric cost estimation. Investment decisions may be based on parameters provided by the method, in terms of equipment sizing and reheat application. As seen in the supplementary firing cases, similar analysis may provide data for peaking plant operation.

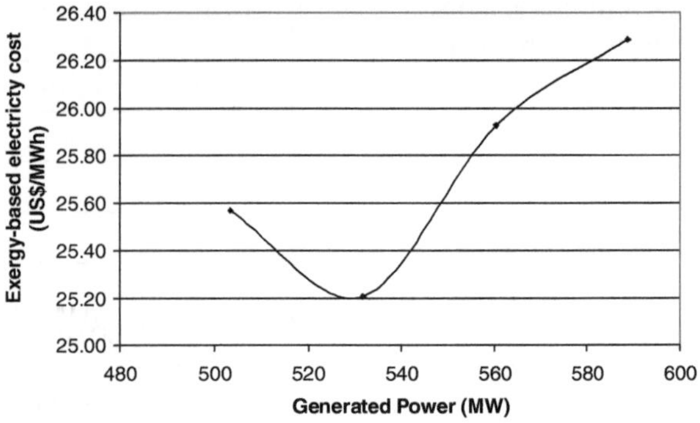

Fig. 3.13 Cost variation according to generated power [5]

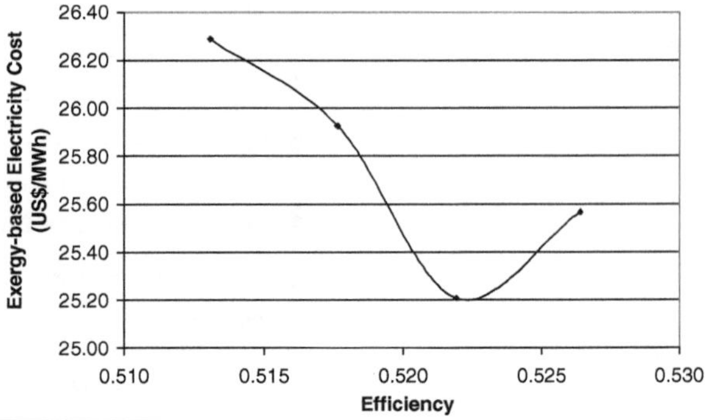

Fig. 3.14 Cost variation according to efficiency [5]

Another possible outcome of using the described method is evaluating the impact of emissions-trading schemes in the cost of electricity. Assuming that a monetary value can be assigned to the exhaust stack gases, the solution of the whole system of equations will then reallocate costs to accurately reflect the costs of emissions.

Additionally, it must be pointed out that the methodology will be also useful in the determination of the steam production cost in cogeneration plants.

3.4 Exergy and Thermoeconomic Evaluation of Cogeneration Plants for a Chemical Industry

3.4.1 Introduction

The projected increase of the natural gas consumption in countries like Brazil has motivated several substitution studies in industrial processes in order to analyze the feasibility of the use of this fuel in utilities plants. Together with these studies, the possibility of adapting these plants to be converted into cogeneration plants is also considered.

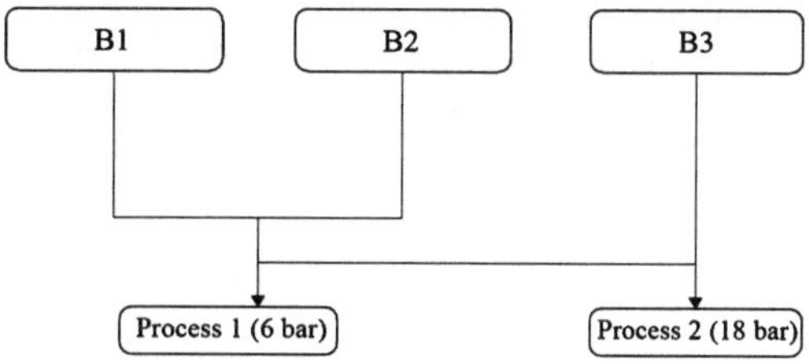

Fig. 3.15 Scheme of the steam distribution line [11]

In the Brazilian Chemical Industrial Sector, 37 % of the energy consumption in 1998 corresponded to steam generation in boilers for heating purposes, with residual fuel oil accounting for 53 % of this consumption [9]. In these industries the average heat-to-power ratio is 1.88 [10].

The study described next presents a thermoeconomic analysis of three cogeneration systems designed to be used in a chemical plant that intends to increase its steam generation capacity and substitute fuel oil by natural gas, to generate electricity and steam to its processes.

The use of exergy and thermoeconomic analysis provides a rational way to evaluate the production costs of these utilities for different technological options, as well as, in different operating conditions.

In this way, three cogeneration systems: a steam cycle with condensation–extraction steam turbine, a gas turbine-based system, and a combined cycle based system, are analyzed in two operating scenarios: in the first one the systems generate steam (10 t/h at 18 bar and 30 t/h at 6 bar) and electricity for the plant (5 MW) and in the second one the systems generate steam (10 t/h at 18 bar and 30 t/h at 6 bar), electricity for the plant (5 MW) and export electricity (12 MW).

3.4.2 Steam and Electricity Demands

The utilities plant of the chemical industry is made up of three steam boilers (B1, B2, and B3), generating steam at two pressure levels, 6 bar (to feed process 2) and 18 bar (to feed process 1). The higher pressure line is connected to the lower pressure one, as shown in Fig. 3.15. According to the Energy Department of the industry, the average monthly consumption of process steam and electricity [11] are:

- electricity: 3,886 MWh
- process steam: 14,942 t

The cost of each one of the utilities considered by the industry [11] is:

- electricity: 68.00 US$/MWh
- process steam: 17.40 US$/t

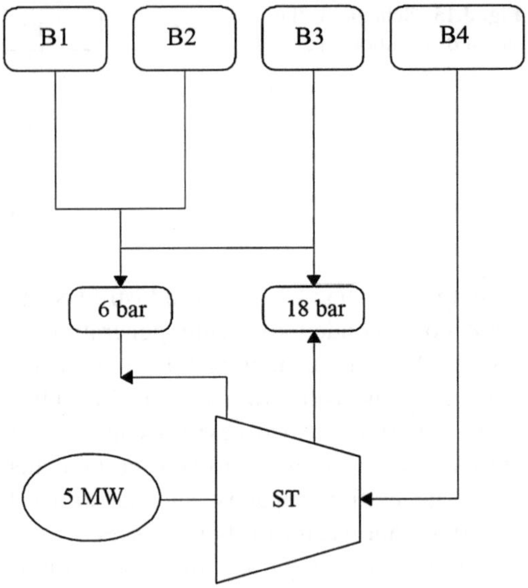

Fig. 3.16 Simplified flow sheet of the utilities plant with condensation–extraction steam turbine [11]

3.4.3 Cogeneration Systems

As mentioned before, the considered cogeneration systems are: a steam cycle with condensation–extraction steam turbine, a gas turbine based system, and a combined cycle based system [11].

The steam turbine based system is composed of a condensation–extraction steam turbine and a high pressure steam generator (B4). The electricity generation capacity is 5 MW. Figure 3.16 shows a simplified flow sheet of this configuration. Steam is generated in the boiler B4 at a pressure of 42 bar and 573 K. This steam is sent to the condensation–extraction steam turbine, where 10 t/h of steam are extracted at 18 bar (process 1) and 30 t/h extracted are at 6 bar (process 2).

The gas turbine-based cycle is made up of a gas turbine of the same capacity as the steam turbine (the combustion chamber outlet temperature is 1,295 K) and a waste heat boiler (B4) that can produce 16.67 t/h of steam at 20 bar. This waste heat boiler must operate with supplementary consumption of natural gas to attain the plant steam demand. This configuration is shown in Fig. 3.17.

Figure 3.18 shows the combined cycle-based cogeneration system. In this configuration the gas turbine-based system is coupled with a steam cycle with a waste heat boiler. The extraction steam conditions and flow rates are the same of the steam turbine-based system. In this system the electricity generation capacity is fixed to 6.3 MW because the steam-based system must produce 40 t/h of steam to supply the processes demand, implying that the industry is able to export 1.3 MW of electricity. This means that the waste heat boiler must consume supplementary fuel to increase the steam production. In Fig. 3.18 it is indicated that the gas

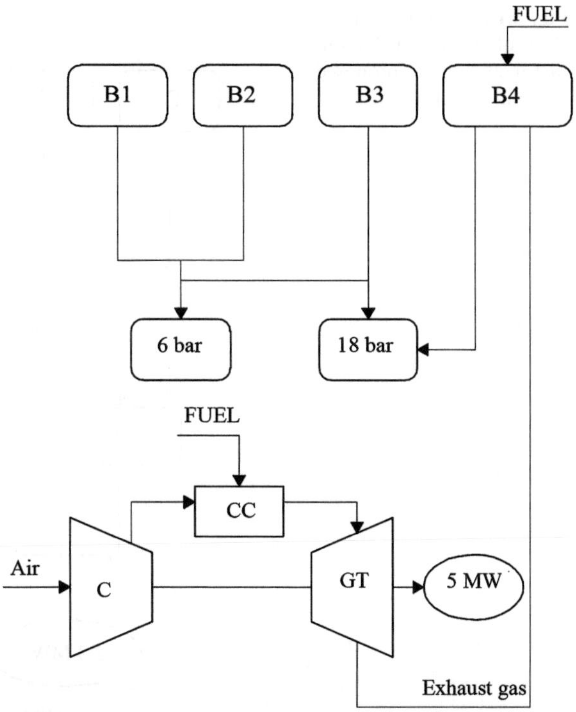

Fig. 3.17 Simplified flow sheet of the utilities plant with the gas turbine-based system [11]

turbine generates 3.0 MW and the steam turbine generates 3.3 MW. This power distribution is obtained by the simulation of the whole system [11].

3.4.4 Exergy Analysis of the Cogeneration Systems

The methodology described in Sect. 3.2 of this chapter (see Table 3.1, Eqs. 3.5–3.8) is applied to analyze the three cogeneration systems, considering the following basic data of each plant:

- thermodynamic reference state: $T_o = 298$ K; $P_o = 1$ bar
- fuel: natural gas (lower heating value = 48,160 kJ/kg)
- gas turbine pressure ratio: 10:1
- isentropic efficiency of the air compressor and gas turbine: 90 %
- thermal efficiency of the gas turbine combustor: 100 %
- gas turbine combustor outlet temperature: 1,295 K
- gas turbine excess air ratio: 275 %
- exhaust gas turbine temperature: 788 K
- steam generation pressure: 42 bar
- steam generation temperature: 573 K
- steam condensation pressure: 0.05 bar

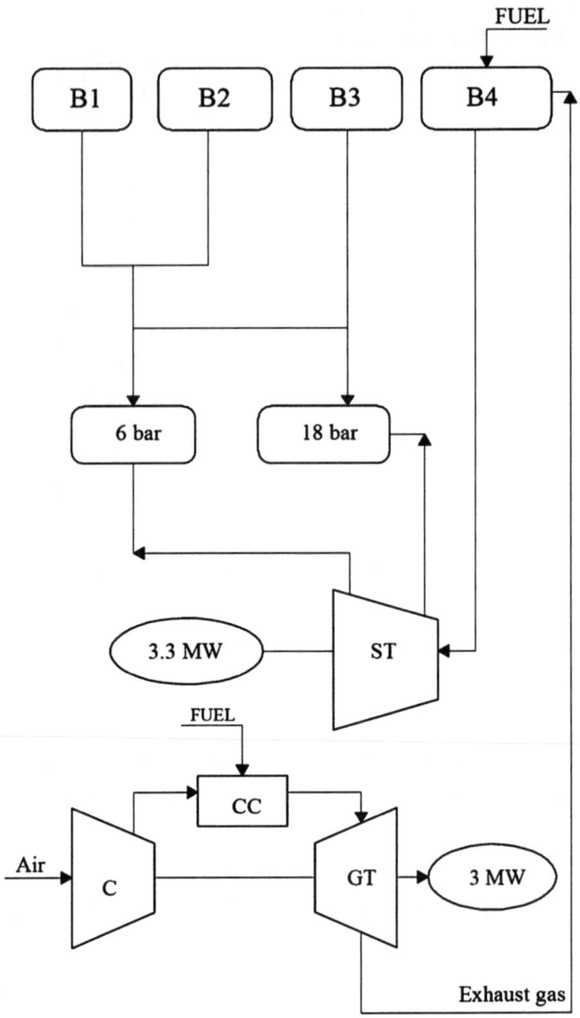

Fig. 3.18 Simplified flow sheet of steam distribution line including the proposed combined cycle [11]

- thermal efficiency of the conventional boiler: 90 %
- thermal efficiency of the heat recovery steam generator: 80 %
- isentropic efficiency of the steam turbine stages: 85 %
- isentropic efficiency of the pumps: 80 %
- mechanical, generator, and transmission efficiency: 95 %
- steam pressure of process 1: 18 bar
- average steam temperature of process 1: 450 K
- steam pressure of process 2: 6 bar
- average steam temperature of process 2: 403 K

The performance behavior of each system was simulated by means of models developed with the aid of the software EES® [12]. Tables 3.10, 3.11, and 3.12

Table 3.10 Parameter f_i and η_{bi} of the components of the steam turbine-based cogeneration system shown in Fig. 3.16 ($W = 5$ MW) [11]

Component	f_i	η_{bi}
Boiler	1.00	0.37
Turbine	0.17	0.75
Process 1	0.05	0.89
Process 2	0.13	0.84
Pumps	0.00	0.75
Deaerator	0.04	0.97
Preheater	0.03	0.97

present the values of f_i and η_{bi} for each component of the three cogeneration systems.

In Table 3.10 it can be seen that the steam turbine and process 2 are the main consumers of fuel exergy. Boiler 4 is the component with the lowest value of η_{bi} due to the heat transfer and combustion irreversibility that take place in this component during the processes of energy conversion.

Table 3.11 summarizes the results obtained with the gas turbine-based cogeneration system. It is interesting to notice the changes in the values of f_i and η_{bi} of the waste heat boiler when it operates with supplementary use of natural gas, indicating a reduction of the exergy efficiency in the steam generation process.

Table 3.12 shows the results of the combined cycle-based system for the second operating scenario ($W = 17$ MW). As the steam turbine must produce 5 MW, the heat recovery steam generator needs to burn natural gas. As a consequence of this operating condition, the value of the exergy efficiency of the waste heat boiler is similar to the values of this component obtained for the steam and gas turbine-based systems.

Table 3.13 presents the overall energy and exergy efficiencies of the cogeneration systems for both operating scenarios. In the first scenario, the combined cycle-based system is the most efficient one based on an exergy analysis. In the second operating scenario, the gas turbine-based system is the most efficient system, in energy and exergy analysis, because it is not necessary to burn supplementary fuel in the waste heat boiler to attain the steam demand in the processes.

3.4.5 Thermoeconomic Analysis of the Cogeneration Systems

In a multi-product plant the determination of the production cost of each utility can be done by the application of utilities cost balances and cost partition methods to the components of the plant. As stated in Chap. 2, in a thermomechanical conversion plant cost balances based on exergy balances provide a rational way to obtain the production costs of the utilities [13].

By applying the cost balance equation to the steam turbine, shown in Fig. 3.16, gives (C_{turb} = steam turbine cost rate, W_e = electric power, hp = high pressure,

Table 3.11 Parameter f_i and η_{bi} of the components of the gas turbine based cogeneration system with and without supplementary burning (sb) of fuel [11]

Component	f_i^{a}	f_{sbi}^{b}	η_{bi}^{a}	$\eta_{b(sb)i}^{b}$
Air compressor	0.40	0.21	0.90	0.90
Combustion chamber	1.36	0.73	0.75	0.75
Turbine	0.74	0.40	0.91	0.91
Waste heat boiler	0.26	0.60	0.51	0.40
Process 1	0.03	0.06	0.89	0.89
Process 2	0.08	0.15	0.84	0.84

[a] $W = 17$ MW
[b] $W = 5$ MW

Table 3.12 Parameter f_i and η_{bi} of the components of the combined cycle based cogeneration system ($W = 17$ MW) [11]

Component	f_i	η_{bi}
Air compressor	0.25	0.90
Combustion chamber	0.84	0.79
Gas turbine	0.49	0.91
Waste heat boiler	0.56	0.38
Steam turbine	0.10	0.79
Process 1	0.03	0.89
Process 2	0.08	0.84
Pumps	0.00	0.74
Deaerator	0.02	0.97
Preheater	0.02	0.98

Table 3.13 Overall energy (η_{eG}) and exergy efficiencies (η_{bG}) of the proposed cogeneration systems for two operating conditions ($W = 5$ MW/$W = 17$ MW) [11]

System configuration	η_{eG}	η_{bG}
Gas turbine	0.80/0.63	0.31/0.36
Steam turbine	0.72/0.43	0.29/0.25
Combined cycle	0.76[a]/0.60	0.32[a]/0.35

[a] $W = 6.3$ MW

e = electricity, $p1$ = steam demanded by process 1, $p2$ = steam demanded by process 2, cd = condenser):

$$c_e W_e + c_{p1} \Delta B_{p1} + c_{p2} \Delta B_{p2} + c_{cd} B_c = c_{hp} B_{hp} + C_{turb} \tag{3.17}$$

In this equation B_{hp}, W_e, ΔB_{p1} and ΔB_p are determined by the exergy analysis of the plant. C_{turb} is known and c_{hp} is obtained by applying the cost balance to boiler 4, where there is only one product (high pressure steam).

To determine the values of c_e, c_{p1}, and c_{p2} it is necessary to consider a cost partition criterion. In this study, the extraction and the equality criteria are used in steam and gas turbines, giving the auxiliary relations shown in Table 3.14.

Table 3.14 Auxiliary relations

Cost partition method	Steam turbine	Gas turbine
Extraction	$c_{hp} = c_{p1} = c_{p2} = c_{cd}$	$c_{gas} = c_{eg}$
Equality	$c_e = c_{p1} = c_{p2} = c_{cd}$	$c_e = c_{eg}$

The compared thermoeconomic analysis of the three cogeneration systems shown in Figs. 3.15, 3.16, and 3.17 is obtained based on the following parameters (the components costs were evaluated for a power generation of 5 MW):

- natural gas cost: 10.40 US$/MWh (3 US$/MBtu);
- capital recovery period: 10 years;
- interest rate: 12 % per annum;
- load factor: 0.80;
- time factor 0.85;
- condensation–extraction steam turbine cost: US$2,500,000;
- conventional boiler cost: US$1,650,000;
- gas turbine cost: US$1,950,000;
- waste heat boiler cost: US$1,100,000;
- auxiliary equipment cost: US$277,000;
- annual operational and maintenance cost: 10 % of the investment cost;
- inflation is not considered.

The equipment cost rate is evaluated according to Eq. 3.18:

$$C_{\text{equip}\,i} = C_{0i}[(n/(1-(1+n)^{-r}) + f_{\text{om}}]/(3600\,N_h f_t f_l) \tag{3.18}$$

Table 3.15 presents the specific production costs of process steam (US$/t) and electricity (US$/MWh), for the three cogeneration systems and using the equality and extraction cost partition methods.

For the combined cycle-based system, the average electricity cost calculated from the values of the electricity cost of the steam and gas turbines is presented. It is interesting to notice that, for this system, the electricity generated by the gas turbine is less expensive than the electricity generated by the steam turbine, in both cost partition methods (28.66 US$/MWh against 52.21 US$/MWh and 46.80 US$/MWh against 72.17 US$/MWh).

Values of Table 3.15 indicate that only the electricity cost of the steam turbine-based system, using the extraction method, is higher than the electricity price paid by the industry (68.00 US$/MWh). In this table it is important to verify that all obtained costs of process steam are lower than the original value considered by the industry.

Another interesting scenario to compare the performance of the systems is the one in which all the three systems are capable to generate more electricity than needed in the industry. In this scenario the company will be able to export electricity to other industries or to the electricity grid. The thermoeconomic analysis is done, in this case, for an electricity generation capacity of 17 MW [10]. In this scenario the gas turbine-based system operates without supplementary use of fuel in the heat recovery steam generator because of the higher capacity of the gas

Table 3.15 Specific production costs of electricity and process steam (mass weighted average value of the two processes) [11]

Method	Equality		Extraction	
System configuration	Electricity (US$/MWh)	Steam (US$/t)	Electricity (US$/MWh)	Steam (US$/t)
Steam turbine	50.26	10.78	70.92	7.73
Gas turbine	29.06	9.70	48.02	7.25
Combined[a] Cycle	40.82	11.26	59.90	7.83

[a] $W = 6.3$ MW

Table 3.16 Specific production costs of electricity and process steam (mass weighted average value of the two processes) considering production of 17 MW [11]

Method	Equality		Extraction	
System configuration	Electricity (US$/MWh)	Steam (US$/t)	Electricity (US$/MWh)	Steam (US$/t)
Steam turbine	59.25	12.71	67.98	8.33
Gas turbine	26.01	12.97	41.88	5.99
Combined cycle	35.75	13.40	47.24	7.48

turbine. The combined cycle-based system needs to burn supplementary fuel in the heat recovery steam generator to attain the steam demand of the processes (in this system the gas turbine generates 12 MW and the steam turbine generates 5 MW).

The new components cost are determined by using some relations presented by Boehm [7] and information given by equipment manufacturers.

Table 3.16 shows the new values of electricity and process steam production costs. The electricity costs obtained for the gas turbine-based system and for the combined cycle-based system are lower than those calculated for the same type of systems in the first scenario. As observed in the first scenario, the steam turbine-based system gives the higher electricity production costs.

3.4.6 Discussion of the Obtained Results

The results given by the thermoeconomic analysis indicate that the three cogeneration systems have attractive performance and production costs of the utilities, which are competitive with the prices paid today by the industry. During the capital recovery period, the system that presents the lowest overall cost rate is the gas turbine one, in both operating scenarios.

Besides the results given by the thermoeconomic analysis some other aspects must be considered to choose the best cogeneration system such as operational flexibility and reliability of the equipment, and environmental impacts from cogeneration systems operation.

3.5 Exergy and Thermoeconomic Evaluation of Utilities Plants for a Dairy Industry

3.5.1 Introduction

The evaluation of cogeneration systems described in this chapter was conducted for the Colombian industrial sector scenario in order to provide a rational utilization of fossil fuels and reduction of electricity demand from the National Interconnected System of the country (SIN). The use of cogeneration systems in the Colombian industrial sector is important because this sector is the major consumer of natural gas, oil, and coal and the second major consumer of electricity in the country.

In the industrial sector, food industry is the major energy consumer. It also has the greater cogeneration potential. Within the food industrial sector, the dairy industry is a very interesting one because it demands various utilities and has a great variety of products and processes. Colombia has important natural gas reserves and the cogeneration systems could optimize its use. In addition to this fact, the substitution of the fuel oil used in the dairy industry by natural gas could bring environmental advantages [14].

The dairy industry, used as a model for the analysis in this study, produces milk, yogurt, cheese, oats, fats, and desserts of different types. The whey, that is a by-product of the processes, is used in the production of some drinks and in the feeding of pigs that are sold alive.

The utilities demand curves of the whole plant were constructed based on the demands of the main equipment and on their hourly tendency. In order to do that, it was made a survey of the processes that take place in every section of the plant and an inventory of the equipment operation in each process. The required utilities were identified and the thermal loads and mass flow rates were all calculated in an hourly basis [14].

3.5.2 Utilities Plant Description

This dairy industry demands electricity, steam, chilled water, compressed air, and potable water (see Figs. 3.19, 3.20).

The energy consumption and demand conditions and the means of production of the utilities for this plant are as follows:

- The plant uses electricity to operate pumps, stirring rods, beaters, mills, slicing machines, and packing machines, heating systems, fans and compressors, among others. The average power demand is 2.89 MW with a peak power demand of 3.16 MW (see Fig. 3.21). The electricity, which is bought from an energy retailer, is obtained from the National Interconnected System at 34.5 kV

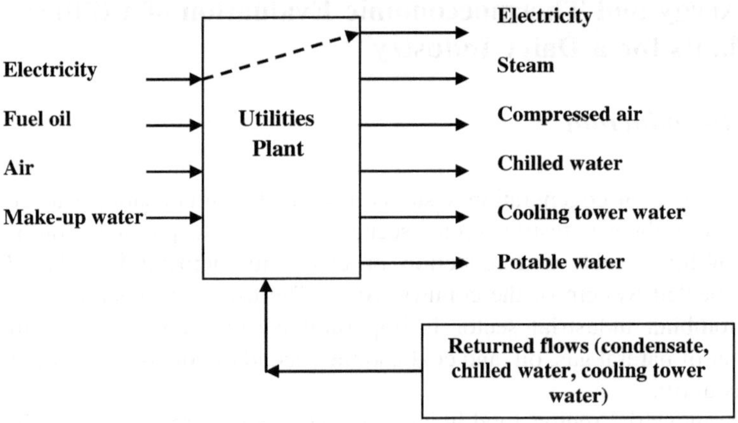

Fig. 3.19 Scheme of the utilities plant [23]

Fig. 3.20 Utilities distribution lines [23]

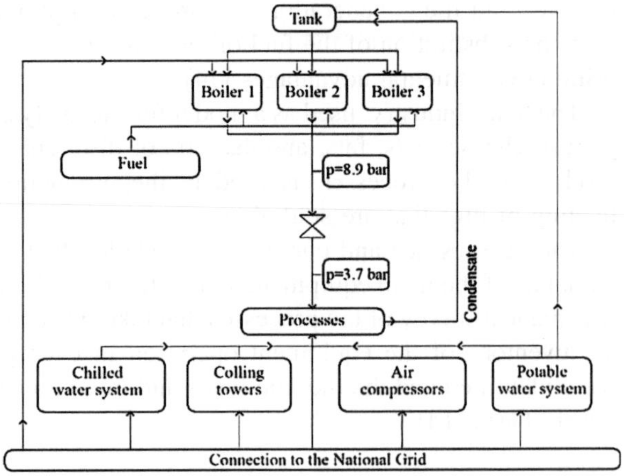

and transformed to 220 and 440 V. The electricity tariff is negotiated. On October/2000, the plant paid 60.56 US$/MWh.

- Steam is used to pasteurize, to sterilize, to heat water, and air and for cleaning purposes. It is also used in packing machines and industrial pots, among others. The majority of the applications use the steam enthalpy of condensation at 369.92 kPa and 414.15 K. The average energy demand of steam is 3.42 MW with a peak demand of 5.18 MW (see Fig. 3.21). This represents an average demand of 1.81 kg/s, with a peak demand of 2.42 kg/s. The plant has three boilers (one in stand-by) that can produce 2.49, 2.15, and 1.36 kg/s of steam at 889.80 kPa and 448.15 K, giving a total steam rate of 6 kg/s.
- Chilled water is employed in pasteurizing machines, cooling machines, production tanks, packing machines, cooling rooms, and in the salt water in which

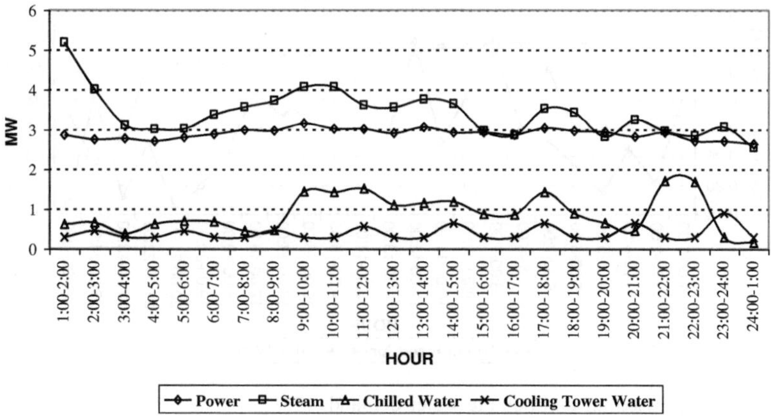

Fig. 3.21 Power, steam, chilled water, and cooling tower water demands [23]

cheese is submersed during manufacture, among others. The average energy demand of chilled water is 893.45 kW with a peak demand of 1705.14 kW (see Fig. 3.21). For the production of chilled water, the plant has an ammonia system composed of three compressors (one of them in stand-by), two evaporative condensers, six expansion valves, and 8 km of pipes. The pipes are submersed in six reservoirs in which water is cooled from 284.18 to 274.15 K, in average. The system has a refrigerating capacity of 1,825 kW.

- Cooling tower water is used in pasteurizing machines and production tanks, among others. It satisfies the cooling needs in processes where the use of chilled water is not required. In other processes, cooling tower water is employed in series with chilled water. The average energy demand of cooling tower water is 389.71 kW with a peak demand of 903.00 kW (see Fig. 3.21). The cooling tower water is produced in a system composed of five cooling towers, which cools water from 308.15 to 296.15 K, in average
- Compressed air is employed in processes, such as pressing, cutting, and packing. The average demand is 0.38 kg/s with a peak demand of 0.45 kg/s (see Fig. 3.22). The plant has a system composed of six compressors that can generate 0.69 kg/s of compressed air at 819.82 kPa, 293.15 K and relative humidity of 85 %.
- Potable water is used in the manufacture of various products, preparation of ferments, and cleaning, among others. The average demand is 15.73 kg/s with a peak demand of 22.10 kg/s (see Fig. 3.22). A system of pumps extracts underground water from four wells. The water is then purified in three parallel treatment plants and stored in tanks. The system can supply 22.69 kg/s of water at 274.94 kPa and 291.15 K.

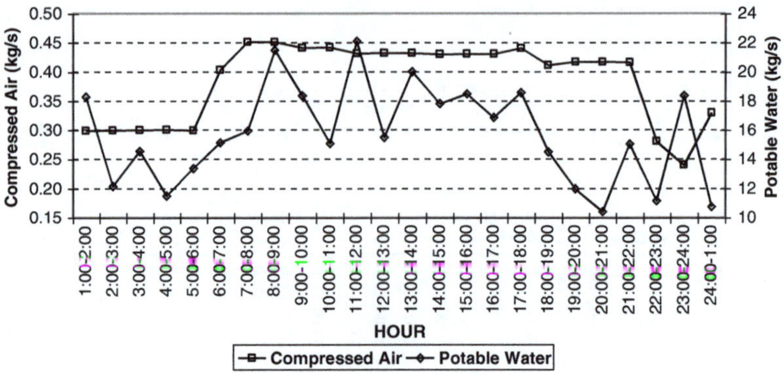

Fig. 3.22 Compressed air and potable water demands [23]

3.5.3 Cogeneration Systems

Figure 3.23 shows a scheme of the cogeneration utilities plant. The cogeneration systems analyzed in this study are: a steam cycle with extraction/condensation steam turbine, a gas turbine based system, and a gas engine-based system.

The steam turbine-based system is composed of an extraction/condensation steam turbine and a high pressure steam generator. The electricity generation capacity is 5 MW. Figure 3.24 shows this configuration. Steam is generated at 4,199 kPa and 596 K. This steam is sent to the extraction/condensation steam turbine and the required steam flow rate, for each hour of the day, is extracted at 400 kPa.

The gas turbine-based system is made up of a gas turbine of the same capacity of the steam turbine (the combustion chamber outlet temperature is 1,295 K) and a waste heat boiler that can produce the required steam flow for each hour at 890 kPa. This configuration is shown in Fig. 3.25.

The gas engine-based system, shown in Fig. 3.26, is made up of a gas engine with an electricity generation capacity of 4.5 MW and a waste heat boiler with a steam generation capacity of 0.75 kg/s at 850 kPa. The remaining steam that is required is generated in the existing boilers.

3.5.4 Comparative Exergy and Thermoeconomic Analysis

The exergy performance of the utilities production systems was quantified by calculating the overall exergy efficiency of the plants [3, 15].

The exergy efficiencies were calculated for the utilities plant in the its original operating condition and with the described cogeneration systems. Their values are presented in Table 3.17. The reference atmosphere has as pressure of 74 kPa, temperature of 291 K, and relative humidity of 75 %.

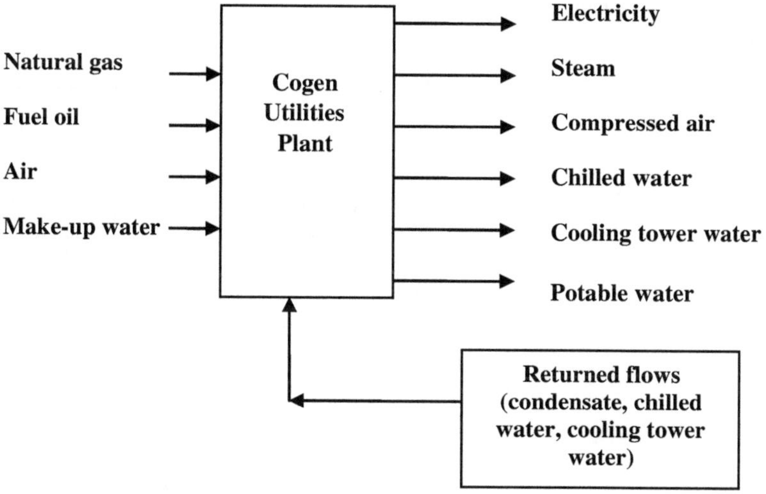

Fig. 3.23 Cogeneration utilities plant [23]

Fig. 3.24 Extraction/condensation steam turbine-based cogeneration system [23]

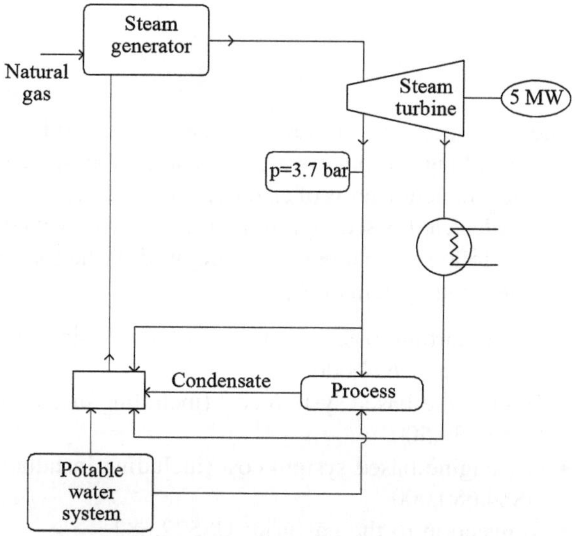

The gas turbine and the gas engine-based systems present the best exergy efficiency values. The exergy efficiency of the steam turbine-based system is smaller than the original one because in the calculation of the exergy efficiency of the original system it was not considered the fuel required to generate the purchased electricity.

The performance behavior of the plant in the original operating condition and for each one of the considered operating scenarios of the cogeneration plants was simulated by modeling all the processes with the aid of the software EES® [12].

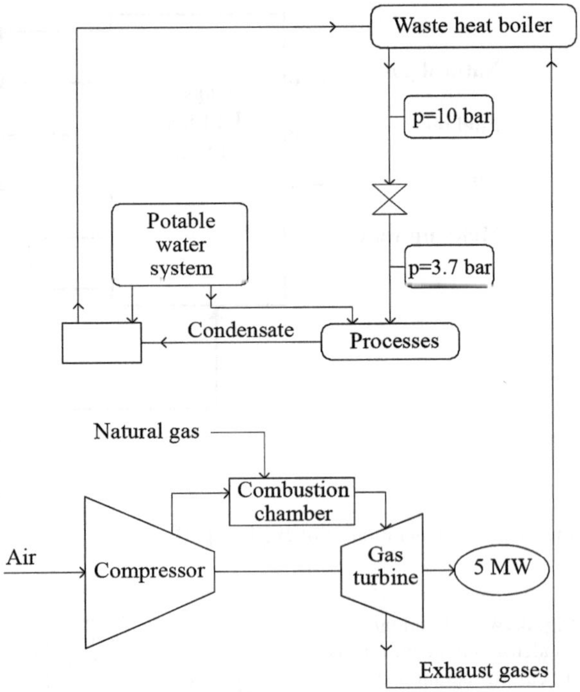

Fig. 3.25 Gas turbine-based cogeneration system [23]

The corresponding production costs of each utility and the costs of the utilities streams along the plant were estimated in an hourly basis.

The production costs of electricity and steam generated in the cogeneration systems were calculated based on programs also implemented in the software EES® [12].

The following parameters were used in the thermoeconomic assessment of the cogeneration systems [14]:

- Steam turbine-based system cost (including installation and commissioning costs): US$5,694,000
- Gas turbine-based system cost (including installation and commissioning costs): US$3,139,500
- Gas engine-based system cost (including installation and commissioning costs): US$4,680,000
- Connection to the gas grid: US$22,682
- Annual operational and maintenance costs: 10 % of investment cost
- Natural gas price: 2.5, 3.5 and 4.5 US$/MMBtu
- Interest rate: 9 % per year
- Capital recovery period: 10 years
- Load factor: 0.89
- Average operation time: 7,796 h/year
- Inflation is not considered

The original natural gas price was 4.5 US$/MMBtu.

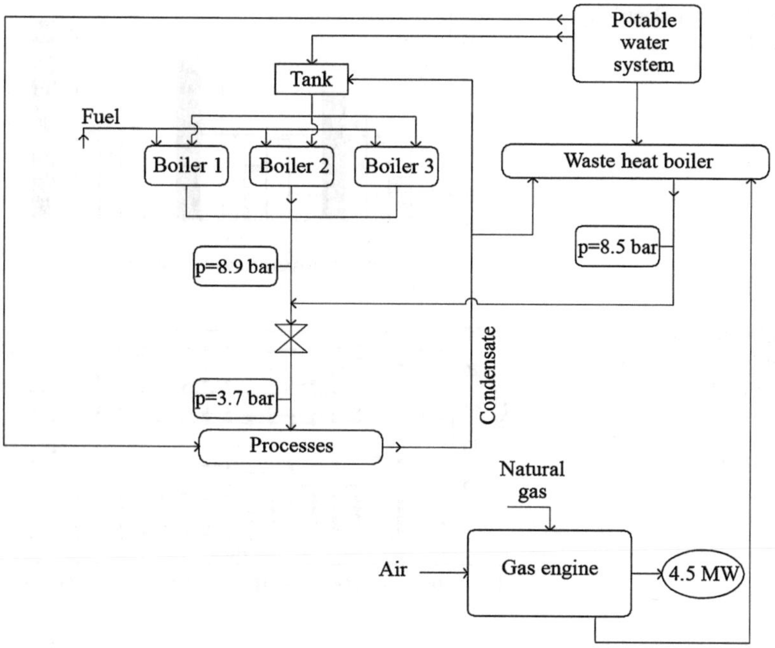

Fig. 3.26 Gas engine-based cogeneration system [23]

Table 3.17 Exergy efficiency of the utilities plant [23]

Production system	Exergy efficiency (%)
Original	24.0
Steam turbine cogeneration	21.3
Gas turbine cogeneration	30.4
Gas engine cogeneration	30.5

It was considered that the capital invested in the utilities production system, presently in use, has already been recovered. However, labor costs have been calculated from information furnished by the accounting department of the dairy industry and included in the analysis.

As used for the cogeneration systems studied for the chemical plant, described in Sect. 3.4, two cost partition methods were also considered here: equality and extraction criteria. In the gas engine, due to the available system information, only the equality method was used. Results for the electricity and the steam production costs are presented in Figs. 3.27 and 3.28.

The obtained results indicate that with natural gas at 4.5 US$/MMBtu, only the gas turbine-based system with the equality method can produce electricity at a lower cost than the price paid to the electricity company in the original operating condition (OOC). With the gas price at 2.5 or 3.5 US$/MMBtu, only the steam turbine-based system produces electricity at a higher cost, as shown in Fig. 3.27.

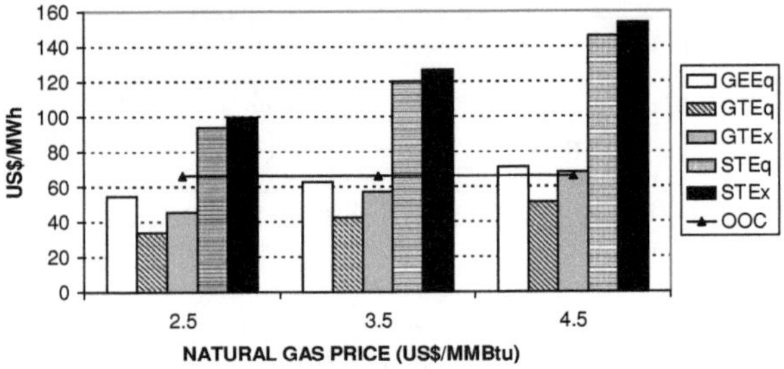

Fig. 3.27 Mean electricity cost [23]

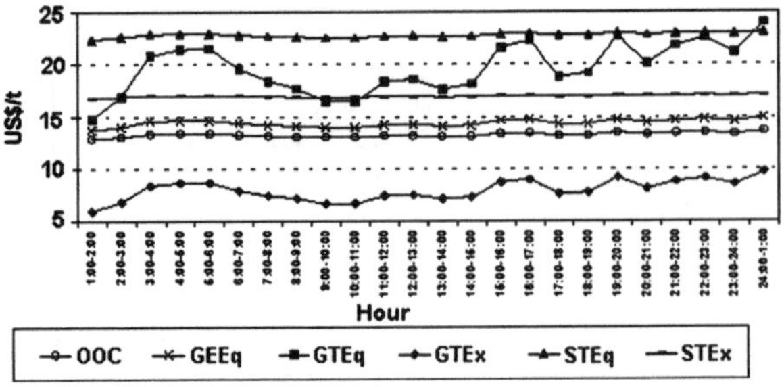

Fig. 3.28 Steam cost for natural gas at 3.5 US$/MMBtu [23]

The steam production costs are presented for the 24 h of the day because, for the gas turbine-based cogeneration system, the quantity of steam generated and consequently the investment cost per unit of mass varies appreciably. With natural gas at 3.5 US$/MMBtu, only the gas turbine-based system with the extraction method can produce steam at a lower cost than the original operating condition (OOC), as shown in Fig. 3.28.

The economic performance of the systems can be assessed by means of the total cost rate (TCR) of the systems, considering the investment cost (including the costs of the equipment, installation, and connection to the gas pipeline, operation and maintenance) and the cost associated to the fuel:

$$\mathrm{TCR} = \mathrm{I} + c_{\mathrm{fuel}} B_{\mathrm{fuel}} \tag{3.19}$$

Table 3.18 presents the TCR for the analyzed cogeneration systems and for the three natural gas prices.

These values can be compared with 268.39 US$/h, corresponding to the TCR in the original operating condition. Only the gas turbine-based system with natural gas at 2.5 US$/MMBtu presents a TCR somewhat lower.

As the considered cogeneration systems produce more electricity than is required in the plant. Then, the surplus could be negotiated in the Colombian Electricity Pool or directly with an electricity retailer or a nonregulated user. Taking the mean price in the Electricity Pool during the year 2000 as a reference, a TCR considering these revenues can be obtained, as indicated in Table 3.19.

Again, only the gas turbine-based system with natural gas at 2.5 US$/MMBtu presents a lower TCR than the reference case.

Table 3.18 Total cost rate (US$/h) [23]

Natural gas price (US$/MMBtu)	Steam turbine	Gas turbine	Gas engine
2.5	411.16	267.05	326.45
3.5	503.41	331.68	368.98
4.5	595.76	396.38	411.54

Table 3.19 Total cost rate selling electricity surplus (US$/h) [23]

Natural gas price (US$/MMBtu)	Steam turbine	Gas turbine	Gas engine
2.5	380.31	236.20	304.22
3.5	472.55	300.83	346.74
4.5	564.90	365.52	389.31

3.5.5 Concluding Remarks

From a thermoeconomic standpoint it is interesting to note the differences of the utilities costs resulting of different cost partition methods. The utilization of these methods evidences the importance of costs of products in a cogeneration plant according to its objectives.

The results show that the gas turbine-based system can be an alternative for the implementation of cogeneration in the Colombian dairy industry.

The results also indicate that the panorama for the dissemination of the cogeneration technology in the sector is not satisfactory due to the high price of natural gas and relative low price of electricity. This situation could be improved with a natural gas tariff policy that stimulates cogenerator with lower natural prices.

3.6 Exergoeconomic Evaluation of Trigeneration Systems

3.6.1 Introduction

A trigeneration system can be defined as a particular type of combined heat and power system that supplies, simultaneously, mechanical or electrical demand as well as heating and cooling demands, by consuming a single energy source. It is a very effective option of utilizing fuels exergy. Trigeneration systems can be divided into two parts: the combined heat and power unit (CHP), which generates electricity and supplies a given heating demand, and the second part, a compression or absorption chiller, which produces refrigerating effect using electricity and/or heat from the CHP unit. Combined heating and power technology has been in use in industrial applications since the end of nineteenth century. However, the rapid development of the technologies involved through the last decades, made easier the application of CHP technology in buildings, hotels, hospitals, schools, community heating, or waste treatment sites. Most recent advances incorporate the use of alternative fuels such as hydrogen or biomass, or the exploitation of excess heat converting it to cooling power, that is used in air conditioning or in various industrial processes [16]. Trigeneration plants have become economically viable due to the commercial spread of absorption chillers [17]. Absorption chillers are generally classified as direct or indirect-fired, and as single, double or triple-effect. In direct-fired units, the heat source can be gas or some other fuel that is burned in the unit. Indirect fired units use steam or some other heat transfer fluid that brings in heat from a separate source, such as a boiler or heat recovered from an industrial process.

Low pressure, steam-driven absorption chillers are available in capacities ranging from 100 to 1,500 TR (ton of refrigeration). Absorption chillers come in two commercially available designs: single-effect and double-effect. Single-effect machines provide a COP of about 0.7 and require about 8.2 kg of steam at 2.0 bar (abs) per TR of cooling capacity. Double-effect machines are about 40 % more efficient, but require higher grade thermal input, using about 4.5 kg of steam at 6.9–10.3 bar (abs) per TR [18].

In short, absorption cooling may fit when a source of free or low-cost heat is available, or if restrictions related to using conventional refrigeration exist. Essentially, the low-cost heat source displaces higher cost electricity in a conventional chiller.

Trigeneration includes various technologies like: gas turbines, steam turbines, combined cycles, internal combustion engines, fuel cells, and Stirling engines. Some works show diverse applications of the trigeneration systems: in supermarkets [19], in the petrochemical industry [20], in the food industry [21], and in hospitals [22].

In the study described in this section, an exergy and exergoeconomic comparison of different trigeneration systems, including a tetra-combined one is carried out in order to calculate the efficiencies and exergy-based cost of electricity, steam, and the exergy transferred to chilled water.

3.6.2 Trigeneration Systems

The analysed trigeneration systems are applied to supply the energy requirements of a dairy industry, taking the same Colombian dairy-industry used in Sect. 3.5 as a case study [23]. The industry energy demands are:

- 2.3 MW of electric power;
- 25 kg/s of chilled water at 5 °C (Evaporator capacity: 525 kW);
- 2 kg/s of saturated steam at 5 bar for process.

Five trigeneration system configurations, shown in Figs. 3.29, 3.30, 3.31, 3.32, and 3.33, are modelled and analysed. These configurations are based on:

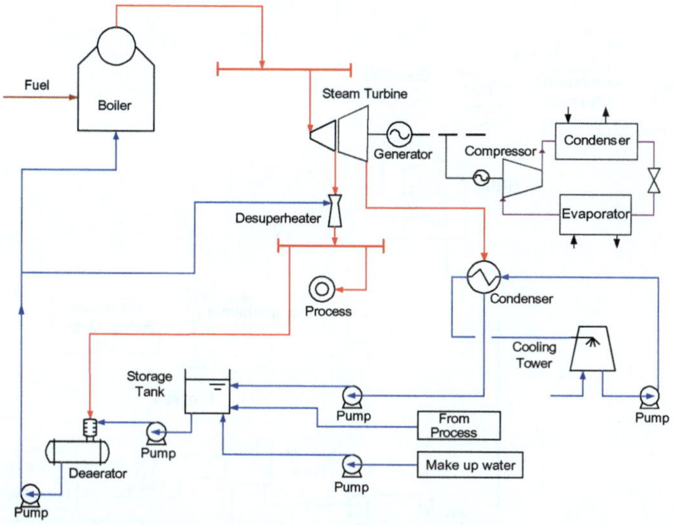

Fig. 3.29 Steam turbine configuration with compression refrigeration system [27]

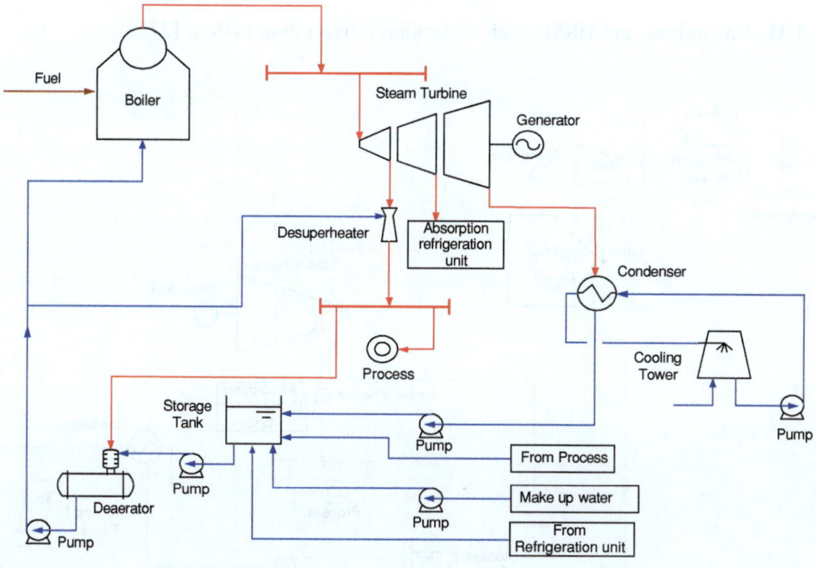

Fig. 3.30 Steam turbine configuration with single effect absorption refrigeration system [27]

- Steam turbine and vapor-compression refrigeration system (Fig. 3.29);
- Steam turbine and single effect absorption refrigeration system (Fig. 3.30);
- Gas turbine and single effect absorption refrigeration system (Fig. 3.31);
- Combined cycle with single effect absorption refrigeration system (Fig. 3.32);
- Tetra combined cycle (Figs. 3.33, 3.34)

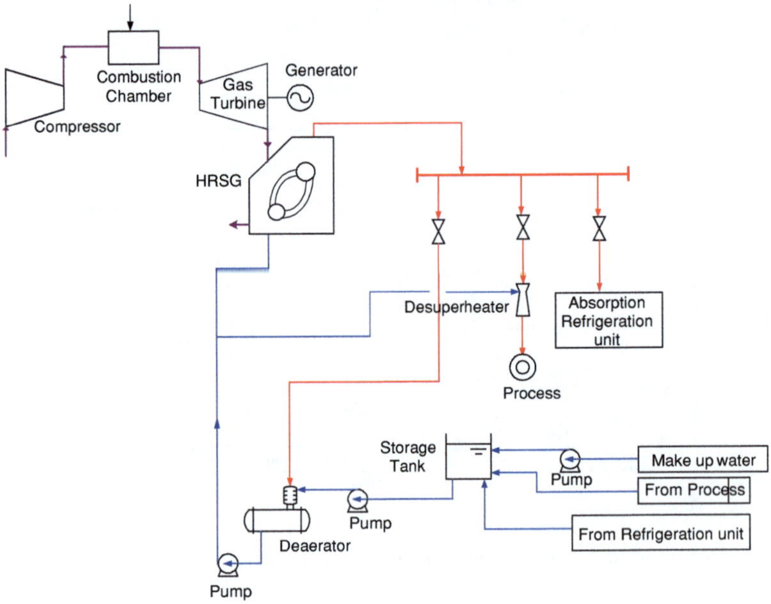

Fig. 3.31 Gas turbine and HRSG with absorption refrigeration system [27]

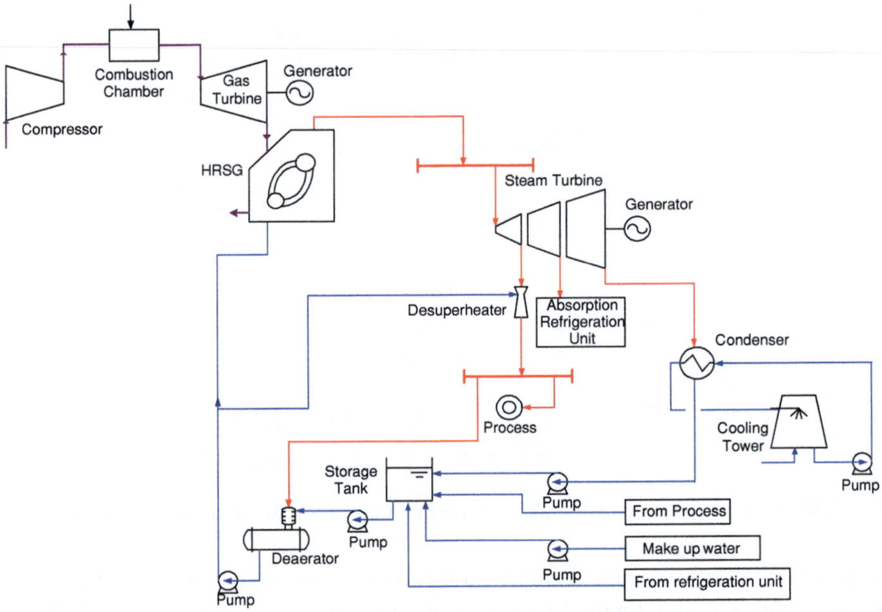

Fig. 3.32 Combined cycle with absorption refrigeration system [27]

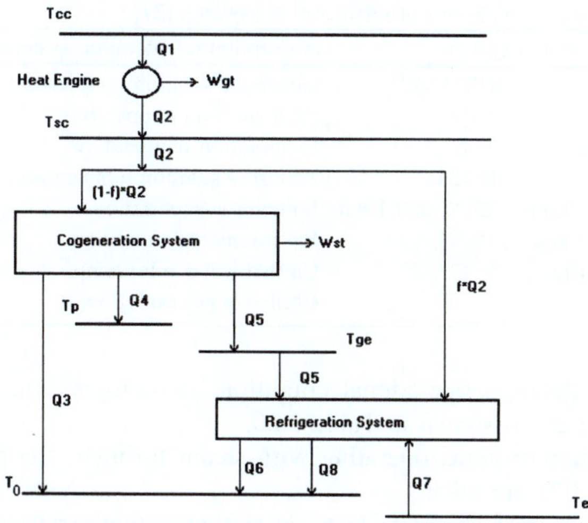

Fig. 3.33 Tetra-combined trigeneration system for electricity production, cooling, and heating purposes [24]

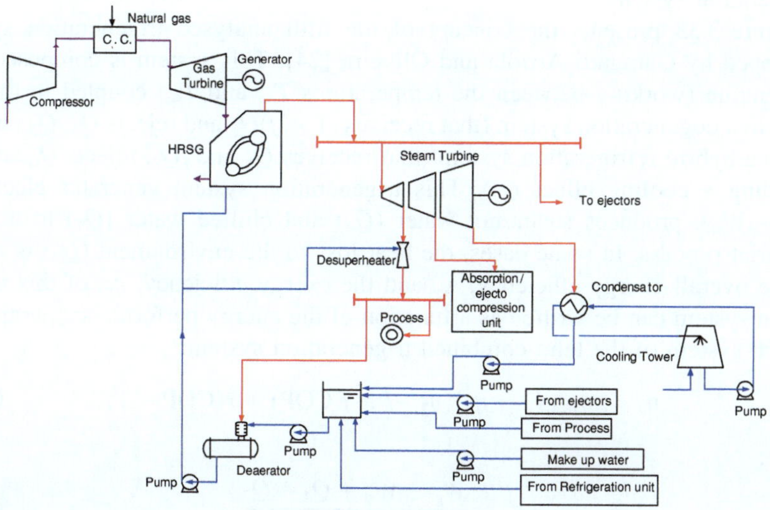

Fig. 3.34 Schematic diagram of the tetra-combined trigeneration system [27]

For the purpose of analysis of single effect water/lithium bromide absorption refrigeration systems, the following assumptions were made: in the generator and in the absorber, the lithium bromide solution is assumed to be in equilibrium; refrigerant (water) at condenser and evaporator exit is saturated; solution pump power and pressure losses in all the heat exchangers, and in the pipelines are negligible. The steam pressure that supplies heat to the generator was taken as 2 bar.

Table 3.20 Operational conditions of refrigeration systems [27]

Compression refrigeration system		Absorption refrigeration system	
Compressor power	107.7 kW	Generator solution temperature.	56–81 °C
Condensation temp.	40 °C	Condensation temperature	40 °C
Evaporation temperature	3 °C	Evaporation temperature	3 °C
Refrigerant	R 134a	Absorber solution temperature	60–31 °C
Environmental condition	25 °C and 1 bar	Solution concentrations	58.2–54.2 %
Chilled water inlet temp.	10 °C	Environmental condition	25 °C and 1 bar
Chilled water outlet temp.	5 °C	Chilled water inlet temp.	10 °C
		Chilled water outlet temp.	5 °C

The data set defining operational conditions of compression and absorption refrigeration systems is shown in Table 3.20.

For trigeneration systems operating with steam turbines, the parameters presented in Table 3.21 are taken.

The parameters used to simulate the trigeneration system based on gas turbine and HRSG with single effect absorption refrigeration system are shown in Table 3.22.

Additionally to parameters presented in Table 3.22, data described in Table 3.23 are considered for the simulation of the combined cycle-based trigeneration system.

Figure 3.33 presents the concept of the fifth analysed trigeneration system developed by Garagatti Arriola and Oliveira [24]. This system is composed of a heat engine (working between the temperatures T_{cc} and T_{sc}) coupled in thermal series to a cogeneration system (that receives $(1-f)Q_2$ and rejects Q_4, Q_5 and Q_3) and to a hybrid refrigeration system (that receives Q_5 and fQ_2, rejects Q_6 and Q_8, providing a cooling effect Q_7). This trigeneration system generates electricity ($W_{gt} + W_{st}$), produces steam/hot water (Q_4) and chilled water (Q_7) to a given industrial process. In some cases, the heat loss to the environment (Q_3) is zero.

The overall energy efficiency, η_e, and the exergy efficiency, η_b, of this trigeneration system can be written as a function of the energy performance parameters of each system of the tetra-combined trigeneration system:

$$\eta_e = \eta_1 + (1-\eta_2)(\eta_2 - f + f\,\mathrm{COP}) + r(\mathrm{COP}-1) \quad (3.20)$$

where:

$$\eta_e = \frac{W_{gt} + W_{st} + Q_4 + Q_7}{Q_1} \quad (3.21)$$

f = the fraction of the rejected heat of the heat engine that is sent to the refrigeration system;

$$\eta_2 = \frac{W_{st} + Q_4 + Q_5}{Q_2(1-f)} \quad (3.22)$$

$$\mathrm{COP} = \frac{Q_7}{(f\,Q_2 + Q_5)} \quad (3.23)$$

Table 3.21 Parameters of trigeneration systems operating under Rankine cycle [27]

Parameter	Value
Boiler efficiency (%, LHV basis)	85
Electric generator efficiency (%)	95
Steam pressure (bar)	42
Steam temperature (°C)	420
Turbine stages isentropic efficiency (%)	78–80
Pump isentropic efficiency (%)	70

Table 3.22 Assumed parameters used to simulate the third trigeneration system [27]

Parameter	Value
Turbine inlet temperature (TIT) (°C)	1,200
Compressor isentropic efficiency (%)	85
Gas turbine isentropic efficiency (%)	87
Saturated steam pressure (bar)	10
HRSG heat losses (%)	2
HRSG Pinch (°C)	10
HRSG Approach (°C)	5
Pump isentropic efficiency (%)	70

Table 3.23 Additional parameters to simulate the combined cycle and absorption refrigeration system based trigeneration system [27]

Parameter	Value
Electric generator efficiency (%)	95
Steam pressure (bar)	42
Steam temperature (°C)	420
Turbine (condensing–extraction) isentropic efficiency (%)	78–80
Pump isentropic efficiency (%)	70

$$r = \frac{Q_5}{Q_1} \tag{3.24}$$

Equation 3.20 shows the influence of the performance parameter of each component, as well as the rejected heat distribution, in the overall performance of the cogeneration system.

The exergy efficiency, η_b, of this trigeneration system is done by Eq. 3.25:

$$\eta_b = \frac{W_{gt} + W_{st} + Q_4\,\theta_4 + Q_7\,\theta_7}{Q_1\,\theta_1} \tag{3.25}$$

The tetra-combined trigeneration system is composed of three subsystems in thermal cascade: gas turbine, a cogeneration system based on a steam cycle and a hybrid absorption ejecto-compression chiller. The expression tetra-combined is derived from the fact of this system to be based on two power cycles (Brayton and Rankine) and two refrigeration technologies (absorption and ejecto-compression).

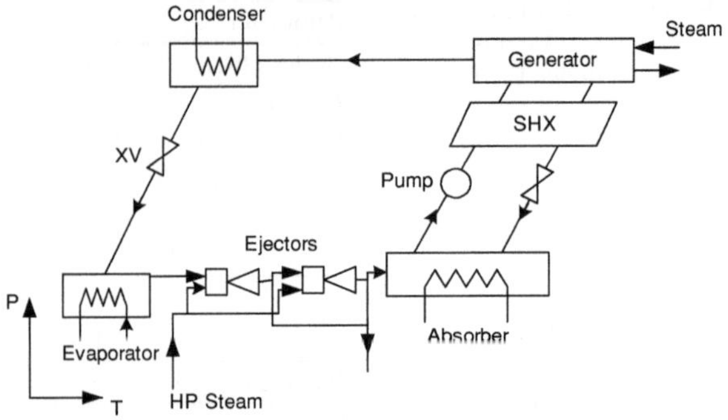

Fig. 3.35 Absorption-ejecto compression refrigeration system Dühring chart schematic [28]

In tetra-combined trigeneration system, the gas turbine produces power and it uses natural gas as energy source. The cogeneration subsystem, based on a steam cycle, uses the rejected gases from the gas turbine to produce superheated steam in a HRSG. The superheated steam feeds an extraction/condensation steam turbine. The steam turbine produces power and has three steam extractions. The first extraction is to feed the ejectors of the hybrid absorption ejecto-compression chiller; the second one is imposed by the process. This process steam is highly superheated and in certain applications saturated steam is needed, so it is necessary to include a desuperheater to take the superheated steam down to the saturated state. In the desuperheater a mixture of superheated steam and water is produced and therefore the saturated steam is obtained. The third steam extraction in turbine is used as heat source to feed the generator of hybrid absorption ejecto-compression chiller. The remaining steam goes out from turbine and enters in the condenser to be recovered like feed water for boiler.

Figure 3.34 shows a schematic diagram of the tetra-combined trigeneration system.

The absorption ejecto-compression refrigeration system had its origin in the work of Oliveira and Le Goff [25]. The system is characterized by having ejectors between the evaporator and absorber. The operation is similar to the single-effect absorption system, with the variation of using ejectors. The number of ejectors depends on the steam pressure decrease required in the evaporator and the steam pressure increase required in the absorber. Each ejector operates with a pressure ratio of about two. The ejector exhaust is discharged to the absorber, causing the absorber pressure to be at a higher level than that in the evaporator. Therefore, the solution within the absorber can be kept away from crystallization when the system is needed to operate with low evaporator temperature or with high absorber temperature such as an air-cooled unit. Figure 3.35 shows a water/lithium-bromide absorption ejecto-compression refrigeration system.

The processes inside the ejector are represented in Fig. 3.36. The refrigerant vapor at low pressure enters the ejector at point 1, and it is drawn by means of the expansion of the high pressure steam at point 2 that produces a vacuum when it flows in the ejector from point 2 and proceeds up to point A, increasing the refrigerant vapor pressure to point 3. This steam at the exit of the first ejector is

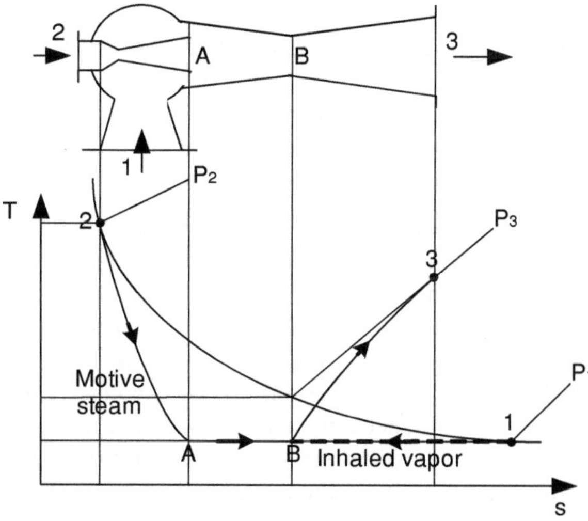

Fig. 3.36 Evolution of motive steam and drawn vapour throughout ejector [29]

used by a second ejector to lift up the refrigerant vapor pressure up to the absorber pressure. Steam then is absorbed by the strong solution coming from the generator, and this diluted solution is sent back to generator by means of the solution pump.

To reduce the motive steam consumption, in each ejector exit there is a mass flow rate deviation that is sent again to the boiler, or used in another process, if the pressure conditions allows.

3.6.3 Modelling and Simulation of Trigeneration Systems

The models presented before were implemented in the software EES® [12], and simulated considering steady-state operation.

For the purpose of analysis of absorption refrigeration systems, the following assumptions are made:

- The analysis is made under steady state conditions;
- The refrigerant at the outlet of the condenser is saturated liquid;
- The refrigerant at the outlet of the evaporator is saturated vapor;
- The outlet temperatures from the absorber and from generators correspond to equilibrium conditions of the mixing and separation, respectively;
- Pressure losses in the pipelines and in heat exchangers are negligible;
- Heat exchanges between the system and surroundings, other than the prescribed at the generator, high temperature generator (in double-effect absorption system), evaporator, condenser, and absorber, are negligible;
- The reference environmental is at 25 °C (T_0) and 1 bar (P_0).

In hybrid absorption ejecto-compression chiller, the following assumptions were considered for ejectors energy balance:

- Adiabatic flow;
- The kinetic energy in Chaps. 1, 2 and 3 of the ejector is negligible;
- 1D flow;
- Steady state condition;
- The saturation pressure at the mixing region of vapors (point A to point B of Fig. 3.36) is constant.

The hybrid absorption ejecto-compression refrigeration system was modeled with two ejectors connected in series flow arrangement and using a pressure ratio of 1.8 for each one.

The production costs were evaluated considering the exergy-based cost for fuel equal to 1 kJ/kJ. For distribution costs in control volumes with more than one product, the equality criterion was adopted (see Chap. 2). That is, each product has the same importance and consequently their exergy-based cost were set equal (i.e. electricity and process steam, in cogeneration systems). Thus, the cost associated to the irreversibility in the control volume is distributed equally among the exergy content of the outlet product flows.

3.6.4 Results

The performance results are presented and discussed for three energy requirements scenarios [26]:

- in the first scenario steam turbine configurations supply the plant requirements (2.3 MW of electric power, 25 kg/s of chilled water at 5 °C, and 2 kg/s of saturated steam at 5 bar for process);
- the second one includes a gas turbine with HRSG and absorption refrigeration system capable to generate 1,800 kW of electricity surplus;
- the third one includes a combined cycle with absorption refrigeration system and the tetra-combined trigeneration system generating 7,500 kW of electricity surplus.

For the described configurations, the energy and exergy efficiencies are calculated according to Eqs. 3.26 and 3.27, and the results are presented in Tables 3.24, 3.25 and 3.26.

$$\eta_e = \frac{W_{plant} + W_{excess} + Q_{process} + Q_{chilledwater}}{m_{fuel}\,\mathrm{LHV}} \tag{3.26}$$

$$\eta_b = \frac{W_{plant} + W_{excess} + \Delta B_{process} + \Delta B_{chilledwater}}{m_{fuel} b_{fuel}} \tag{3.27}$$

where W_{plant} is the electric dairy demand, W_{excess} is the excess generated electricity that can be sold to the grid, $Q_{process}$ is the heating process demand, $Q_{chilledwater}$ is the cooling demand, m_{fuel} is the fuel mass flow rate and LHV is the lower heating

Table 3.24 Energy and exergy efficiency of trigeneration systems based on steam turbine [28]

Trigeneration system	η_e (%)	η_b (%)
Steam turbine with compression refrigeration system	56.88	26.64
Steam turbine with absorption refrigeration system	57.37	26.87

Table 3.25 Energy and exergy efficiency of trigeneration systems with 1,800 kW excess electricity [28]

Trigeneration system	η_e (%)	η_b (%)
Steam turbine with compression refrigeration system	44.89	25.23
Steam turbine with absorption refrigeration system	45.28	25.45
Gas turbine and HRSG with absorption refrigeration system	79.12	44.65

Table 3.26 Energy and exergy efficiency of trigeneration systems with 7,500 kW excess electricity [28]

Trigeneration system	η_e (%)	η_b (%)
Steam turbine with compression refrigeration system	33.53	23.90
Steam turbine with absorption refrigeration system	33.82	24.11
Combined cycle with absorption refrigeration system	65.45	46.21
Tetra-combined cycle	65.79	46.45

value, $\Delta B_{\text{process}}$ is the exergy flow rate variation of process steam, $\Delta B_{\text{chilledwater}}$ is the exergy flow rate variation of chilled water, and b_{fuel} is the fuel specific exergy.

In order to show the results of exergoeconomic study for the trigeneration systems, the fuel exergy rate, exergy destruction rate, and exergy-based costs for each configuration were calculated and discussed next.

Figure 3.37 shows fuel exergy rate (kW) of each studied configuration for the analyzed scenarios: without and with electricity surplus.

As Fig. 3.37 shows, fuel exergy rate for configuration using steam turbines systems is very similar, in all cases, with a small advantage for the cycle with absorption refrigeration system. In the second scenario, gas turbine with HRSG and absorption refrigeration system presents a reduction in fuel exergy rate, around 44 %, with respect to steam turbine configurations.

For 7,500 kW excess electricity generation, combined cycle with absorption refrigeration system and tetra-combined cycle present important reduction (around 48 %) of fuel exergy rate with respect to steam turbine with compression and absorption refrigeration systems, respectively.

Figure 3.38 shows the exergy destruction rate caused by systems operation for all studied cases.

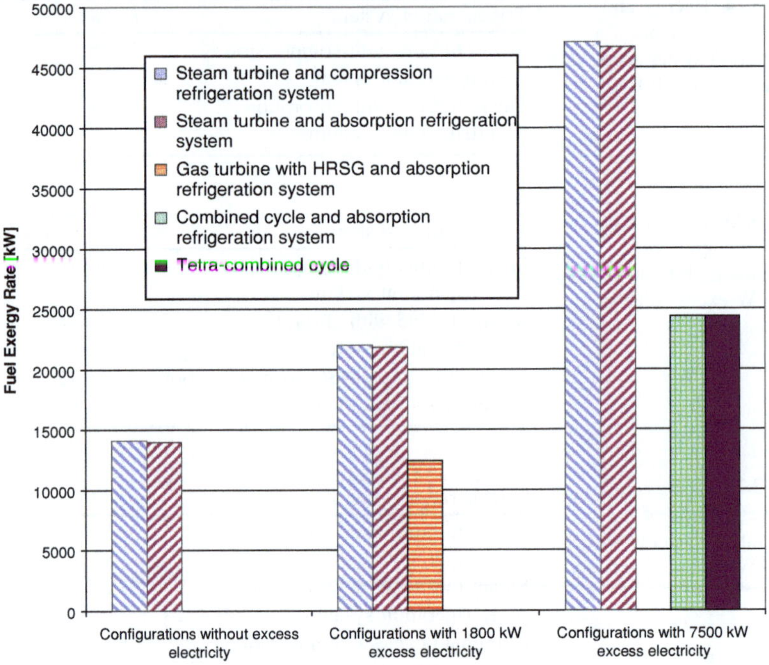

Fig. 3.37 Fuel exergy rate comparison for the analyzed systems [28]

A comparative analysis of the studied systems, shown in Fig. 3.38 is similar to the one presented in Fig. 3.37. The exergy destroyed rate by the system with steam turbine with compression refrigeration system is slightly higher than the configuration with steam turbine and absorption refrigeration system, for all analyzed cases. For 1,800 kW of electricity surplus, gas turbine with HRSG and absorption refrigeration system presents a reduction in exergy destruction of around 63 % with respect to steam turbine configurations. For 7,500 kW of electricity surplus, the exergy destroyed by tetra-combined cycle is lightly lower than combined cycle with absorption refrigeration system and also a reduction of exergy destruction around of 71 % with respect to steam turbine with compression and absorption refrigeration systems.

Exergy-based costs (kJ/kJ) of electricity, process steam and exergy transferred to chilled water for the studied configurations in the three operating scenarios are summarized in Fig. 3.39.

In Fig. 3.39 it can be concluded that without electricity surplus, the exergy-based cost of electricity and process steam are similar in steam turbine configurations, being slightly higher in the system with compression refrigeration system. The comparison for chilled water cost between these two configurations is interesting, since in the first case, exergy based cost is 11.3 kJ/kJ, whereas with absorption cycle it is 19.55 kJ/kJ. The reason for this difference is due to the values

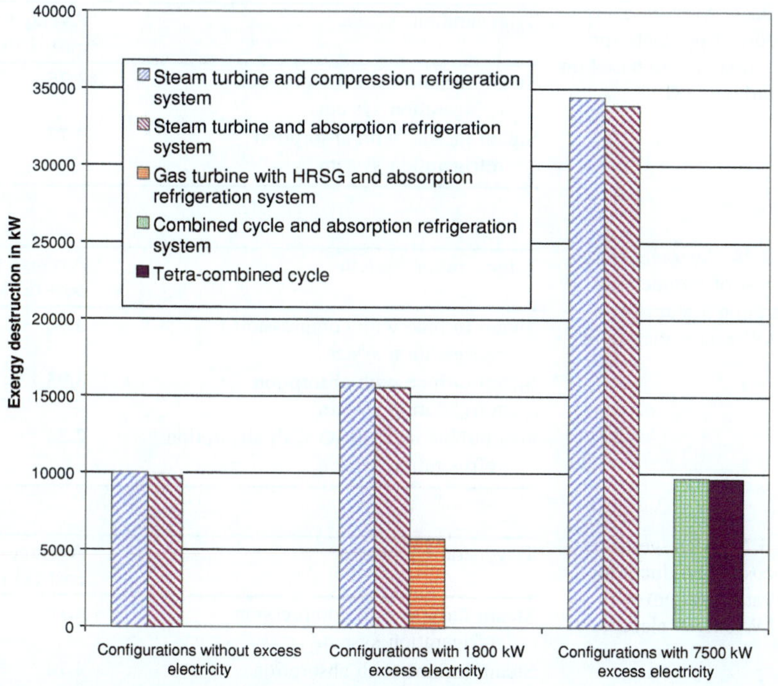

Fig. 3.38 Exergy destruction for the different systems [28]

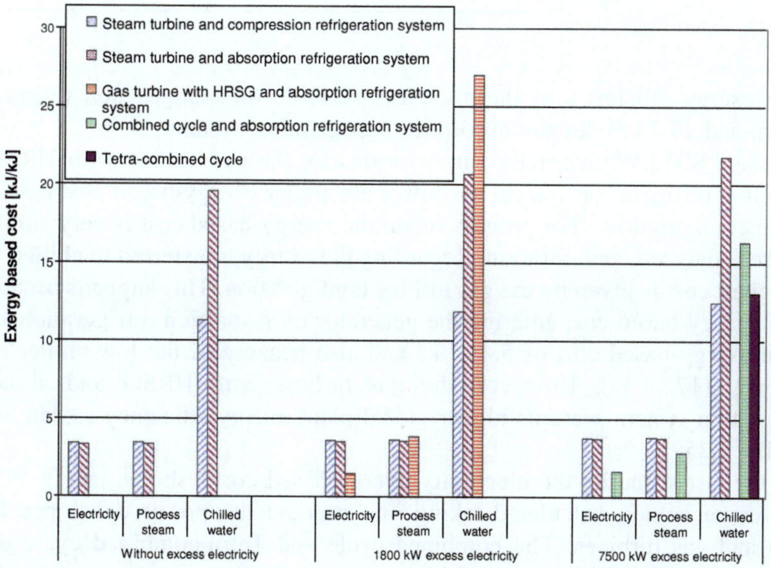

Fig. 3.39 Exergy-based costs (kJ/kJ) of electricity, process steam and chilled water for the studied configurations [28]

Table 3.27 Average exergy-based cost of products for trigeneration system based on steam turbine [28]

Trigeneration system	Average exergy cost (kJ/kJ)
Steam turbine with compression refrigeration system	3.75
Steam turbine with absorption refrigeration system	3.72

Table 3.28 Average exergy-based cost of products for trigeneration system with 1,800 kW excess electricity [28]

Trigeneration system	Average exergy cost (kJ/kJ)
Steam turbine with compression refrigeration system	3.96
Steam turbine with absorption refrigeration system	3.93
Gas turbine and HRSG with absorption refrigeration system	2.24

Table 3.29 Average exergy-based cost of products for trigeneration system with 7,500 kW excess electricity [28]

Trigeneration system	Average exergy cost (kJ/kJ)
Steam turbine with compression refrigeration system	4.18
Steam turbine with absorption refrigeration system	4.14
Combined cycle with absorption refrigeration system	2.17
Tetra-combined cycle	2.15

of the exergy efficiency of these cycles: 30.41 % for compression refrigeration system, and 17.14 % for the absorption refrigeration system.

In the 1,800 kW electricity surplus scenario, the gas turbine with HRSG and absorption refrigeration system has lower electricity exergy-based cost than steam turbine configurations. For process steam the exergy-based cost is very similar for the three analyzed configuration. Regarding the exergy transferred to chilled water, the highest cost is given by the gas turbine configuration. This happens because the steam exergy-based cost entering the generator of absorption refrigeration system has an exergy-based cost of 5.4 kJ/kJ and also because of the low chiller exergy efficiency (17.14 %). However, the gas turbine with HRSG and absorption refrigeration system presents higher global-plant exergy efficiency as can be seen in Table 3.25.

In the last scenario, the electricity exergy-based costs, shown in Fig. 3.39 are the average values, calculated taking into account the power developed by the steam and gas turbines. The combined cycle and Tetra-combined cycle present considerable reduction in electricity and process steam exergy-based costs in comparison with steam turbine configuration. In this scenario, the biggest reduction in the exergy-based cost of chilled water is presented in the Tetra-combined cycle, as consequence of having an exergy efficiency in the hybrid absorption ejecto-compression chiller of 21.24 %, in contrast with the exergy efficiency of the single effect absorption chiller of combined cycle, 17.14 %, showing an advantage of including ejectors between the evaporator and absorber.

Tables 3.27, 3.28 and 3.29 show the average exergy-based cost of products for the different studied configurations.

In general, observing the impact in the formation of energy conversion costs for the proposed configurations, the minor impact in exergy-based costs of products (electricity, process steam and exergy transferred to chilled water) is for the Tetra-combined cycle, as it is possible to appreciate looking at the average exergy-based costs of the different configurations on Tables 3.27, 3.28 and 3.29.

Nevertheless, to choose an alternative or another, it is also necessary to take into consideration technical and financial aspects, since better exergy-based costs are directly linked to the efficiency of the energy conversion processes.

3.6.5 Concluding Remarks

Trigeneration represents a quite interesting alternative of producing electricity and reducing the production costs of utilities. In this comparative study, an analysis of different trigeneration systems was done, including a tetra-combined system, by means of the use of exergoeconomic analysis to quantify its energy and exergy efficiency and the impact in the production of electricity, process steam and chilled water for air conditioning purposes. The preliminary performance results of the studied trigeneration systems show the viability of tetra-combined cycle. This system has higher exergy efficiency than the single effect absorption chiller. This configuration also presents a bigger impact in the exergy-based cost of chilled water in comparison with other analyzed systems.

References

1. Szargut J, David RM, Steward F (1988) Exergy analysis of thermal, chemical, and metallurgical processes. Hemisphere Publishing, New York
2. Bejan A (1988) Advanced engineering thermodynamics. Wiley, New York
3. Oliveira S Jr, Van Hombeeck H (1997) Exergy analysis of petroleum separation processes in offshore platforms. Energ Convers Manage 38:1577–1584
4. Beyer J (1970) Strukturuntersuchungen-notwendiger Bestandteil der effekivitatsanalyse von warmeverbrauchersysteme. Energieanwendung 19:358–361
5. Borelli SJS, Oliveira S Jr (2008) Exergy based method for analysing the composition of the electricity cost generated in gas-fired combined cycle plants. Energy 33:153–162
6. Means RS (2002) Mechanical cost data. RS Means Company, Inc., EUA—2002
7. Boehm RF (1987) Design analysis of thermal systems. Wiley, New York
8. GATECYCLE software, v.5.5.1 (2003) General Electric Power Systems Inc.

9. Means RS (1998) Ministry of mines and energy. In: National energy balance, Brasilia (In Portuguese)
10. Tolmasquim M et al (1999) Evaluation of technical and economic potentials and difficulties identification to the use of cogeneration in selected sectors in Brazil. In: PROCEL/COOPE, Rio de Janeiro (In Portuguese)
11. Teixeira MS, Oliveira S Jr (2001) Thermoeconomic evaluation of cogeneration systems for a chemical plant. Int J Thermodyn 4:157–163
12. Klein SA (2011) Engineering equation solver—EES, F-Chart Software, www.fChart.com
13. Cespedes JFP, Oliveira S Jr (1995) Cogeneration in the Brazilian tertiary sector: exergetic and thermoeconomic analysis. In: Proceedings of the 8th international conference on efficiency, costs, optimization, simulation and environmental impact of energy systems, Istanbul
14. Larrazábal ML (2001) Thermoeconomic analysis of the use of cogeneration with natural gas in the Colombian dairy industry. Masters on Energy Dissertation, University of São Paulo, São Paulo, Brazil (In Portuguese)
15. Tsatsaronis G (1995) On the efficiency of energy systems In: Proceedings of the 8th international conference on efficiency, costs, optimization, simulation and environmental impact of energy systems, Istanbul
16. Kavvadias KC, Tosios AP, Maroulis ZB (2010) Design of a combined heating, cooling and power system: sizing, operation strategy selection and parametric analysis. Energy Convers Manag 51:833–845
17. Cardona E, Piacentino A (2003) A methodology for sizing a trigeneration plant in Mediterranean areas. Appl Therm Eng 23:1665–1680
18. Absorption Chillers (2010) Available at: http://www.absorptionchillers.com/. Cited Jan 2010
19. Maidment GG, Tozer RM, Missenden JF (2001) Combined cooling, heat and power (CCHP) in Supermarkets. In: Heat powered cycles conference, conservatoire national des arts et métiers, Paris
20. Colonna P, Gabrielli S (2003) Industrial trigeneration using ammonia-water absorption refrigeration systems (AAR). Appl Therm Eng 23:381–396
21. Bassols J, Kuckelkorn B, Langreck J et al (2002) Trigeneration in the food industry. Appl Therm Eng 22:595–602
22. Ziher D, Poredos A (2006) Economics of trigenerationsystem in a hospital. Appl Therm Eng 26:680–687
23. Larrazábal ML, Oliveira S Jr (2002) Thermoeconomic evaluation of cogeneration systems for a dairy industry. In: Proceedings of the 18th international conference on efficiency, costs, optimization, simulation and environmental impact of energy systems, Berlin
24. Garagatti Arriola DW, Oliveira S Jr (2001) Tetra-combined cogeneration system. Exergy and thermoeconomic analysis. In: Proceedings of the congress HPC'01—cooling, heating and power systems, Paris
25. Oliveira S Jr, Le Goff P (1993) Hybrid systems absorption-compression to upgrade industrial waste heat. In: In: Proceedings of the energy systems and ecology proceedings of the international conference, Krakow
26. Burbano JC, Pellegrini LF, Oliveira S Jr (2009) Comparative exergoeconomic analysis of trigeneration systems for a dairy industry. In: Proceedings of the 22nd international conference on efficiency, costs, optimization, simulation and environmental impact of energy systems, Foz de Iguaçú
27. Burbano JC (2011) Exergoeconomic optimization of tetra-combined trigeneration system. Ph.D. Thesis Polytechnic School of the University of São Paulo, São Paulo, Brazil (In Portuguese)
28. Burbano JC, Pellegrini LF, Oliveira S Jr (2010) Exergoeconomic analysis of tetra-VCombined trigeneration systems. In: Proceedings of the 23rd international conference on efficiency, costs, optimization, simulation and environmental impact of energy systems, Lausanne

29. Oliveira S Jr (1991) Upgrading industrial thermal effluents—Exergetic, entropic and economic analysis. Ph.D. Thesis. Polytechnic National Institute of Lorraine, Nancy (In French)
30. Borelli SJS (2005) Method for the analysis of the composition of electricity costs in combined cycle thermoelectric power plants. Master Dissertation, Institute of Eletro-technical and Energy, University of Sao Paulo, São Paulo, Brazil (In Portuguese)
31. Gas Turbine World Handbook 2004–2005, Volume 24—Pequot Publications, 2005

Chapter 4
Exergy Evaluation of Petroleum Production and Refining Processes

Symbols

AC	Air compressor
b	Specific exergy (kJ/kg)
bpd	Barrels per day
B	Exergy rate/flow rate of a substance/product (kW)
C_i	Net monetary rate value, cost rate of a substance, product or equipment (US$/year, US$/s)
c_i	Exergy specific cost of a substance or product (kJ/kJ, US$/kJ, US$/kWh, US$/t)
f_i	Relation between the exergy consumed in the module, i and in the whole plant
f_a	Capital recovery factor
f_l	Load factor
f_omf	Operational and maintenance fixed cost factor
f_omv	Operational and maintenance variable cost factor
h, H	Specific enthalpy (kJ/kg), enthalpy rate (kW)
ΔH	Enthalpy variation between reactants and combustion products (kJ/kg)
I	Cost (US$)
m	Mass flow rate (kg/s)
P	Pressure (bar)
P_0	Environment pressure (kPa)
Q	Heat transfer rate (kW)
R	Gas constant (kJ/kg K)
s, S	Specific entropy (kJ/kgK), entropy rate (kW/K)
ΔS	Entropy variation between reactants and combustion products (kJ/kgK)
T_0	Environment temperature (°C)
W	Power (kW)
x	Weight factor

S. de Oliveira Jr., *Exergy*, Green Energy and Technology,
DOI: 10.1007/978-1-4471-4165-5_4,

Greek Letters

α, β, γ	Stoichiometric coefficients
α	Relation between chemical exergy and lower heating value
Δ	Variation
η	Efficiency
θ	Carnot factor $(1 - T_o/T)$
ρ_m	Specific gas consumption

Subscripts

aeq	Annual equipment cost
air	Air
av	Average
b	Exergy
bb	Exergy based
bo	Boiler
bo, fu	Boiler fuel
bt	Time based
cgbo, fu	Fuel gas for boiler
cgtg	Fuel gas for gas turbine
ch	Chemical
ci	Turbine condenser
cm	Compression module
cold	Cold fluid
condensed	Condensed
comp	Compressor
comp, i	Compressor inlet
comp, o	Compressor outlet
cw	Cooling tower
dest	Destroyed
e	Electricity
eg	Exhaust gas
electric	Electric
eq	Equipment
equip	Equipment
extraction	Extraction
f	Furnace
fw	Feed water
fu, fuel	Fuel
g	Gas
gas	Gas
g, bo	Gas sent to the boiler
g, gt	Gas sent to the gas turbine
gs	Gas at separator outlet
ger	Generated

$g0$	Reference for the calculation of gas exergy
gs	Gas at the exit of the separator
gt	Gas turbine
gt, fu	Gas turbine fuel
h	Heating, heater
he	Heat exchanger
he, i	Heat exchanger inlet
he, o	Heat exchanger outlet
hot	Hot fluid
i	Indicates a flow or a module, inlet, component i
in	Inlet
$i0$	Reference for the calculation of substance i exergy
m	Number of carbon atoms
mb	Mass based
mechanical	Mechanical
min	Minimum
n	Number of hydrogen molecules
o	Oil, reference for the calculation of exergy
o0	Reference for the calculation of oil exergy
os	Oil at the exit of the separator
out	Outlet
overall	Considering the whole plant
p	Petroleum
p_m	Pumping module
P_0	Reference for the calculation of petroleum exergy
prod	Product
pu	Pump
pu, i	Pump inlet
pu, o	Pump outlet
rec	Heat recovery system
s	Outlet
sep	Separator, separation process
steam	Steam
t	Turbine
tb	time basis
valv, tg	Gas turbine fuel expansion valve
valv, bo	Boiler fuel expansion valve
w, water	Water
w0	Reference for the calculation of water exergy
wp	Water pump

Superscripts

0	standard
wf	Without supplementary fuel
sf	With supplementary fuel

Abbreviations

B	Boiler
DEA	Deaerator
E	Electricity
FCC	Fluidized catalytic cracking
FG	Fuel gas
FO	Fuel oil
GT	Gas turbine
HDT	Hydro-treatment
HP	High pressure
HT	Heat transfer
IP	Intermediate pressure
LHV	Lower heating value
LP	Low pressure
LPG	Liquefied petroleum gas
NG	Natural gas
MP	Mechanical Power
PSA	Pressure swing adsorption
RB	Recovery boiler
ST	Steam
T	Steam turbine
US$	American dollar
V	Valve
W	Water
WTP	Water treatment process

4.1 Introduction

Petroleum is the most important energy source in the World. In 2008, it represented 33.2 % of the primary energy offer in the World [1]. According to EIA-US [2], the global conventional petroleum-derived liquid fuels consumption tends to increase from 81.85×10^6 barrels per day (bpd) in 2010 to 97.7×10^6 bpd in 2035. A challenge as important as the substitution of the petroleum by the so-called renewable energy sources is the efficient and rational use of petroleum reserves.

The decrease in the availability of light crudes pushes the petroleum companies to search the black gold into deeper regions of the sea, increasing the production risks and the probability of environmental impacts.

The efficient and environmentally acceptable production, processing, and use of petroleum reserves are very important fields to be explored by the exergy analysis of energy conversion processes that take place from the well to the atmosphere.

Thus, in the next sections the method of the exergy and thermoeconomic analysis applied to petroleum processing will be described and some examples of analysis, including processes that take place in offshore platforms and refineries, will be presented and discussed.

4.2 Exergy Analysis of Petroleum Separation Processes in Offshore Platforms

4.2.1 Introduction

In an offshore platform, petroleum is separated into oil, gas, and water. This plant consumes exergy in order to heat the petroleum, compress the natural gas, and pump the oil to the coast.

The fuel utilized in the offshore platform is normally part of the natural gas produced. This gas is consumed in gas turbines (GT) or engines, to generate electricity for the plant and to drive gas compressors, and in furnaces to heat the petroleum before the separation process. The heating load can be partially supplied by the exergy of turbine (or gas engine) exhaust gases. The recovery of prime movers exhaust gases for heating purposes characterizes the offshore plant as a typical cogeneration plant.

Figure 4.1 shows a scheme of the energy conversion processes that take place in an offshore platform. It can be seen two modules of GTs, one that drives a gas compressor and one that is coupled to an electric generator, one booster compressor (driven by an electric motor), a set of two pumps (driven by electric motors), a heat recovery system, a furnace, and a separator. The heat load required by the separator is supplied by the furnace and by the turbine exhaust gases that are sent to the heat recovery system. Figure 4.1 shows also the flows of petroleum (sent to the separator), water (discarded from the plant), oil (sent to the pumps), and gas (that is partially consumed in the plant).

As natural gas is used as the plant fuel, inefficient energy conversion processes imply additional petroleum consumption for the plant operation and more CO_2 sent to the atmosphere. This fact shows the importance of developing the exergy analysis of the platform.

4.2.2 Exergy Analysis of an Offshore Primary Petroleum Processing Plant

Figure 4.2 shows the mass flow rates, enthalpy, entropy, and exergy rates during the operation of an offshore platform. Considering that it operates in steady state conditions and that there are no heat interactions with the environment, the

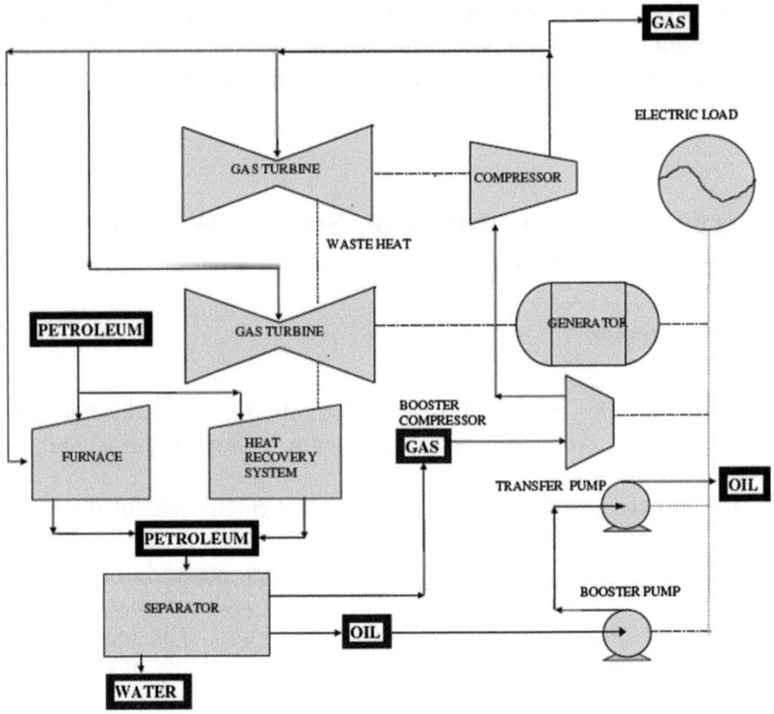

Fig. 4.1 Schematic representation of a petroleum primary processing plant [3]

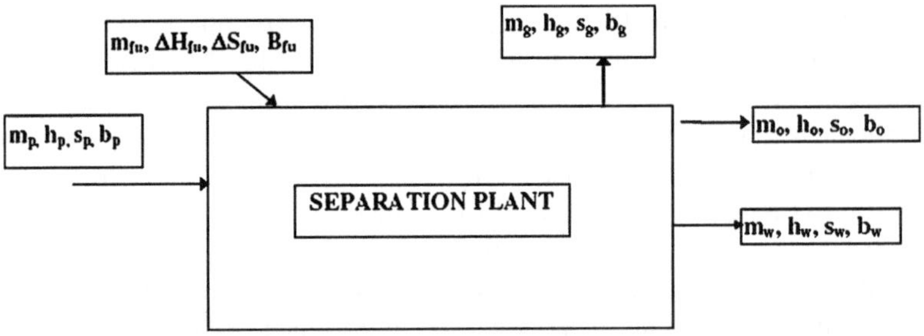

Fig. 4.2 Mass flow rate, energy, entropy, and exergy rates in an offshore platform [3]

balances of mass (restricted to the petroleum separation), energy, entropy, and exergy are given by Eqs. 4.1–4.4:

$$m_o + m_g + m_w = m_p \tag{4.1}$$

$$m_p h_p + m_{fu} \Delta H_{fu} = m_o h_o + m_g h_g + m_w h_w \tag{4.2}$$

$$m_o s_o + m_g s_g + m_w s_w - (m_p s_p + m_{fu} \Delta S_{fu}) = S_{ger} \tag{4.3}$$

$$m_p b_p + B_{fu} = m_o b_o + m_g b_g + m_w b_w + B_{dest} \tag{4.4}$$

Overall enthalpy and exergy of petroleum separation (considering the processes of separation, compression, and pumping that take place in the platform) are defined, respectively, by Eqs. 4.5 and 4.6:

$$H_{overall} = m_o h_o + m_g h_g + m_w h_w - m_p h_p \tag{4.5}$$

$$B_{overall} = m_o b_o + m_g b_g + m_w b_w - m_p b_p \tag{4.6}$$

Writing Eq. (4.4) in terms of specific enthalpies and entropies:

$$\begin{aligned} m_p\left[h_p - h_{p0} - T_0(s_p - s_{p0})\right] + B_{fu} &= m_o\left[h_o - h_{o0} - T_0(s_o - s_{o0})\right] \\ &+ m_g\left[h_g - h_{g0} - T_0(s_g - s_{g0})\right] + m_w\left[h_w - h_{w0} - T_0(s_w - s_{w0})\right] + B_{dest} \end{aligned} \tag{4.7}$$

It is possible to obtain the following expression:

$$\begin{aligned} B_{fu} = {} & m_o(h_o - T_0 s_o) + m_g(h_g - T_0 s_g) + m_w(h_w - T_0 s_w) - m_p(h_p - T_0 s_p) \\ & - \left[m_o(h_{o0} - T_0 s_{o0}) + m_g(h_{g0} - T_0 s_{g0}) + m_w(h_{w0} - T_0 s_{w0}) - m_p(h_{p0} - T_0 s_{p0})\right] \\ & + B_{dest} \end{aligned}$$

As the plant operates in steady state conditions, as shown in Chap. 2:

$$\sum_i m_i(h_{i0} - T_0 s_{i0}) = 0 \tag{4.8}$$

and then, Eq. 4.7 is reduced to:

$$\begin{aligned} B_{fu} = {} & m_o(h_o - T_0 s_o) + m_g(h_g - T_0 s_g) + m_w(h_w - T_0 s_w) - m_p(h_p - T_0 s_p) \\ & + B_{dest} \end{aligned} \tag{4.9}$$

With Eq. 4.9, it is possible to evaluate the thermodynamic performance of the separation, compression, and pumping processes that take place in the offshore platform. The exergy efficiency of these processes can be calculated by means of the following expression:

$$\eta_b = \frac{\text{Useful exergy effect}}{\text{Consumed exergy}} \tag{4.10}$$

For the plant presented in Fig. 4.2, the global exergy efficiency is:

$$[\eta_b]_{overall} = \frac{\Delta B_{overall}}{B_{fu}} \tag{4.11}$$

$$[\eta_b]_{overall} = \frac{m_o(h_o - T_0 s_o) + m_g(h_g - T_0 s_g) + m_w(h_w - T_0 s_w) - m_p(h_p - T_0 s_p)}{B_{fu}}$$

The exergy efficiency of the main modules that compose the separation plant: separator (including the heat recovery system and the furnace), compressor, and pumping, are given by Eqs. 4.12–4.14:

$$[\eta_b]_{sep} = \frac{m_o(h_{os} - T_0 s_{os}) + m_g(h_{gs} - T_0 s_{gs}) + m_w(h_w - T_0 s_w) - m_p(h_p - T_0 s_p)}{Q_h\left(1 - \frac{T_0}{T}\right)} \quad (4.12)$$

$$[\eta_b]_{comp} = \frac{m_g(b_{comp,o} - b_{comp,i})}{W_{comp}} \quad (4.13)$$

$$[\eta_b]_{pu} = \frac{m_o(b_{pu,o} - b_{pu,i})}{W_{pu}} \quad (4.14)$$

By introducing the factor f_i, defined as the relation between the exergy consumed by each module and the exergy consumed by the plant, and presented in Eqs. 4.15–4.17, respectively for the separator, compressor, and pumping modules, it is possible to relate the global efficiency $[\eta_b]_{global}$ with the exergy efficiencies of each module, as shown in Eq. 4.18.

$$f_{sep} = \frac{Q_h\left(1 - \frac{T_0}{T}\right)}{B_{fu}} \quad (4.15)$$

$$f_{comp} = \frac{W_{comp}}{B_{fu}} \quad (4.16)$$

$$f_{pu} = \frac{W_{pu}}{B_{fu}} \quad (4.17)$$

$$[\eta_b]_{overall} = [\eta_b]_{sep} f_{sep} + [\eta_b]_{comp} f_{comp} + [\eta_b]_{pu} f_{pu} \quad (4.18)$$

The $[\eta_b]_{overall}$ can also be written as:

$$[\eta_b]_{overall} = \Sigma([\eta_b]_i f_i) \quad (4.19)$$

It must be pointed out that:

$$f_{sep} + f_{comp} + f_{pu} < 1 \quad (4.20)$$

due to the irreversible processes of heating, compression, and pumping.

In this way the importance of each module, in the overall exergy efficiency can be characterized, allowing the evaluation of the influence of the consumed exergy and of the efficiency of the energy conversion processes of each module.

Considering that gas compressors and oil pumps are driven by GTs with thermal efficiency η_t, that the heat recovery system has a thermal efficiency η_{rec}, and that the furnace has a thermal efficiency η_f, the consumed exergy is done by:

$$B_{fu} = B_t + B_f \quad (4.21)$$

The exergy rates B_t and B_f are given, respectively, by Eqs. 4.22 and 4.23:

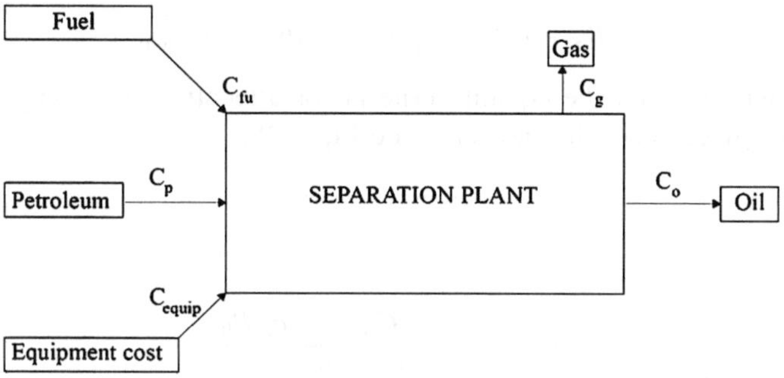

Fig. 4.3 Inputs and products of an offshore platform

$$B_{\mathrm{t}} = \frac{1}{\eta_{\mathrm{t}}}\left(\sum\left(W_{\mathrm{comp}}+W_{\mathrm{pu}}\right)\right) \tag{4.22}$$

$$B_{\mathrm{f}} = \frac{\theta_{\mathrm{f}}}{\eta_{\mathrm{f}}}\left(Q_{\mathrm{h}} - \eta_{\mathrm{rec}}\left(\frac{1}{\eta_{\mathrm{t}}} - 1\right)\sum\left(W_{\mathrm{comp}}+W_{\mathrm{pu}}\right)\right) \tag{4.23}$$

where the heat transfer rate in the furnace, neglecting turbine heat losses, is:

$$Q_{\mathrm{f}} = \left(Q_{\mathrm{h}} - \eta_{\mathrm{rec}}\left(\frac{1}{\eta_{\mathrm{t}}} - 1\right)\sum\left(W_{\mathrm{comp}}+W_{\mathrm{pu}}\right)\right) \tag{4.24}$$

The mass flow rate of gas consumed in the turbines and in the furnace is:

$$m_{\mathrm{fu}} = \frac{B_{\mathrm{fu}}}{b_{\mathrm{ch}}} \tag{4.25}$$

The calculation of m_{fu} allows establishing the relation between the mass flow rates of fuel gas and of the produced gas in the platform:

$$\rho_m = m_{\mathrm{fu}}/\, m_{\mathrm{g}} \tag{4.26}$$

This parameter, the specific gas consumption, can be considered as another platform performance parameter.

4.2.3 Thermoeconomic Analysis of an Offshore Platform

Figure 4.3 shows the cost balance for a separation plant, identifying the inputs, petroleum, fuel, and capital cost, and the products, oil, and gas. The water that is separated from the petroleum is considered as waste of the plant.

The cost balance of the separation plant is given by Eq. 4.27.

$$c_o B_o + c B_g = c_p B_p + c_{fu} B_{fu} + C_{equip} \tag{4.27}$$

Taking into account the equality criteria for allocating the exergy-based production cost of gas and oil, one can write Eq. 4.28:

$$c_g = c_o = c \tag{4.28}$$

with

$$c = \frac{c_p B_p + C_{equip} + c_{fu} B_{fu}}{B_g + B_o} \tag{4.29}$$

A deeper analysis of the cost formation processes in the offshore platform requires the application of cost balances for every component of the plant. The set of cost balances for the utilities plant (see Fig. 4.1) is composed of balances applied to GTs, boilers, separators, gas compressors, and pumps. These cost balances will give the exergy specific costs of heating the petroleum for separation purposes, pumping the oil, and compressing the gas, as presented next.

4.2.3.1 Gas Turbine

Taking the GT as the control volume, the cost balance gives:

$$c_{gt,fu} B_{gt,fu} + C_{gt} = c_e W_e + c_{eg} B_{eg} \tag{4.30}$$

As the GT generates two products, mechanical/electrical power and the exergy associated to the exhaust gas, it is necessary the adoption of a cost allocation criterion in order to be possible the determination of c_e and c_{eg}. Two criteria can be chosen, the equality and extraction one:

$$c_e = c_{eg} \text{ (equality criterion)} \tag{4.31}$$

or

$$c_{gt,\,fu} = c_{eg} \text{ (extraction criterion)} \tag{4.32}$$

The adoption of the extraction criterion results in Eq. 4.33:

$$c_e = \left[c_{gt,fu}\left(B_{gt,fu} - B_{eg}\right) + C_{gt}\right] / W_e \tag{4.33}$$

4.2.3.2 Boiler

In an offshore platform, the boilers (indicated in Fig. 4.1 as furnace and heat recovery system) can operate without (when the thermal load is supplied by the GT exhaust gases) and with supplementary fuel consumption to supply the required heat load to heat the petroleum that will undertake the separation operation. The cost balances for both conditions are described by Eqs. 4.34 and 4.35:

- without supplementary gas consumption:

$$c_{eg}B_{eg} + W_{wp}^{wf}c_e + B_{fw}^{wf}c_{fa} + C_{bo}^{wf} = B_{he,i}^{wf}c_{he,\,i}^{wf} \tag{4.34}$$

- with supplementary gas consumption:

$$c_{g,bo,\,fu}B_{g,\,bo} + B_{fw}^{sf}c_{fw} + W_{wp}^{sf}c_e + c_{eg}B_{eg} + C_{bo}^{sf} = B_{he,\,i}^{sf}c_{he,\,i}^{sf} \tag{4.35}$$

The average feed water cost at the inlet of the heat exchanger used to increase the petroleum temperature (see Fig. 4.1) can be determined by Eq. 4.36:

$$c_{he,av} = \left(B_{he,\,i}^{sf}c_{he,\,i}^{sf} + B_{he,\,i}^{wf}c_{he,\,i}^{wf}\right)\Big/\left(B_{he,\,i}^{sf} + B_{he,\,i}^{wf}\right) \tag{4.36}$$

4.2.3.3 Separator

The cost balance for the separator includes the inlet and outlet flows of the hot water (see Fig. 4.1):

$$B_pc_p + C_{sep} + \left(B_{he,\,i}^{sf} + B_{he,\,i}^{wf}\right)c_{he,\,av} = B_{gs}c_{gs} + B_{os}c_{os} + B_{he,\,o}c_{he,\,o} \tag{4.37}$$

As there are three unknowns in Eq. 4.37 (c_{gs}, c_{os}, and $c_{he,o}$) two cost partition criteria are used. They are given by Eqs. 4.38 and 4.39:

- equality criterion for the products of the separator, taking into account that the water that is sent back to the sea (after being properly treated) is valueless:

$$c_{os} = c_{gs} \tag{4.38}$$

- extraction criterion for the hot water used to heat the petroleum:

$$c_{he,av} = c_{he,o} = c_{he} \tag{4.39}$$

These two criteria allow rewriting Eq. 4.37:

$$B_pc_p + C_{sep} + \left(B_{he,\,i}^{sf} + B_{he,\,i}^{wf} - B_{he,\,o}\right)c_{he} = \left(B_{gs} + B_{os}\right)c_{gs} \tag{4.40}$$

4.2.3.4 Hot Water Pump

The cost balance for the boiler and the waste heat boiler hot water pump is:

$$B_{he,\,o}c_{he} + C_{wp} + W_{wp}c_e = B_{fw}c_{fw} \tag{4.41}$$

4.2.3.5 Gas Compression Module

The cost balance applied to the gas compression module gives the gas production cost:

$$B_{gs}c_{gs} + C_{cm} + W_{cm}c_e = B_g c_g \tag{4.42}$$

4.2.3.6 Oil Pumping Module

The cost balance applied to the oil pumping module gives the oil production cost:

$$B_{os}c_{os} + C_{pm} + W_{pm}c_e = B_o c_o \tag{4.43}$$

4.2.3.7 Gas Expansion Valves

The gas cost change when it is expanded up to the GTs and boilers operating pressures is obtained by Eq. 4.44 (for GTs) and 4.45 (for boilers):

$$B_{g,\,gt}c_g + C_{valv,\,gt} = B_{gt,\,fu}c_{gt,\,fu} \tag{4.44}$$

$$B_{g,\,bo}c_g + C_{valv,\,bo} = B_{bo,\,fu}c_{bo,\,fu} \tag{4.45}$$

The solution of this set of equations gives the values of c_i for each analyzed control volume, including the products of the offshore platform: c_o and c_g.

4.2.4 Exergy Evaluation of an Offshore Petroleum Separation Plant

The described exergy analysis approach is employed to evaluate the quality of the energy conversion processes in an offshore platform. The analyzed offshore plant is composed by the following modules, presented in Fig. 4.4:

- separation, with a petroleum heater and two three-phase separators;
- compression, with four compression stages and five sets of gas cooling and liquid separation;
- pumping, with two pumps.

There is one inlet flow of petroleum (section FEED), one outlet flow of gas (section GAS), one outlet flow of oil (section OIL), and two outlet flows of water (Sections 1A and 2A). Table 4.1 shows the mole fractions of components of the inlet flow of petroleum and the outlet flows of oil and gas. Table 4.2 shows important variables in some sections of the plant.

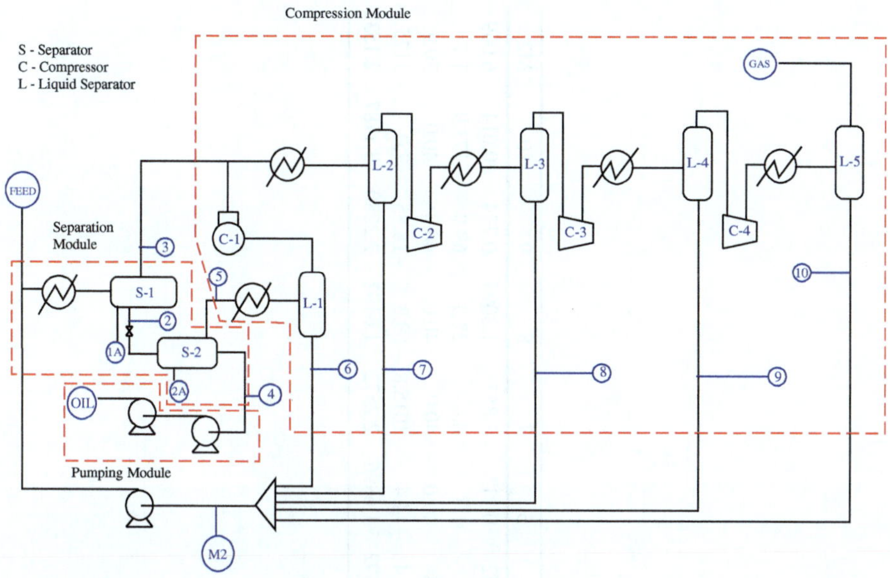

Fig. 4.4 Scheme of the analyzed offshore platform [3]

Table 4.1 Mole fractions of inlet and outlet platform flows [3]

Section/component	Feed	Gas	Oil
Methane	0.1255	0.8468	0.0037
Ethane	0.0092	0.0613	0.0022
Propane	0.0066	0.0421	0.0061
i-Butane	0.0016	0.0094	0.0035
n-Butane	0.0034	0.0190	0.0097
i-Pentane	0.0014	0.0062	0.0080
n-Pentane	0.0018	0.0072	0.0121
n-Hexane	0.0031	0.0056	0.0370
n-Heptane	0.0034	0.0014	0.0521
n-Octane	0.0042	0.0002	0.0682
n-Nonane	0.0066	–	0.1078
n-Decano	0.0019	–	0.0310
n-C11	0.0039	–	0.0637
C12+	0.0363	–	0.5932
H_2O	0.7912	0.0008	0.0018

The prime movers of compressors and pumps are GTs (that drive these machines directly and by means of an electric motor) that are equipped with heat recovery systems in order to heat the petroleum before the separation operation. The heating load is complemented with the use of a furnace that consumes fuel gas.

The analysis of the plant was developed utilizing the process simulator HYSIM, version C 2.50 [4], in order to obtain the thermodynamic properties of the

Table 4.2 Flow characteristics [3]

Section/variable	Feed	Gas	Oil	1A	2A	3	4	5	6	7	8	9	10	M2
m (t/h)	450.000	29.013	285.068	135.917	–	34.230	285.068	0.852	0.115	3.857	1.3095	0.774	0.014	6.069
p (bar)	10.78	173.87	68.65	9.3	–	9.3	2.2	2.2	1.7	8.6	22.9	69.9	173.9	1.7
T (°C)	7.4	40.0	92.7	90.0	–	90.0	89.9	89.8	40.0	40.0	40.0	40.0	40.0	30.9
H (kW)	−67343.3	2930.1	16006.6	−61659.4	–	6083.6	15240.1	126.4	−14.4	−1015.8	−69.3	−18.85	−7.0	−1125.4
s (kJ/kgK)	2.1178	6.6590	1.6933	3.8345	–	8.1598	1.6810	5.7892	1.6896	2.2572	1.6192	2.2280	3.1787	2.1532

Table 4.3 Energy inputs of the modules/equipment [3]

Energy inputs	(kW)
Heating load	2,8190.6
Power of compressor 1	40.8
Power of compressor 2	1,493.3
Power of compressor 3	1,604.5
Power of compressor 4	1,139.7
Pumping power	835.3

Table 4.4 Exergy inputs of the platform [3]

Exergy inputs	(kW)
Turbine demand	20,942.8
Furnace demand	24,492.1
Total demand	45,434.9

Table 4.5 Performance parameters of the modules of the platform [3]

Parameters/module	Separation	Compression	Pumping
η^a	0.222	0.480	0.621
f	0.186	0.094	0.018
$(\eta\ f)$	0.041	0.045	0.011

[a] $T_o = 25$ °C, $\theta = 0.3$, $(B_{ch})_{fu} = 50{,}953$ kJ/kg

substances considered in the analysis, and to give the energy balances for each module. The Peng–Robinson equation of state [5] was chosen for the characterization of the behavior of petroleum and its derivatives. The performance parameters of some components of the plant are listed bellow:

- polytropic efficiency of compression: 0.75
- mechanical efficiency of compressors: 0.90
- transmission efficiency: 0.95
- pump efficiency: 0.75
- GT thermal efficiency: 0.30
- furnace thermal efficiency: 0.85
- heat recovery system thermal efficiency: 0.60

The calculation of the chemical exergy of fuel gas was based on data presented by Kotas [6].

Tables 4.3, 4.4, 4.5 and 4.6 present performance parameters of the modules and of the whole offshore plant.

The results of Tables 4.5 and 4.6 point out:

- The importance of the petroleum heating operation in the exergy consumption of the plant. As the separation temperature was fixed on 90 °C (due to the petroleum characteristics), the heating load required an exergy consumption in the furnace 17 % higher than that of turbine, even with the use of heat recovery system.

Table 4.6 Performance parameters of the offshore platform [3]

Overall separation exergy (kW)	4,407.2
Overall exergetic efficiency	0.097
Combustion gas consumption (kg/h)	3,210
Specific gas consumption	0.111

- Heating and compression operations are the main exergy consumers of the plant.
- The exergy efficiency of the separation module presents the lowest value of the plant, due to the difference between the separation temperature and the combustion/exhaust gases temperature.
- The low $\eta_{b\ \mathrm{overall}}$ value is due to the high heating load required by the plant and because mechanical power is generated in an internal combustion machine with an exergy efficiency lower than 30 %.

These results highlight the importance of the exergy consumption in heating operations that precede the separation of petroleum, showing the relevance of the utilization of the exergy associated to the exhaust gases of GTs, as well as the correct choice of the separation temperature.

4.2.5 Exergo-Economic Comparison of Petroleum Primary Processing Artificial Lift Systems

4.2.5.1 Introduction

A procedure to determine the electricity and heat demands as well as the production specific costs for an offshore platform operating with and without the aid of artificial lift systems is described in this section in order to compare their operational performance. The considered artificial lift systems are a subsea multiphase pump (SMPS) and the gas lift system (GL).

The specific costs are determined in an exergy basis to allow a comparison of different configurations in a rational way. Moreover, with the exergy approach, it is possible to identify the main sources of irreversibilities (inefficiencies) and directly compare similar processes that occur in petroleum boosting systems. The evaluation of the performance of the SMPS among other alternatives is necessary since there are several intrinsic advantages and disadvantages associated to this boosting technology. This must be assessed in order to establish the best operational conditions and scenarios for the best performance.

4.2.5.2 Twin-Screw Multiphase Pump

The twin-screw multiphase pump is a positive displacement device with an axial flow through the screws. In this kind of pump, the rotation and meshing of the

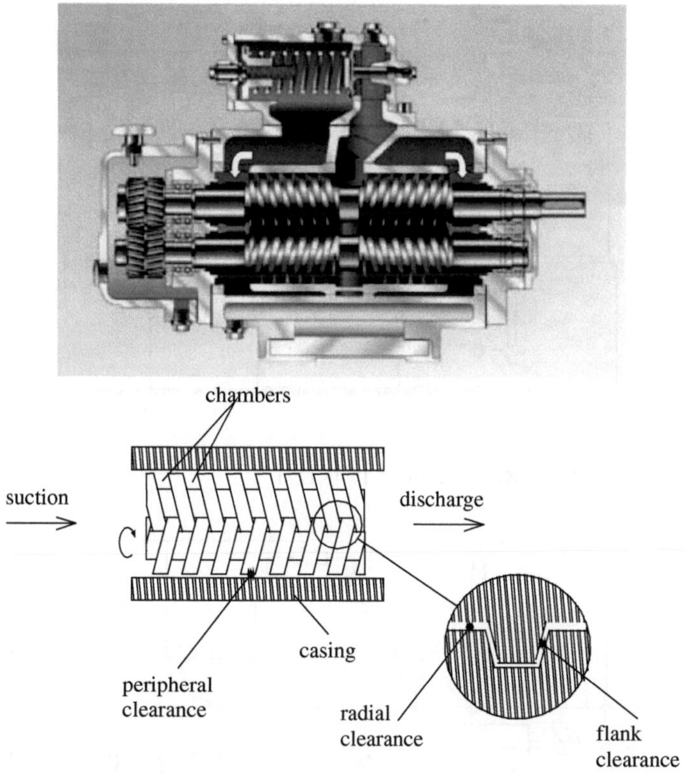

Fig. 4.5 Twin-screw multiphase pump [10]

screws create one or more cavities which move continuously from the suction to the discharge region (Fig. 4.5). Therefore, any type of product (liquid, gas, or solid) that can be introduced into the cavities will be carried along to the discharge. This characteristic allows the pump to transport multiphase fluids with gas volumetric fraction (GVF) varying from 0 to 95 % (100 % if 5 % of liquid can be recirculated). By simultaneously "pumping" gas and liquid, the multiphase pump could minimize facilities in deepwater and reduce operational costs, as cited by Caetano et al. [7]. For example, PETROBRAS has a subsea system (SBMS-500) based on a twin-screw multiphase pump, which will be installed in the Campos Basin (see [8]). Its objective is to increase well production through energy transfer to the unprocessed fluids. The use of twin-screw pumps is seen as a feasible option to make possible exploitation in ultra deep waters. The cost of a multiphase equipment [9] can reach about 70 % of the cost of a conventional equipment but the multiphase pump has lower efficiency (30–50 %) compared to the conventional pumps (60–70 %) and compressors (70–90 %). Figure 4.6 shows a scheme of an offshore platform operating with a SMPS.

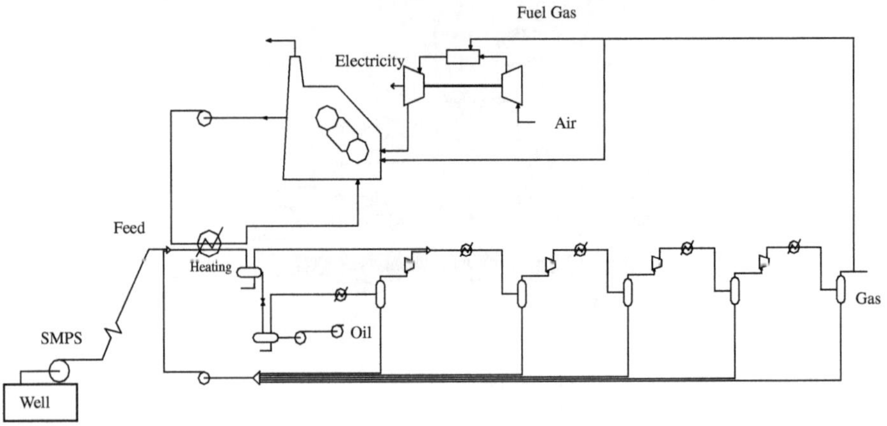

Fig. 4.6 Platform/SMPS schematic process plant [11]

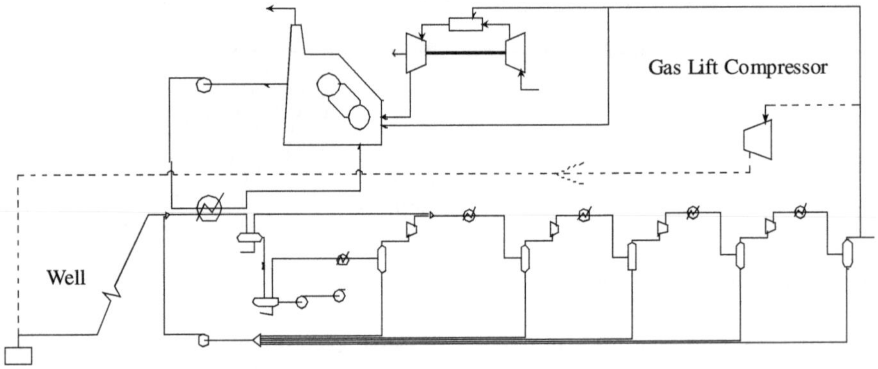

Fig. 4.7 Platform/GL schematic process plant [9]

4.2.5.3 Gas Lift System

The GL is the most common artificial lift method used in subsea petroleum boosting. In this system, part of the gas separated in the platform is recompressed with a gas lift compressor and returns to the well where it is downhole injected in the production column, as presented in Fig. 4.7. By using this gas injection, the petroleum specific weight is reduced, the pressure in the well head is lowered increasing the production.

4.2.5.4 Exergoeconomic Analisys

The process plant of the offshore platform used as a basis to all comparisons in this study is similar to the one represented in Figs. 4.1 and 4.3. The plant is fed with petroleum boosted from the well. This petroleum is mixed with the recirculated liquid and goes to the heat exchanger. The heated mixture proceeds to the separators where it is divided into oil, gas, and water. Oil and gas are exported to the coast through pumps and compressors and part of the gas is separated to be used as fuel in a cogeneration system. The turbine is responsible for supplying electricity to pumps and compressors and the rejected gases are used in the boiler to heat the water of the heat exchanger. Additional fuel is burned in the boiler when the rejected gases cannot supply the required energy. The efficiencies of the considered platform components in the simulations are as follows:

- isentropic efficiency of compression (η_{cp}): 0.75;
- mechanical efficiency of pump and compressors (η_m): 0.90;
- transmission efficiency (η_{tr}): 0.95;
- isentropic efficiency of conventional pumps (η_p): 0.75;
- thermal efficiency of GTs (η_{tg}): 0.30;
- thermal efficiency of boiler (η_{bo}): 0.80;
- heat recovery system thermal efficiency (η_{rec}): 0.60.

The physical and chemical exergies were calculated with the aid of software Hysys. Process v2.1 [12]. The chemical exergy, following Rivero et al. [13], is calculated through Eq. 4.46:

$$b_{ch} = \sum_{i=1}^{C} \left(b^0_{ch,i} + RT_0 x_i \ln(x_i) \right) \tag{4.46}$$

Where b_{ch}, b^0_{ch}, T_0, R, and x_i stand respectively for chemical exergy, standard chemical exergy, reference temperature, universal gas constant, and molar fraction of component i.

Equation 4.46 would be valid for ideal mixtures only. However, as shown by Rivero et al. [13], eventhough petroleum mixtures cannot be considered ideal, the calculations with this equation do not cause significative errors due to the huge standard chemical exergies of the petroleum components. The standard exergy for the light components of the petroleum is tabulated in Kotas [6]. The composition of the heavy fraction of petroleum is unknown. In this case, the heavy fraction is divided in pseudo-components and the standard chemical exergy of these components have to be calculated based on the lower heating value (LHV):

$$\alpha = \frac{b^0_{ch}}{\text{LHV}} \tag{4.47}$$

Where α can be calculated according to Eq. 4.48, where c, h, o, s are the mass fractions of, respectively, C, H, O, and S.

$$\alpha = 1.0401 + 0.1728\frac{h}{c} + 0.0432\frac{o}{c} + 0.2169\frac{s}{c}\left(1 - 2.0628\frac{h}{c}\right) \tag{4.48}$$

Table 4.7 Specific exergy cost of fee, oil, gas, electricity and heat [11]

Stream	Exergy flow rate (MW)	η_{tg}		
		0.30	0.20	0.15
Feed	4134.5	1.000	1.000	1.000
Oil	3753.4	1.006	1.007	1.007
Gas	329.2	1.025	1.027	1.028
Electricity	6.3	1.932	2.085	2.140
Heat	6.0	5.125	5.166	5.362

In this study, the contaminants where not taken into account and the value h/c for heavy fractions is generally about 0.1. Therefore, its influence over α is less than 2 %. Then, it was considered that α has a constant value of 1.0401. The heating value can be estimated using the equations suggested by Guthrie [14].

The determination of the performance parameters and products exergy-based costs for the studied configurations are done by using the method described in Sects. 4.2.2 and 4.2.3. The platform cogeneration system (GT, boiler, heat exchanger, and water pump) can be simulated apart with the software EES® [15].

The mean logarithmic temperature of the heat transfer in the heat exchanger is taken as $T_{\mathrm{mlt}} = 154\ ^{\circ}\mathrm{C}$, which is based on measured temperatures.

As the costs of electricity and rejects are unknown in the turbine, a partition criterion–equality or extraction—must be adopted to allow calculations.

With these considerations, and having all values of exergy rates and flow rates calculated, all the exergy-based costs of the process can be determined. This is an iterative calculation because the cost of the produced gas in the platform (c_{gas}) must be used to calculate the costs in the cogeneration system.

Table 4.7 shows results of the influence of the GT thermal efficiency in the production costs. The equality criterion is utilized because both electricity and rejects are used in the platform processes. The obtained costs show a small increase between the feed and the discharge (oil and gas). This behavior is due to the high chemical exergy values of the currents. Compared to such values, the destroyed exergy in the equipment of the process is almost negligible and the increments are small. Only in the cogeneration system, the cost increase is higher owing to the elevated irreversibility of the combustion processes.

4.2.5.5 Comparative Study

Three systems and four different cases were compared: standalone offshore platform, offshore platform operating with a GL, and an offshore platform operating with a SMPS under two different operational conditions. The GL process plant used in the simulations is presented in Fig. 4.7. The GL was considered as a compressor that injects the outlet gas in the petroleum flow downhole in the well.

The SMPS is placed at the seabed, in the production line between the well and the platform. The SMPS has a broad range of operational conditions and two limit

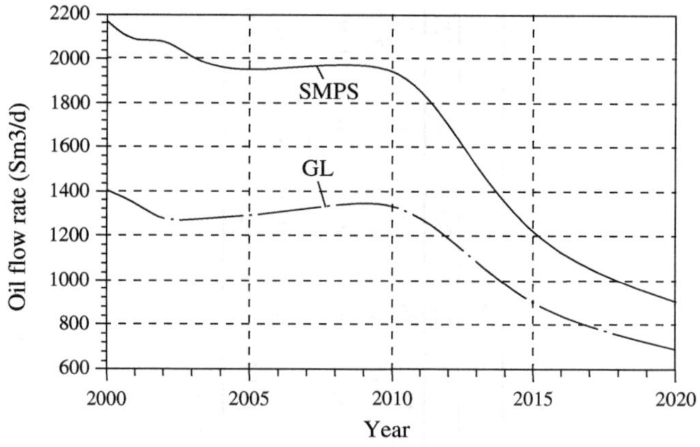

Fig. 4.8 Volumetric oil flow rate: GL and SMPS-60 [11]

situations were chosen. The first one denoted here as SMPS-00, is the SMPS working in order to give the same oil production obtained when the GL is used (dashed line in Fig. 4.8). Due to the low flow rate, the multiphase pump operates far bellow its maximum capacity. In the second one, denoted as SMPS-60, the pump operates with the maximum allowable differential pressure (60 bar) with a higher flow rate (continuous line in Fig. 4.8). Table 4.8 summarizes the operational conditions for the four cases. All these values refer to the production line position where the SMPS will be installed.

Based on a well production data with the use of artificial lift methods [11], the well operating with SMPS or GL could be simulated. The SMPS discharge conditions and absorbed power are calculated by using a thermodynamic model for twin-screw multiphase pumps [10, 16].

The SMPS electrical motor efficiency was set to 0.8 during calculations. This low value is due to the canned conception adopted for subsea operation. The GL compressor adiabatic efficiency was taken as 0.75. Heat exchange between ducts and environment was not considered, nor was the electricity transmission efficiency of the SMPS energy cable. In this study, no operational limits were considered for any equipment involved in the petroleum processing. This means that maximum or minimum rotation, power, and flow rate are ignored. For all the simulations, some parameters were considered fixed:

- Platform feed pressure: 21.6 bar g;
- Gas pressure at platform outlet: 172.6 bar g;
- Oil pressure at platform outlet: 67.6 bar g;
- Well outlet flow's specific cost of: 1.0 $/kWh.

The specific costs were calculated for the years 2000, 2001, 2010, and 2020. These years were chosen because the oil is considered the main petroleum product and, as shown in Fig. 4.8, the oil production has two distinct tendencies: one,

Table 4.8 SMPS and GL operating conditions [11]

	Platform				GL				SMPS-00				SMPS-60			
Year	2000	2001	2010	2020	2000	2001	2010	2020	2000	2001	2010	2020	2000	2001	2010	2020
P (bar_g)	58.3	63.1	63.4	72.5	68.0	67.3	70.7	74.1	51.9	62.0	53.0	43.7	29.0	35.0	29.5	30.0
V (m^3/h)	106.4	165.7	81.7	16.7	180.0	187.5	165.9	167.1	146.5	173.9	128.3	109.6	347.3	438.1	276.1	171.8
GVF	0.53	0.65	0.40	0.21	0.65	0.68	0.59	0.59	0.57	0.66	0.47	0.38	0.73	0.79	0.65	0.49
V_{gl} (Sm^3/h)	–	–	–	–	4167	1667	4167	5833	–	–	–	–	–	–	–	–
ΔP (bar)	–	–	–	–	–	–	–	–	13.6	2.5	18.8	37	60.0	60.0	60.0	60.0
W (kW)	–	–	–	–	–	–	–	–	83.1	33.5	99.1	174.3	645.1	822.0	519.1	346.0

P pressure, V volumetric flow rate, V_{gl} gas lift flow rate, ΔP differential pressure, W SMPS consumed electricity

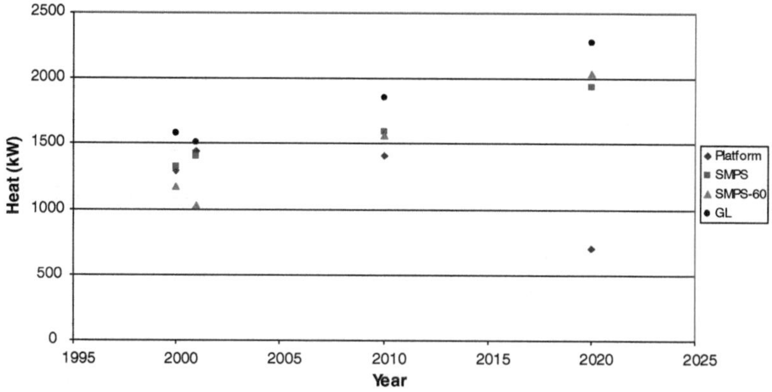

Fig. 4.9 Heat rate demands [11]

between 2000 and 2010, where the production is kept almost constant, and the other, between 2010 and 2020, where there is a linear reduction in the oil flow. The year 2001 was simulated because it predicted a sudden increase in gas void fraction of the oil (GVF) in the first years which differs from the tendency of continuous decrease of the following years. With this increase, the multiphase pump has to deal with a higher volumetric flow rate and, in the GL case, there is a lower necessity of gas lift. As a consequence, the behavior of the systems suffers a considerable change and it is taken into account in the comparisons.

4.2.5.6 Comparative Results

Figures 4.9 and 4.10 show the heating load and electricity demands consumed in each case. These results show that heat load demand with the GL is slightly higher. This is due to the higher gas flow rate, a consequence of the gas lift recirculation. The difference is somewhat lower in 2001, because there is a decrease in the necessity of gas lift, and rises a little again in the following years. The heat load demand with the SMPS-60, which should be the highest values due to the high oil and gas flow rates, has a reduction in the first years owing to the higher shaft speed and friction power inside the pump. In 2020, the volumetric flow decreases with the consequent reduction in the shaft speed and friction power and the heat consumption raises.

The electricity demand of the GL is driven mainly by the compressors consumption, which is large and approximately constant due to the gas lift. Despite its variation, the electric power of the pumps is a small part of total power and its influence is not significant. The electricity consumption of the SMPS goes along with the gas volumetric flow rate, which explains, in both cases, the initial increase followed by a reduction of the consumed electricity. With the SMPS-60, however, the electricity consumption is much higher. In this case, there is a considerable

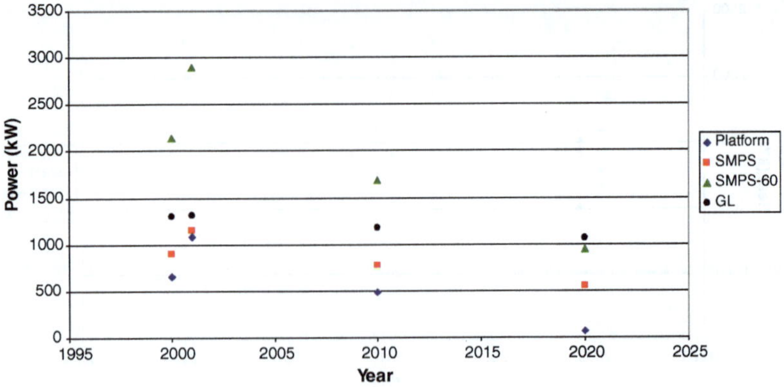

Fig. 4.10 Power demands [11]

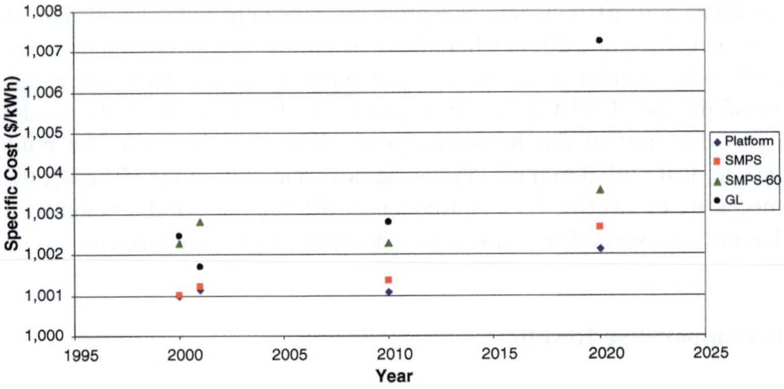

Fig. 4.11 Feed specific costs [11]

increase in the mass flow rate of liquid and gas owing to the reduction in the well head pressure. The increase in the volumetric flow rate, however, is much higher because of the gas expansion. The large volumetric flow rate, together with the elevated differential pressure, causes higher power consumption as seen in Fig. 4.10. Due to the lower production, the platform has in general a lower consumption in pumps and compressors.

Figures 4.11 and 4.12 show the feed and oil specific costs, in exergy basis, calculated for the four studied cases. The standalone platform has always the lowest feed costs because there is no material or energy injection in the production column. However, when the oil cost is analyzed this behavior is not repeated. Only in 2001, the platform has the best cost because the well productivity is high even without auxiliary methods. In the other years, the oil cost for the SMPS has lower values than the standalone platform. The higher feed cost is attenuated during the petroleum processing because SMPS provides an increase in petroleum

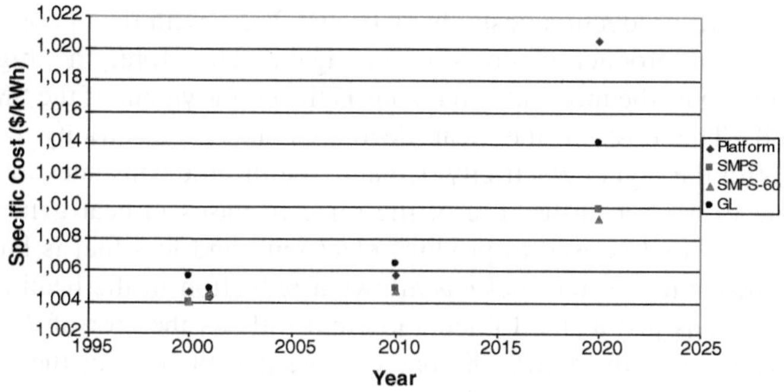

Fig. 4.12 Oil specific costs [11]

production. The higher exergy flow rate, a consequence of the production increase, reduces the impact of energy expenses in the platform over the cost formation. In 2020, because of the extremely low production, the oil cost for the platform is the highest.

Comparing the GL and SMPS-00 cases, which have the same oil production, the resulting specific costs show that the SMPS-00 allows lower feed and oil specific costs. The main advantage of this system, when compared to the GL, is that there is not gas recirculation. The absence of this recirculation provides a reduction in the electric power consumed by the compressors, so the total electric power used in the platform is lower when the SMPS-00 is employed instead of the GL. With the GL, part of the produced gas will be burned to generate electricity that will be used to compress the gas again. The compressed gas is, then, re-injected downhole in the well in order to reduce the specific weight of the petroleum. In this case, besides the losses caused by successive conversions of energy, part of the gas is being processed (separated and compressed) in a redundant way since this gas, after previous processing, is recompressed and mixed again with petroleum. Related to this aspect, the SMPS-00 has an advantage because it has no material (gas) recirculation.

The electric power generated in the plant is transferred directly to the petroleum flow through the multiphase pump without gas recirculation. Therefore, there is less gas to be processed by the separators and compressors, which causes a reduction in the consumption of heat and electricity. Another advantage of the SMPS-00 is its subsea operating position. This is a high pressure location and, compared to surface operations, allows the multiphase pump to work with lower volumetric flow rate and, with lower energy consumption. These advantages, however, depend clearly on the amount of recirculating lift gas. In 2001, for instance, there is an increase in GVF—due to reservoir characteristics and production conditions—and a consequent lower necessity of lift gas. Within this scenario, the multiphase pump performance decreases as well as the impact of gas recirculation in the GL. The consequence is a reduction in the difference between

SMPS-00 and GL production costs. In 2010 and 2020, with the opposite situation, the difference in production costs raises again. Therefore, the higher is the necessity of lift gas, the more advantageous is the employment of the SMPS for the studied well. The cost of the heat decreases as the electricity consumption increases. With the higher electricity demand, the turbine consumes more fuel and it is possible to have a further use of the rejected gases in heat generation. This means that more fuel is used to produce electricity and less fuel is burned in the boiler. The better use of the fuel gas allows a reduction in the heating cost. The electricity cost has just a small fluctuation and follows the cost of the fuel gas.

By analyzing only the SMPS-00, one can see that the cost of the produced oil enhances continuously. In 2001, the small cost increment is due to performance degradation in the multiphase pump owing to high GVF. In 2010 and 2020, the increment occurs mainly due to the decreasing oil flow. The total (water and oil) liquid flow rate is kept almost constant. As Fig. 4.8 shows only the oil flow rate, the increasing flow of water causes a reduction in oil production and in the exergy associated with the petroleum flow. Therefore, the influence of electricity and heat costs over the stream is stronger. Physically, this means that part of the energy inputs are consumed to process a substance with low value (water) which will not be part of the products. In the heating process, for instance, the water is heated, since it is part of the petroleum, and thrown away later on. The cost of the gas is lower with higher GVF because, in this case, the percentage of dissolved gas, which is more difficult to separate, is lower.

When the GL is considered, the gas and oil costs have a tendency to follow the magnitude of the lift gas flow. The higher is this flow, the larger are the costs. The lowest cost is obtained in 2001 when the necessity of lift gas is marginal. Before 2010 the lift gas flow raises again and, besides, the amount of water increases continuously as in the SMPS case. These two factors cause the costs to increase again.

The SMPS-60 has, in general the same behavior of the SMPS-00. However, as shown in the previous item, there is an increase mainly in the electricity consumption, which modifies the use of the energy inputs.

When the costs for the two conditions (SMPS-00 and SMPS-60) of the SMPS are compared (Figs. 4.11, 4.12), it is possible to see that the platform feed cost increases in the SMPS-60 case. Several parameters contribute to this increase: higher viscous losses due to higher shaft speeds, higher backflow rates due to higher differential pressures, and higher GVFs, which cause higher recirculation inside the multiphase pump. The lower multiphase pump efficiency causes higher feed costs. However, when the oil costs are compared, it is possible to realize that the SMPS-60 has lower costs despite of the higher feed values. Due to the high electricity consumption, all the heat required by the process is obtained from the turbine reject and there is no necessity of additional combustion in the boiler. This means that the gas is burned only to generate electricity (and increase production) and not to heat the petroleum.

4.2.5.7 Concluding Remarks

The application of the exergy and exergoeconomic analysis to compare the performance of artificial lift systems proved to be a reliable tool to highlight the advantages and disadvantages of each system on the same basis: the quality of the energy conversion processes that take place in the primary petroleum processing.

The developed comparisons between GL and SMPS presented in this comparative study indicate that the employment of the SMPS could bring significant benefits in the offshore oil and gas production. When the costs of the produced oil are compared, the SMPS has always the lowest values, which means that the operational flexibility provided by the multiphase pump can be fully profited. With this flexibility it is possible to manage the pump so the gas is burned only to generate electricity and increase the production while the heat is obtained entirely from the rejected gases. Such flexibility does not exist with the GL.

The advantage of the SMPS is much more evident when the GVF values are not so high. In this case, the multiphase pump performance increases and the need of lift gas is higher. When the GVF is high, the performance of the SMPS could be improved by installing the SMPS in a high pressure position (nearer the well head). This could reduce the volumetric flow and the power consumption (not considering the electricity transmission efficiency).

The cost values can change according to the platform configuration since the processes involved in petroleum production will define the costs. For different process plants it is necessary to evaluate the methods again. The production costs when the platform operates below its capacity would be higher by taking into account equipment and operational costs. With these fixed costs a lower production would have higher production costs. In theory, this would increase the advantage of the SMPS over lower productivity methods because it is possible to use the SMPS to fulfill the platform capacity. An exact evaluation, however, must take into account all the equipment and operational costs.

4.3 Exergy and Thermoeconomic Analysis of a Petroleum Refinery Utilities Plant

4.3.1 Introduction

Oil refining industry worldwide is an activity with high fossil fuel consumption and consequently high CO_2 emissions. Approximately 7–15 % of the crude oil input is used by the refinery processes [17]. As stated previously, the global oil consumption tends to increase from 86×10^6 barrels per day (bpd) in 2010 to 111×10^6 bpd in 2035 [2], while energy consumption per refining unit tends to increase 11 % during the same period. This increment in energy consumption in refining processes can be explained by the rising demand for high quality products and decreasing availability of light crudes [18].

In the utilities plant of a petroleum refinery a huge consumption of exergy occurs to supply the demands of power and steam required by the refining processes. Therefore, the application of the exergy analysis to point out the main causes of inefficiencies in the utilities plant as well as the thermoeconomic approach to evaluate how much money each process unit should pay for or receive from utilities plant, due to the exchange of fuels' and utilities' flows, can provide useful information to operate and optimize the refinery processes. This section describes an exergy and thermoeconomic analysis of a petroleum refinery utilities plant in order to determine the operating performance of its main components and the production cost of the utilities.

4.3.2 Refinery Description

The studied refinery processes about 360,000 bpd of acid crude, which has a density of 26 API. Its production scheme can be described as one between a cracking and coking typical scheme. It maximizes the cuts of diesel and kerosene, and it is composed of atmospheric and vacuum distillation, fluid catalytic cracking (FCC), delayed coking, hydrotreating (HDT), hydrogen generation, sulfur recovery, and other auxiliary units such as sour water treatment and storage/transportation units.

The final products of the refinery are: liquefied petroleum gas (LPG), gasoline, diesel, kerosene, naphtha, aromatic residue, asphalt, coke, and sulfur.

The utilities plant of a refinery comprises several components with many interfaces to the refining process. Therefore, for the sake of simplicity, the similar components and process units were aggregated to form the synthesis plant shown in Fig. 4.13.

The utilities plant is a typical polygeneration plant since several products are generated. The analyzed plant has five different products:

- Electricity: generated by gas and steam turbines used mainly to drive the electric motors (sections 37 and 53).
- Mechanical power: mainly used to drive pumps, compressors, and blowers etc. (sections 54, 55, 56, and 57).
- Steam: provided in three different grades: 90 bar (section 7), 13 bar (sections 2, 8, 16, 22, 29 and 31), and 3 bar (sections 3, 9 and 30), and used in all process units, providing heat wherever it is needed, it is also used to reduce the partial pressure of mixture components such as in distillation process, being separated afterwards by condensation and resulting in sour water.
- Compressed water: provided at three different pressure levels: 120 bar, 28 bar, and 1 bar, and used in several process units to generate steam while transferring heat, cleaning, and others (sections 89, 93 and 112).
- Compressed Air: for service and instrumentation (section 78).

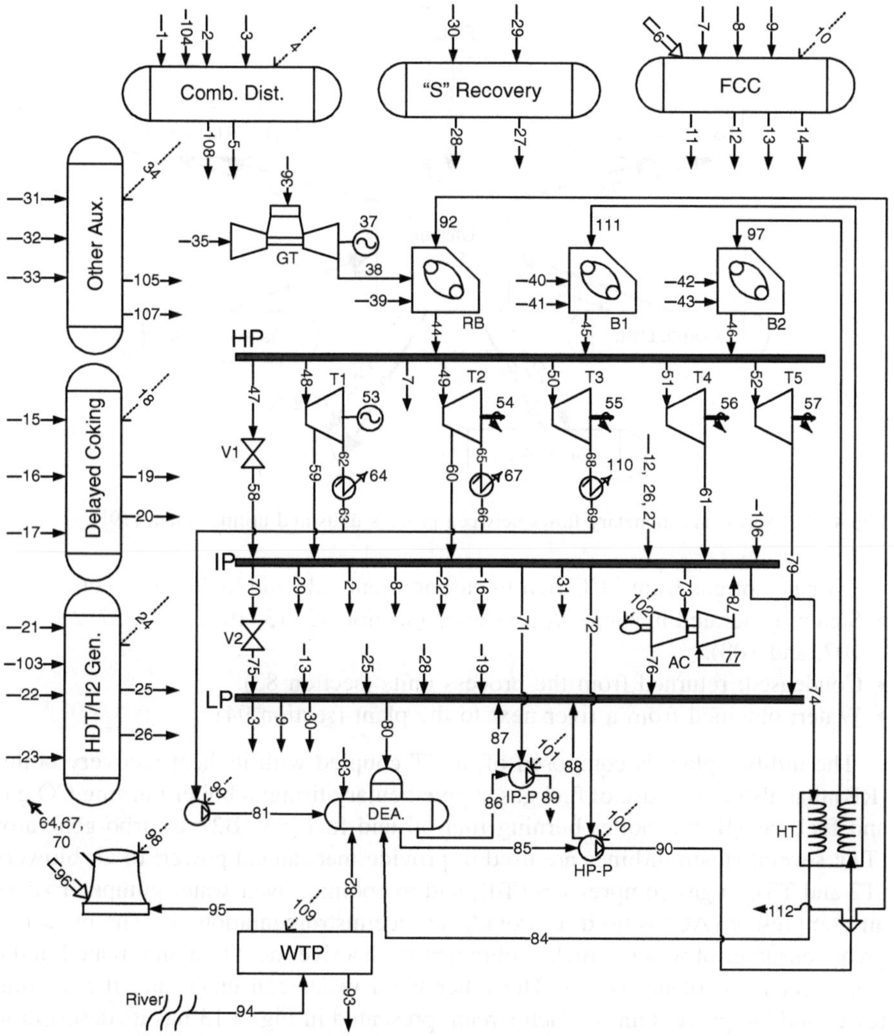

Fig. 4.13 Utilities plant simplified scheme and process units [19]

In order to generate the above listed products the polygeneration plant makes use of several different fuels (exergy sources) produced internally and externally to the refinery, as shown by Fig. 4.14 where are indicated the net exergy and cost flows among the utilities plant and the processes:

- Natural gas: produced externally and bought to fuel the GTs (it is also used in H_2 generation for HDT process) (sections 36 and 21).
- Fuel gas: produced by FCC unit, delayed coking unit, and some other units in minor quantities (sections 14, 20 and 105).
- Fuel oil: produced by combined distillation (section 5).

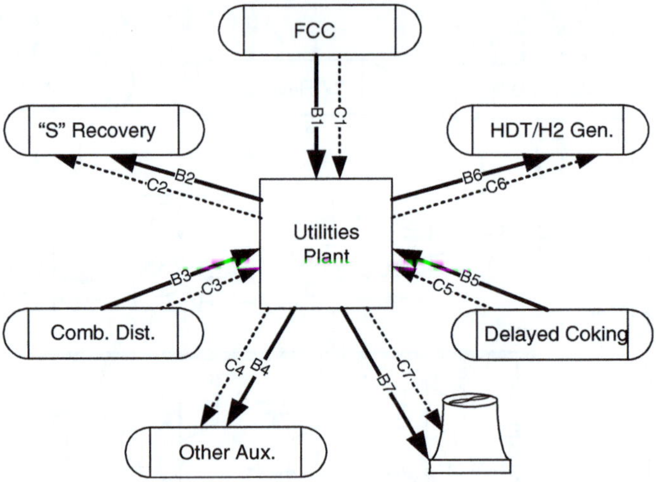

Fig. 4.14 Exergy and monetary flows between process units and utilities plant [19]

- CO gas: hot gas from FCC rich in carbon monoxide and hydrogen (section 11).
- Steam: generated in some process units (section 12, 13, 19, 25, 26, 27, 28, 106, 107, and 108).
- Condensed: returned from the process units (section 83).
- Water: obtained from a river next to the plant (section 94).

The utilities plant is composed of: a GT coupled with its heat recovery boiler (RB) that also makes use of fuel gas supplementary firing; a boiler burning CO gas and fuel gas (B1); a boiler burning fuel oil and fuel gas (B2); a turbo-generator (T1); several steam turbines are used to provide mechanical power: to air blowers (T2 and T3), to gas compressors (T4), and to cooling tower water pumps (T5); an air compressor (AC) is used to provide service/instrumentation air. The plant is a great consumer of water, which is obtained from a river next to it and treated in the water treatment plant (WTP). The other main water consumers are the cooling tower and the process units. Each stream presented in Fig. 4.13 has its description and exergy rates and costs detailed in Table 4.9.

4.3.3 Exergy Analysis and Thermoeconomic Approach

Exergy analysis is used to highlight the components of the utilities plant responsible for the main exergy destructions in addition to the information about the exergy, from the utilities plant, consumed by each refinery process unit.

Through the use of thermoeconomic approach, the exergy of the utilities plant fuels is rationally allocated to the plant streams, thus, creating a picture of the fuels' exergy distribution to the whole refinery. Furthermore, the monetary costs

Table 4.9 Description, exergy, and costs of the streams shown in Fig. 4.13 [19]

Stream	Description	Exergy rate/flow rate (kW)	c (US$/kJ)	C (US$/s)	c_b (kJ/kJ)
1	NG	18,636	5.177E−06	0.10	1.00
2	IP ST	29,447	2.018E−05	0.59	2.38
3	LP ST	5,768	2.075E−05	0.12	2.45
4	E	11,022	2.119E−05	0.23	3.06
5	FO	118,903	8.404E−06	1.00	1.00
6	M P	26,765	3.590E−05	0.96	4.26
7	HP ST	17	2.015E−05	0.00	2.38
8	IP ST	50,841	2.018E−05	1.03	2.38
9	LP ST	14,403	2.075E−05	0.30	2.45
10	E	4,375	2.119E−05	0.09	3.06
11	CO	126,571	8.819E−06	1.12	1.00
12	IP ST	24,844	2.018E−05	0.50	2.38
13	LP ST	4,787	2.075E−05	0.10	2.45
14	FG	236,140	8.819E−06	2.08	1.00
15	NG	1,215	5.177E−06	0.01	1.00
16	IP ST	21,300	2.018E−05	0.43	2.38
17	HP W	127	2.359E−05	0.00	2.89
18	E	12,706	2.119E−05	0.27	3.06
19	LP ST	3,305	2.075E−05	0.07	2.45
20	FG	215,352	8.819E−06	1.90	1.00
21	NG	206,328	5.177E−06	1.07	1.00
22	IP ST	14,648	2.018E−05	0.30	2.38
23	HP W	4,255	2.359E−05	0.10	2.89
24	E	7,397	2.119E−05	0.16	3.06
25	LP ST	13,998	2.075E−05	0.29	2.45
26	IP ST	11	2.018E−05	0.00	2.38
27	IP ST	5,777	2.018E−05	0.12	2.38
28	LP ST	46	2.075E−05	0.00	2.45
29	IP ST	0	2.018E−05	0.47	2.38
30	LP ST	279	2.075E−05	0.01	2.45
31	IP ST	399	2.018E−05	0.01	2.38
32	NG	6,014	5.177E−06	0.03	1.00
33	IP W	5,067	2.272E−05	0.12	2.78
34	E	10,820	2.119E−05	0.33	3.06
35	AIR	0	0.000E + 00	0.00	0.00
36	NG	110,920	5.177E−06	0.57	1.00
37	E	37,796	1.138E−05	0.43	2.20
38	FLUE G	27,844	5.177E−06	0.14	1.00
39	FG	92,941	8.819E−06	0.82	1.00
40	CO	126,571	8.819E−06	1.12	1.00
41	FG	53,677	8.819E−06	0.47	1.00
42	FO	118,903	8.404E−06	1.00	1.00
43	FG	60,196	8.819E−06	0.53	1.00
44	HP ST	51,271	2.100E−05	1.08	2.63
45	HP ST	96,795	1.873E−05	1.81	2.15

(continued)

Table 4.9 (continued)

Stream	Description	Exergy rate/flow rate (kW)	c (US$/kJ)	C (US$/s)	c_b (kJ/kJ)
46	HP ST	81,231	2.129E−05	1.73	2.51
47	HP ST	517	2.015E−05	0.01	2.38
48	HP ST	95,325	2.015E−05	1.92	2.38
49	HP ST	38,082	2.015E−05	0.77	2.38
50	HP ST	18,153	2.015E−05	0.37	2.38
51	HP ST	63,289	2.015E−05	1.28	2.38
52	HP ST	13,914	2.015E−05	0.28	2.38
53	E	21,359	3.854E−05	0.82	4.58
54	M P	5,757	5.996E−05	0.35	7.13
55	M P	9,396	3.855E−05	0.36	4.61
56	M P	13,717	2.666E−05	0.37	3.15
57	M P	4,719	2.816E−05	0.13	3.33
58	IP ST	368	2.825E−05	0.01	3.34
59	IP ST	54,284	2.015E−05	1.09	2.38
60	IP ST	20,862	2.015E−05	0.42	2.38
61	IP ST	45,135	2.015E−05	0.91	2.38
62	W-ST	4,556	–	–	–
63	W	828	2.015E−05	0.02	2.38
64	HEAT	3,727	–	–	–
65	W-ST	2,078	–	–	–
66	W	381	2.015E−05	0.01	2.38
67	HEAT	1,698	–	–	–
68	W-ST	4,275	–	–	–
69	W	783	2.015E−05	0.02	2.38
70	IP ST	1,722	2.018E−05	0.03	2.38
71	IP ST	0	2.018E−05	0.00	2.38
72	IP ST	7,780	2.018E−05	0.16	2.38
73	IP ST	1,532	2.018E−05	0.03	2.38
74	IP ST	1,017	2.018E−05	0.02	2.38
75	LP ST	1,270	2.735E−05	0.03	3.23
76	LP ST	1,130	2.018E−05	0.02	2.38
77	AIR	0	0.000E + 00	0.00	0.00
78	AIR	861	3.475E−05	0.03	4.77
79	LP ST	7,319	2.015E−05	0.15	2.38
80	LP ST	17,265	2.075E−05	0.36	2.45
81	W	2,003	2.021E−05	0.04	2.39
82	W	2,181	6.787E−06	0.01	1.84
83	W	12,401	2.025E−05	0.25	2.39
84	W	203	2.018E−05	0.00	2.38
85	W	24,772	2.248E−05	0.56	2.73
86	W	4,975	2.248E−05	0.11	2.73
87	LP ST	0	2.018E−05	0.00	2.38
88	LP ST	5,738	2.018E−05	0.12	2.38
89	IP W	5,067	2.272E−05	0.12	2.78
90	HP W	26,894	2.359E−05	0.63	2.89

(continued)

Table 4.9 (continued)

Stream	Description	Exergy rate/flow rate (kW)	c (US$/kJ)	C (US$/s)	c_b (kJ/kJ)
91	HP W	17,473	2.359E−05	0.41	2.89
92	HP W	5,032	2.359E−05	0.12	2.89
93	W	61	6.787E−06	0.00	1.84
94	W	15,657	9.412E−07	0.01	1.00
95	W	12,942	6.787E−06	0.09	1.84
96	M P	6,824	3.590E−05	0.25	4.26
97	HP W	8,623	2.375E−05	0.20	2.91
98	E	609	2.119E−05	0.01	3.06
99	E	16	2.119E−05	0.00	3.06
100	E	2,500	2.119E−05	0.05	3.06
101	E	154	2.119E−05	0.00	3.06
102	E	1,030	2.119E−05	0.02	3.06
103	FG	48,837	8.819E−06	0.43	1.00
104	FG	201,571	8.819E−06	1.78	1.00
105	FG	0	8.819E−06	0.05	1.00
106	IP ST	530	2.100E−05	0.01	2.63
107	LP ST	0	2.075E−05	0.00	2.45
108	LP ST	0	2.075E−05	0.00	2.45
109	E	4,351	2.119E−05	0.09	3.06
110	HEAT	3,492	–	–	–
111	HP W	9,419	2.375E−05	0.22	2.91
112	HP W	4,389	2.359E−05	0.10	2.89

associated with the utilities plant are allocated to the exergy of streams so that the process units will be charged from a thermodynamic point of view, for utilities consumption. This is a part of the analysis of the cost formation process of the products of the refinery. This approach, clearly, has a strong environmental meaning since destruction of exergy can be directly associated with more fuel consumption.

The exergy performance and the exergy-based production cost of the petroleum derivatives are not only quantitative information about the quality of the energy conversion processes, but also are indicators to aid the optimization of the refinery operating conditions and to compare the performance of different utilities and refinery plants.

The analysis is based on thermodynamic data obtained from the Plant Information system (PI) and sent to a Microsoft Excel Spread Sheet. In this spread sheet the thermodynamic properties: enthalpy, entropy, and quality as well as the flow are calculated.

Next, the exergy of each stream, exergy efficiency of the components in addition to the exergy destruction are figured. The set of linear equations, used for costs and fuels exergy distribution among the streams, are solved to calculate the unitary exergy cost of the flows (kJ/kJ), unitary cost (US$/kJ), and the total cost rate of each streams (US$/s).

The exergy efficiency (η_b) of every component of the synthesis utilities plant is presented in Table 4.10. In this table, it is also included the usual expressions of the energy efficiencies (η_e) of these components.

In the steam turbine efficiencies the condensers were aggregated to the steam turbine as suggested by Lozano and Valero [20] and Lazzaretto and Tsatsaronis [21]. Thus, the exergy spent in cooling tower to destroy the exergy present in the thermal load of a given condenser (the power of pumps, fans, and the exergy rate of the make-up water) was also allocated to the respective steam turbine, as shown in Table 4.10. The parameter x is the weight factor calculated as the ratio between the exergy transferred by each condenser and the overall exergy transferred in the cooling water circuit of the cooling tower. Neglecting the differences between the average logarithmic temperature of the cooling water circuit, before and after each turbine condenser (T_{ci}) and the average logarithmic temperature of the condensing water circuit before and after the cooling tower (T_{cw}), the determination of x can be simplified by calculating the ratio between the heat rate rejected by each condenser and the overall thermal load of the cooling tower system, as shown in Eq. 4.49.

$$x = \frac{Q_i\left(1-\frac{T_0}{T_{ci}}\right)}{\left(1-\frac{T_0}{T_{cw}}\right)\sum_{i} Q_i} \approx \frac{Q_i}{\sum_i Q_i} \tag{4.49}$$

In the thermoeconomic approach adopted in this analysis, as the utilities plant receives steam from several process units and distributes this steam together with the steam produced back to the units, it was assumed that the received steam has the same unitary exergy cost of the same grade steam produced by utilities plant. Besides, the unitary exergy cost of the condensed returning from process units was considered to be the average unitary exergy cost of the steam that goes to the units (three different grades). The fuel principle suggested in Lazzaretto et al. [21] is used in the steam turbines. Thus, the unitary exergy cost of the extraction and exhaust/condensed is equal to the unitary exergy cost of the live steam entering the turbine.

4.3.4 Results

The energy and exergy efficiency for each component of utilities plant are indicated in Fig. 4.15. The reference temperature and pressure are $T_0 = 298.15$ and $P_0 = 101.325$ kPa, respectively.

The calculated overall utilities plant exergy efficiency is 42 % while the overall energy efficiency is 74 %.

The water treatment process (WTP) has a technical function rather than energetic one, thus, it spends energy/exergy to produce a non-energetic/exergetic

Table 4.10 Energy performance parameters and exergy efficiencies of the utilities plant components [19]

Component	Energy performance parameters	Exergy efficiencies
Gas turbine	$\eta_e = \frac{W_{Electric}}{m_{Fuel} LHV}$	$\eta_b = \frac{W_{Electric}}{m_{Fuel} b_{Fuel} - m_{Exhaust} b_{Exhaust}}$
Steam turbines	$\eta_e = \frac{W_{Electric/Mechanical}}{H_{In} - H_{Extraction} - H_{Condensed} + x(H_{H_2O} + E_{Electric} + E_{Mechanical})}$	$\eta_b = \frac{W_{Electric/Mechanical}}{B_{In} - B_{Extraction} - B_{Condensed} + x(B_{H_2O} + E_{Electric} + E_{Mechanical})}$
Pumps[a]	$\eta_e = \frac{\sum\left(\frac{m_{Water} \Delta P_{Water}}{\rho_{Water}}\right)}{\sum[W_{Electric} + (H_{In\_Steam} - H_{Out\_Steam})]}$	$\eta_b = \frac{\sum(B_{Out\_Water} - B_{In\_Water})}{\sum[W_{Electric} + (B_{In\_Steam} - B_{Out\_Steam})]}$
Boilers	$\eta_e = \frac{H_{Steam} - H_{Water}}{m_{Fuel1} LHV_{Fuel1} + m_{Fuel2} LHV_{Fuel2}}$	$\eta_b = \frac{B_{Steam} - B_{Water}}{m_{Fuel1} b_{Fuel1} + m_{Fuel2} b_{Fuel2}}$
Heat exchanger	$\varepsilon = \frac{C_{Cold}(T_{Cold\_In} - T_{Cold\_Out})}{C_{Min}(T_{Hot\_In} - T_{Cold\_In})}$	$\eta_b = \frac{B_{Cold\_Out} - B_{Cold\_In}}{B_{Hot\_In} - B_{Hot\_Out}}$
Valves	$\eta_e = \frac{H_{Out}}{H_{In}}$	$\eta_b = \frac{B_{Out}}{B_{In}}$
Deaerator	$\eta_e = \frac{\sum H_{Out}}{\sum H_{In}}$	$\eta_b = \frac{\sum B_{Out}}{\sum B_{In}}$
Air Compressor[b]	$\eta_{isoT} = \frac{\sum\left[mRT_0 \ln\left(\frac{P_{Out\_Air}}{P_0}\right)\right]}{\sum[W_{Electric} + (H_{In\_Steam} - H_{Out\_Steam})]}$	$\eta_b = \frac{\sum(B_{Out\_Air} - B_{In\_Air})}{\sum[W_{Electric} + (B_{In\_Steam} - B_{Out\_Steam})]}$
WTP	$\eta_e = \frac{H_{Out\_Water} - H_{In\_Water}}{W_{Electric}}$	$\eta_b = \frac{B_{Out\_Water} - B_{In\_Water}}{W_{Electric}}$

[a] This efficiency is calculated for group of pumps driven by electricity and steam turbines such as in Fig. 4.14.
[b] This efficiency is calculated for group of compressor driven by electricity and steam turbines such as in Fig. 4.14

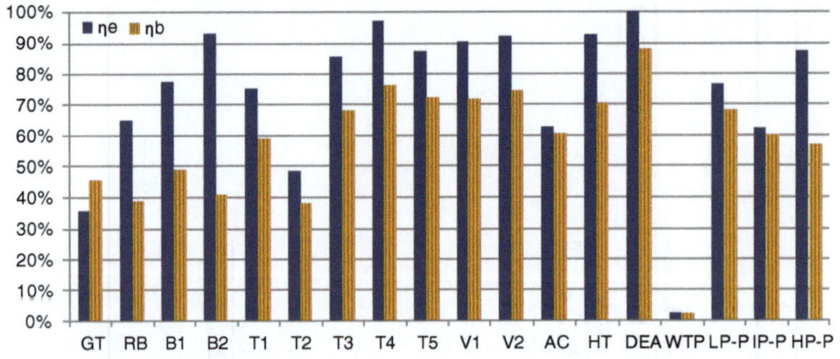

Fig. 4.15 Energy and exergy efficiency for the utilities plant components [19]

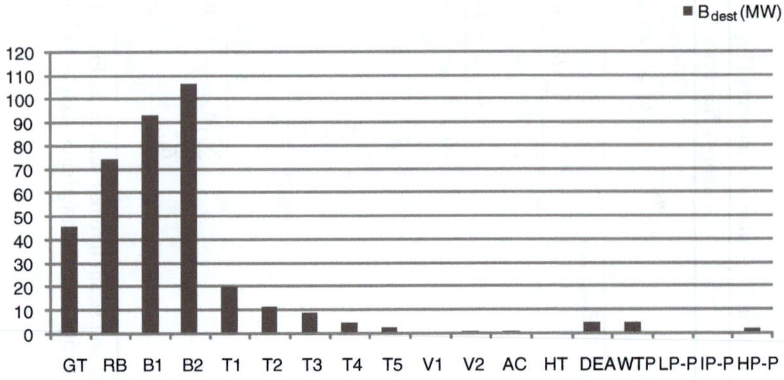

Fig. 4.16 Exergy destroyed rate in each component of utilities plant [19]

product (clean water), and therefore this process has a very low energy and exergy efficiency.

Figure 4.16 provides an enlightening picture of the components responsible for exergy destruction in the utilities plant. The boilers (RB, B1 and B2) together with the GT are responsible for 85 % of the exergy destroyed in utilities plant. The five steam turbines (T1, T2, T3, T4 and T5) together (taking into consideration the heat sent by their condensers to be dissipated in the cooling tower) are responsible for only 12 % of all exergy destroyed, while the others 3 % are mainly due to deaerator (DEA) and WTP. It is worth noting that only 9 % of the exergy dissipated in cooling tower comes from the utilities condensers while 91 % comes from process units.

The unitary exergy cost (kJ/kJ) of main flows leaving and entering the utilities plant can be seen in Fig. 4.17. In order to evaluate the fuel distribution through the streams of the plant, the unitary exergy cost of the fuels (NG, FG, CO, FO and river water) was considered equal to one. The unitary exergy cost of the steam increases from high pressure to low pressure due to the two valves (V1 and V2). These two valves use more than one unit of exergy of higher pressure steam to produce one unit of exergy of lower pressure steam.

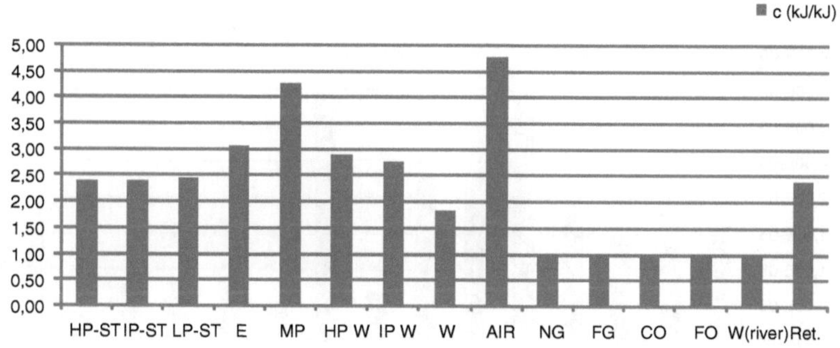

Fig. 4.17 Unitary exergy cost of main input and output flows of utilities plant [19]

Table 4.11 Fuel costs in mass and exergy basis [19]

Substance	Cost (US$/t)	Cost (US$/kJ)
River water	0.05	9.41E−07
FO	375.04	8.40E−06
FG	423.12	8.82E−06
NG	255.63	5.18E−06

Table 4.12 Average monetary unit cost of the products of the utilities plant [19]

Stream	c (US$/t or US$/MWh)
HP-ST	29
IP-ST	21
LP-ST	16
E	76.273[a]
MP	129.256[a]
Compressed air	7

[a] The electricity and mechanical power costs are given per MWh

The adopted fuel costs used in the thermoeconomic analysis are listed in Table 4.11. The plant was considered already amortized.

Using the listed fuel costs, each utilities plant stream cost was calculated (see Fig. 4.18 for the main inputs and outputs monetary unit costs). Table 4.12 shows the average costs in mass and exergy basis (US$/kg and US$/MWh) for the products of the utilities plant.

According to the adopted thermoeconomic approach, shown in Fig. 4.14, the units should be charged or awarded by a given net monetary rate value, C_i, calculated as in Eq. 4.50, for the interaction between the utilities plant and the sulfur recovery system:

$$C_2 = C_{27} + C_{28} - C_{29} - C_{30} \tag{4.50}$$

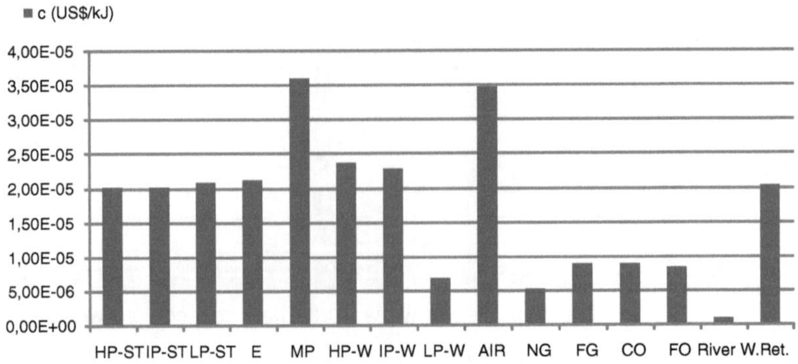

Fig. 4.18 Monetary unit cost of the main flows entering and leaving the utilities plant [19]

Table 4.13 Net exergy rate and net economic rate values [19]

Process unit	Net exergy rate (MW)	Net monetary rate (US$/s)
FCC (1)	166.64	0.28
Sulfur recovery (2)	−17.46	−0.35
Combined distillation (3)	72.67	0.05
Others (4)	−18.70	−0.44
Delayed coking (5)	66.60	0.23
HDT/H_2 (6)	−12.29	−0.26
Cooling tower (7)	−18.53	−0.31

The values of C_i and B_i (net exergy rate value) of the interactions shown in Fig. 4.14 are summarized in Table 4.13. The negative values of Table 4.13 indicate that the process unit should pay the calculated value for utilities plant. It is important to note that FCC and Delayed Coking units are awarded because they provide fuel gas and CO gas for utilities plant, while the Combined Distillation unit is awarded since it provides fuel oil. In addition to the interactions utilities plant-processes, the utilities plant is charged with the costs of external inputs such as natural gas and river water. For that reason, the utilities plant-processes interactions have an overall negative value of 0.80 US$/s.

4.3.4.1 Concluding Remarks

The exergy analysis was used to pinpoint the components responsible for the main exergy destructions in the utilities plant of a refinery. It highlights that the GT together with the boiler are responsible for 85 % of total exergy destruction while the overall plant exergy efficiency is 42 %. The thermoeconomic approach uses the efficiency of exergy conversion in each component as basis for cost distribution among the plant streams, penalizing the streams produced in a non-efficient way, such as: mechanical power, compressed air, compressed water, and electricity.

Figure 4.14 provides an overview of exergy and monetary rates' interactions between utilities plant and process units. It is important to note that the directions of exergy and monetary flows can be opposed to each other. This may happen if the exergy provided by a process unit has a higher cost than a greater amount of exergy provided by utilities plant, and vice versa. The cooling tower serves utilities plant and process units. The 9 % of cooling tower costs (exergy and monetary) was allocated to the utilities plant condensers and then to the respective steam turbines products. The remaining costs of cooling tower can be distributed to the other process units using the same criteria employed in utilities plant: quantity of exergy transferred by each unit to be dissipated.

4.4 Petroleum Refinery Hydrogen Production Unit: Exergy and Production Cost Evaluation

4.4.1 Introduction

Hydrogen is normally found combined with other elements, such as oxygen in water, carbon in hydrocarbons, and the majority of organic composites. Among fossil fuels, natural gas is the main raw material used to produce hydrogen (48 % of all raw materials, according to Argonne National Lab. [22]). Due to its great chemical activity and consequent easiness of reaction, pure hydrogen is often used in industrial processes, such as the ammonia production and oil refineries to purify several products and fuels. According to Ball and Wietschel [23], about 50 % of the hydrogen use is in the ammonia production while slightly less than 40 % in petroleum processing.

One of the main important hydrogen production route is the methane steam reforming one, responsible for 48 % of the primary energy sources for hydrogen production [24].

The hydrogen production unit analyzed in this chapter has to supply 550,000 Nm^3 of hydrogen per day to purify diesel oil. Based on a synthesis plant of a petroleum refinery hydrogen production unit, the exergy efficiency of each component and of the overall plant are calculated. The hydrogen production cost is determined by means of a thermoeconomic analysis in which the equality cost partition method is employed, including capital and operational costs, in order to determine the production cost of hydrogen and other products of the plant [25].

4.4.2 Methane Reforming Process

Hydrogen can be obtained using the methane reforming process. In this process [26], steam (H_2O) reacts with natural gas composed of a mixture of CH_4 (89 %),

C_2H_6, (9 %), C_3H_8, C_4H_{10}, N_2, and CO_2. The product of this reaction is a mixture of hydrogen (H_2), carbon monoxide (CO), carbon dioxide (CO_2), and steam (H_2O), according to Eq. 4.51 (for each hydrocarbon):

$$\begin{aligned}\alpha C_mH_{2n} + \beta H_2O \rightarrow [(m+n)\alpha + \gamma]H_2 + (m\alpha - \gamma)CO + \gamma CO_2 \\ + (\beta - m\alpha - \gamma)H_2O\end{aligned} \tag{4.51}$$

with

$$0 \leq \gamma \leq m\alpha \tag{4.52}$$

The stoichiometric coefficient γ of Eq. 4.51 depends on the chemical equilibrium of the reaction.

The design of a hydrogen production plant is complex. A great number of equipment, valves, pipes, and connections are necessary to build the plant and guarantee its production. Due to this great number of information and variables, it is necessary to develop a synthesis plant to describe the main functions of the real process, as presented in Fig. 4.19.

Before entering the Reformer, the feed (node 100) is compressed in the Feed Compressor (node 110). From the Feed Compressor, natural gas is heated in the Feed Preheater (from node 110 to node 141), mixed with recycled hydrogen (node 145) and sent to the Reformer Feed Preheat Coil (node 150). The gas leaving the Feed Preheater Coil (node 160) goes to the Desulphurizer.

The outlet flow from the Desulphurizer (node 170) is mixed with process steam (node 590) and sent to the Reformer Mix Feed Preheater Coil, located in the reformer convection section. Then it goes to the catalyst tubes (node 220) located in the Reformer at 2.8 MPa and 460 °C.

Hydrogen is produced in the reforming section by the reaction of hydrocarbons with steam in the presence of a catalyst. As the reforming reaction is strongly endothermic and the heat required is at very high temperature, the reforming catalyst is placed in vertical tubes installed inside the Reformer radiant section.

The Reformer consists of a single top-fired radiant cell (the burners are located in the roof of radiant cell) and a vertical convection box. Flue gases leave the radiant bottom.

The combustion air is injected in the burners by the Forced Draft Fan (node 330) and is preheated in the Combustion Air Preheater located in the convection section of the Reformer. The fuel used by these burners comes from a second stream from the feed (node 310).

In the convection section the flue gas, in addition to preheating the air, exchanges heat in the following coils: Steam Generator Coil, Mix Feed Preheat Coil, Feed Preheat Coil, and Steam Superheater Coil.

The reformed gas leaves the catalyst tubes (node 230) at 840 °C and 2.58 MPa and goes to the Waste Heat Exchanger. The reformer effluent temperature is controlled by the burners' control system.

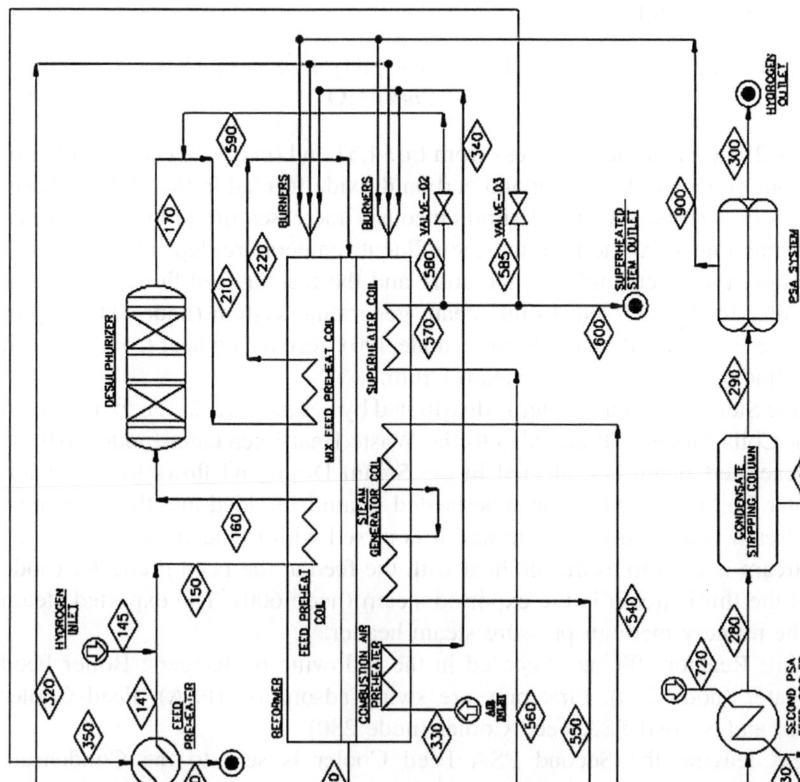

Fig. 4.19 Synthesis plant [25]

The reformer effluent is cooled in the Waste Heat Boiler and goes to the Shift Reactor (node 240). In the Shift Reactor CO reacts with steam to form hydrogen and CO_2, as shown in the Eq. 4.53:

$$(m\alpha - \gamma)\text{CO} + (\beta - m\alpha - \gamma)\text{H}_2\text{O} \rightarrow (m\alpha - \gamma)\text{H}_2 + (m\alpha - \gamma)\text{CO}_2 + (\beta - 2m\alpha)\text{H}_2\text{O} \tag{4.53}$$

where $(\beta - 2\ m\alpha)$ is the steam excess from Eq. 4.51 and $(m\alpha - \gamma)$ is the number of mols of both molecular hydrogen and carbon dioxide formed in the Shift Reactor.

This reaction reduces CO and simultaneously increases the production of H_2. The shift reaction is exothermic and the effluent temperature depends on the CO concentration, the reactor inlet temperature and the reactor feed flow.

The boiler feed water sent to the steam generating system (node 480) is preheated in the Boiler Feed Water Heater (node 490), recovering heat from the Shift Reactor effluent, and goes to the Steam Drum.

From the Steam Drum the water is distributed by natural circulation to the Steam Generator Coils (node 540) and also to the Waste Heat Exchanger (node 510).

The generated steam is collected in the Steam Drum and flows to the Steam Superheater Coil (node 560). The superheated steam is divided into three streams: the first stream is the process steam and it is mixed with the feed (node 580), the second stream is used to exchange heat with the feed at the Feed Preheater (node 350), and the third stream is the exported steam (node 600). The exported steam goes to the refinery medium pressure steam header.

The Shift Reactor effluent is cooled in the following exchangers: Boiler Feed Water Heater (node 260), First pressure swing adsorption (PSA), Feed Cooler (node 270) and Second PSA Feed Cooler (node 280).

The gas leaving the Second PSA Feed Cooler is sent to the Condensate Stripping Column, where the process condensate is removed (node 400), and then flows to the PSA System (node 290). The final hydrogen purification (99.90 % vol.) is done in the PSA system. The PSA System generates two streams: the purge gas, which is used as a fuel in the Reformer (node 900), and the hydrogen product (node 300).

The design, installation, maintenance, and operation of every component of the plant and each chemical element that composes the process have a direct impact on the hydrogen production cost. Each of them is difficult to evaluate but proper evaluation during the design and construction of a new plant can mean the difference between profits and losses.

4.4.3 Exergy Analysis of the Plant

The exergy analysis of the natural gas reforming plant was developed using the synthesis plant shown in Fig. 4.19 considering the composition, mass flow rate, pressure, and temperature indicated in each one of the sections of that figure

Table 4.14 Main inlet and outlet fluid characteristics [25]

Section		Product	Mass flow rate (kg/h)	Temperature (°C)	Pressure (kPa)	Physical exergy (MW)	Chemical exergy (MW)	Total (MW)
Inlet flows	100	Natural gas	7,910.00	40	2465.3	0.91	103.63	104.54
	145	Hydrogen	37.00	40	3085.9	0.03	0.96	0.99
	310	Natural gas	662.00	40	2465.3	0.07	8.42	8.49
	330	Dry air	78,564.0	25	101.3	0.00	0.09	0.09
	480	Compressed water (8.4 MPa)	41,739.00	145	8473.8	1.08	0.61	1.69
	720	Compressed water to second PSA feed cooler	52,400.00	25	465.8	0.01	0.74	0.75
Outlet flows	300	Hydrogen	2,004.00	44	2091.1	2.17	68.05	70.22
	360	Saturated water (1.3 Mpa)	1,223.00	192	1322.7	0.04	0.01	0.05
	400	Liquid water from stripping column	13,177.00	40	2189.5	0.01	0.19	0.20
	600	Superheated steam (3.1 MPa)	17,011.00	436	3056.3	6.11	0.25	6.36
	730	Compressed water from second PSA feed cooler	52,400.00	25	465.8	0.01	0.74	0.75
	920	Combustion products	97,657.00	359	99.3	3.66	1.31	4.97

assuming steady state operating conditions. Thermodynamic and transport properties of every flow were determined by means of the software EES® [15]. Table 4.14 presents the main inlet and outlet fluid characteristics of the synthesis plant. All the gases and gas mixtures were treated as ideal gases because their compressibility factors are close to one (in node 100 the compressibility factor is 0.958; in node 145 it is 1.018 and in node 150 it is 0.995 [27]).

Based on the synthesis plant and with the thermodynamic and transport properties of the fluids in each section, it is possible to make calculations of mass, energy, and exergy balances of each component of the synthesis plant, as well as its respective exergy efficiency. In these balances, it is assumed that there are no heat losses to the environment in any analyzed component, and $P_0 = 101.3$ kPa and $T_0 = 25$ °C were used as the reference environment pressure and temperature.

Table 4.15 Exergy destruction rate in each component of the process [25]

Component	Exergy destruction rate (kW)	(%)
Feed compressor	20.55	0.06
Feed preheater	110.10	0.30
Mixer-01	47.07	0.13
Mixer-02	1,024.07	2.77
Waste heat exchanger	2,885.86	7.82
Shift reactor	197.68	0.54
Boiler feed water heater	759.27	2.06
First PSA feed cooler	4,890.60	13.25
Second PSA feed cooler	410.08	1.11
PSA system	867.01	2.35
Valve-01	45.17	0.12
Valve-02	35.20	0.10
Valve-03	27.11	0.07
Valve-04	35.27	0.10
Steam drum	3.62	0.01
Reformer	25,559.05	69.23
Overall	36,917.70	100.00

The atmospheric composition was considered to be the same as proposed by Szargut et al. [28]. These conditions represent the real average values of pressure, temperature, and atmospheric composition of the environment in which the studied hydrogen production unit will be installed. Based on these definitions and on the process data, the exergy flow rate in each point of the system was calculated and, consequently, it was possible to evaluate the exergy destruction rate in each component. These values are shown in Table 4.15.

The exergy efficiencies for the main components and for the overall plant were calculated using the following equations:

- Feed compressor:

$$\eta_b = \frac{B_{110} - B_{100}}{W} \tag{4.54}$$

- Feed preheater:

$$\eta_b = \frac{B_{141} - B_{110}}{B_{350} - B_{360}} \tag{4.55}$$

- Waste heat exchanger:

$$\eta_b = \frac{B_{520} - B_{510}}{B_{230} - B_{240}} \tag{4.56}$$

Table 4.16 Exergy efficiency of the main components [25]

Component	Exergy efficiency (%)
Feed compressor	84.95
Feed preheater	51.85
Waste heat exchanger	63.05
Boiler feed water heater	73.48
Reformer	46.40
Overall plant	66.60

– Boiler feed water heater:

$$\eta_b = \frac{B_{490} - B_{480}}{B_{250} - B_{260}} \tag{4.57}$$

– Reformer:

$$\eta_b = \frac{\left[\begin{array}{l}(B_{160} - B_{150}) + (B_{220} - B_{210}) + (B_{570} - B_{560}) \\ + (B_{550} - B_{540}) + (B_{340} - B_{330}) + (B_{230} - B_{220})\end{array}\right]}{(B_{900} + B_{340} + B_{320})} \tag{4.58}$$

– Overall plant:

$$\eta_b = \frac{(B_{300} + B_{360} + B_{400} + B_{600} + B_{730})}{(B_{100} + B_{145} + B_{310} + B_{330} + B_{480} + B_{720})} \tag{4.59}$$

The exergy efficiencies calculated for the main components are presented in Table 4.16.

The results shown in Table 4.15 indicate the influence of the reformer efficiency on the overall efficiency of the plant, due to reaction and heat transfer irreversibilities.

Using the same approach utilized in the exergy analysis, the energy efficiency of the plant was evaluated, giving an overall value of 81.7 %. This value was calculated differently as proposed by Lutz et al. [29]. It was calculated considering not only the hydrogen flow rate, but also every secondary flow like high-pressure steam and water, since these secondary products are used in another refinery production process.

4.4.4 Thermoeconomic Analysis

Data concerning equipment costs, engineering costs, construction, and erection costs have been obtained through some commercial proposals and interviews with the professionals involved in the construction of a real hydrogen production plant [25] and are presented in Table 4.17.

Table 4.17 Direct and indirect costs by equipment and annual levelized costs [25]

Component	I_{eq} (US$) (2003)	C_{aeq} (US$/year)
Feed compressor	11,200,000.00	2,688,000.00
Feed preheater	142,000.00	34,080.00
Mixer-01	5,000.00	1,200.00
Desulphurizer	3,570,000.00	856,800.00
Mixer-02	5,000.00	1,200.00
Waste heat exchanger	715,000.00	171,600.00
Shift reactor	1,785,000.00	428,400.00
Boiler feed water heater	2,143,000.00	514,320.00
First PSA feed cooler	1,650,000.00	396,000.00
Second PSA feed cooler	640,000.00	153,600.00
Condensate stripping column	22,800.00	5,472.00
PSA system	6,411,000.00	1,538,640.00
Valve-01	13,000.00	3,120.00
Valve-02	13,000.00	3,120.00
Valve-03	13,000.00	3,120.00
Valve-04	13,000.00	3,120.00
Demixer	5,000.00	1,200.00
Steam drum	85,000.00	20,400.00
Reformer	32,140,000.00	7,713,600.00
Total	60,570,800.00	14,536,992.00

To continue the economic analysis it was necessary to distribute the equipment costs throughout time and, for this reason, several economic variables have been defined as follows:

- Operational and maintenance fixed cost factor (f_{omf}): 6 % of the total investment;
- Operational and maintenance variable cost factor (f_{omv}): 2 % of the total investment;
- Load factor (f_l): 100 %
- Annual operation time (t_{op}): 8,400 h/year
- Annual interest rate (i): 15 %
- Capital recovery period (n): 20 years
- Capital recovery factor: 16 %
- Natural gas cost: $c_{100} = 2.20$ US$/GJ
- Compressed water @85 bar: $c_{480} = 1.10$ US$/t

The levelized annual cost (C_{aeq}) for each component of the plant (Table 4.17) was determined by using Eq. 4.60 and annual operation time:

$$C_{aeq} = I_{eq} \left(f_a + f_{omf} + f_l \, f_{omv} \right) \tag{4.60}$$

The combination of the costs balances of the main components of the synthesis plant gives Eq. 4.61

Table 4.18 Considerations for determination of the exergy-based production costs [25]

Hypothesis	Justification
$c_{920} = 0\,\text{US}\,\$/\text{kJ}$	Flue gases are discharged into the atmosphere
$c_{330} = 0\,\text{US}\,\$/\text{kJ}$	Combustion air is taken from the atmosphere
$c_{145} = 1.5c_{300}$	Hydrogen that is added to process (node 145) was evaluated taking into account a correction factor that considers transportation, storage and compression costs
$c_{480} = c_{720} = 1.10\,\text{US}\,\$/\text{t}$	Compressed liquid water cost in Sect. 720 is considered equal to compressed liquid water @ 85 bar
$c_{300} = c_{360} = c_{400} = c_{600} = c_{730} = c_{\text{prod}}$	As the flows in sections. 300, 360, 400, 600, and 730 are utilized in other refinery processes, the equality cost partition criteria is taken as an auxiliary relation to determine the production costs of these five mass flows

Table 4.19 Costs of the products of the plant [25]

Product	c_{bb} (US$/GJ)	c_{mb} (US$/t)	c_{tb} (US$/h)
Hydrogen	9.75	1,185.86	2,463.61
Superheated steam (3.1 MPa)	9.75	12.39	223.19
Saturated water (1.3 MPa)	9.75	1.95	1.67
Liquid water from condensate stripping column	9.75	0.52	7.27
Compressed water from second PSA feed cooler	9.75	0.52	27.53

$$c_{100}B_{100} + c_{145}B_{145} + c_{310}B_{310} + c_{330}B_{330} + c_{480}B_{480} + c_{720}B_{720} + \Sigma C_{eq} \\ = c_{300}B_{300} + c_{360}B_{360} + c_{400}B_{400} + c_{600}B_{600} + c_{730}B_{730} + c_{920}B_{920} \tag{4.61}$$

In order to determine the production costs for every one of the outlet flows some hypothesis were taken into consideration, as summarized in Table 4.18

With the aforementioned considerations, Eq. 4.61 can be solved together with equations presented in Table 4.18, allowing the determination of the value of c_{prod}.

Table 4.19 shows the calculated costs of the products of the studied plant on an exergy basis (c_{bb}), mass basis (c_{mb}), and time basis (c_{tb}).

The approach employed to determine the production costs of the hydrogen production plant can also be utilized to estimate the new cost of the products if any change or improvement in the operating conditions is added. Consider that instead of discharging the reformer flue gases into the atmosphere, at 359 °C and with a significant amount of exergy, they were used to preheat any refinery flow. For example, if an additional heater was considered in the system with a similar cost to the first PSA cooler that uses the flue gases flow with an exergy efficiency of 70 and 10 % pressure loss between inlet and outlet (in this new operating condition, the flue gases leave the heater at 120 °C), the new hydrogen exergy-based cost will be 9.67 US$/GJ, or 2,443.39 US$/h. This simplified calculus shows that it is possible to obtain a reduction of 0.82 % in the hydrogen production cost, or an economy of 169,800.00 US$/year.

4.4.5 Concluding Remarks

The exergy and thermoeconomic analysis of the hydrogen production by the steam reforming route is a useful approach in the determination of the hydrogen production cost. Even with the simplifications used in the evaluation, this approach can also be applied to other hydrogen production processes based on the different primary energy sources such as electrolysis, coal gasification, and oil/naphtha reforming, in order to perform a comparative production cost analysis.

Taking this analysis of the hydrogen production process in a larger context, the values of both the exergy efficiency for the overall plant (66.60 %) and the hydrogen production cost (1.18 US$/kg), although obtained in a refinery plant, show clearly the thermodynamics performance limitations for reforming natural gas to produce hydrogen for fuel cells-based systems.

References

1. Key world energy statistics. In: International energy agency 2010. Available at http://www.iea.org. Cited 05 Feb 2011
2. Annual Outlook 2010. In: EIA—US energy information administration. Available via DIALOG http://www.eia.gov. Cited 03 Oct 2011
3. Oliveira Jr S, Van Hombeeck M (1997) Exergy analysis of petroleum separation processes in offshore platforms. Energ Convers Manage 38:1577–1584
4. HYSIM User's Guide. Version C1.50. [S.l.]: Hyprotech, 1991
5. Peng DY, Robinson DB (1976) A two-constant equation of state. Ind Eng Chem Fundam 15:59–64
6. Kotas TJ (1995) The exergy method of thermal plant analysis. Krieger Publishing Company, Florida
7. Caetano EF, Silva RM, da Silva RG et al. (1997) Petrobras and bornemmann pumpen cooperation on multiphase flow pumping. In: Proceedings of the 29th annual offshore technology conference, Houston
8. Baruzzi JOA, Caetano EF, Fagundes Netto JR (2001) Production forecast for a subsea multiphase pumping system in the marlim field. In: Proceedings of the SPE annual technical conference and exhibition, New Orleans
9. Gonzáles R, Guevara E (1995) Economic field development in venezuela heavy oil fields using multiphase pumping technology. In: Proceedings SPE international oil symposium, Richardson
10. Nakashima CY (2005) Modeling and simulation of a twin-screw multiphase pump. Ph.D. Thesis, Polytechnic School of the University of São Paulo, São Paulo (in Portuguese)
11. Nakashima CY, Oliveira S Jr, Caetano EF (2004) Subsea multiphase pumping system X gas lift: an exergo-economic comparison. Therm Eng 3:107–114, ISSN 1676-1790
12. Hyprotech (1996) Hysys version 1.1—Reference vol 1 e 2, Hyprotech Ltd

13. Rivero R, Rendón C, Monroy L (1998) Calculation of the exergy of crude oil mixtures. In: Proceedings of the 11th international conference on efficiency, costs, optimization, simulation and environmental impact of energy systems, Nancy
14. Guthrie VB (1960) Petroleum products handbook. McGraw Hill Book Company, London
15. Klein SA (2011) Engineering equation solver—EES, F-Chart Software. www.fChart.com
16. Nakashima CY, Oliveira S Jr, Caetano EF (2002) Thermodynamic model of a twin-screw multiphase pump. In: Proceedings of ASME engineering technology conference on energy, Houston
17. Szklo A, Schaeffer R (2007) Fuel specification, energy consumption and CO_2 emission in oil refineries. Energy 32:1075–1092
18. Williams B (2003) Future energy supply-5. Refining adjustments. Oil Gas J 32:20–40
19. Mendes da Silva JA, Pinto CP, Rucker C, Oliveira Jr S (2011) Evaluation of a refinery utilities plant through exergy analysis and its cost distribution through thermoeconomy. In: Proceedings of the 24th international conference on efficiency, costs, optimization, simulation and environmental impact of energy systems, Novi Sad
20. Lozano MA, Valero A (1993) Theory of exergetic cost. Energy 18:939–960
21. Lazzaretto A, Tsatsaronis G (2006) SPECO: a systematic and general methodology for calculating efficiencies and costs in thermal systems. Energy 31:1257–1289
22. U.S. Department of Energy (2009) Assessing current, near-term, and long-term U.S. hydrogen markets. In: Argonne national laboratory, U.S. Department of Energy. Available via DIALOG. http://www.dis.anl.gov/news/HydrogenMarkets.html. Cited 17 Mar 2009
23. Ball M, Wietschel M (2009) The future of hydrogen—opportunities and challenges. Int J Hydrogen Energ 34:615–627
24. Ewan BCR, Allen RWK (2005) A figure of merit assessment of the routes to hydrogen. Int J Hydrogen Energ 30:809–819
25. Cruz FE, Oliveira Jr S (2008) Petroleum refinery hydrogen production unit: exergy and production cost evaluation. Int J Thermodyn 11:187–193
26. Docekal T (1986) Hydrogen production from hydrocarbons. Int J Hydrogen Energ 11:709–714
27. Cruz FE (2004) Thermoeconomic evaluation of a petroleum refinery hydrogen production unit. Undergraduate conclusion project. Polytechnic School of the University of São Paulo, São Paulo (in Portuguese)
28. Szargut J, David RM, Steward F (1988) Exergy analysis of thermal, chemical, and metallurgical processes. Hemisphere Publishing, New York
29. Lutz AE, Bradshaw RW, Keller JO et al (2003) Thermodynamic analysis of hydrogen production by steam reforming. Int J Hydrogen Energ 28:159–167

Chapter 5
Chemical Processes Analysis and Improvement

Symbols

a	Constant of Eqs. 5.31, 5.39 and 5.40
A	Parameter of Eq. 5.42
b	Constant of Eqs. 5.31 and 5.39
b	Specific exergy (kJ/kmol)
BL	Column liquid side draw (kmol/s)
c	Constant of Eq. 5.31 and 5.39
c_p	Specific heat at constant pressure (kJ/kmol K)
c_v	Specific heat at constant volume (kJ/kmol K)
d	Constant of Eq. 5.31
e	Constant of Eq. 5.31
F	Stage feed flow rate (kmol/s)
G	Specific Gibbs free energy (kJ/kmol)
h	Specific enthalpy (kJ/kmol)
I	Irreversibility/exergy destroyed rate (kW)
k	$=\frac{c_p}{c_v}$
K	Equilibrium ratio
L	Liquid flow rate (kmol/s)
P	Pressure (kPa)
Q	Heat duty or heat rate (kW)
R	Universal gas constant (8.314 kJ/kmol K)
s	Specific entropy (kJ/kmol K)
T	Temperature (K)
UA	Product of the overall heat transfer coefficient by the heat transfer area of the heat exchanger (kW/K)
V	Vapor flow rate (kmol/s)
x	Liquid phase concentration (kmol/kmol)
y	vapor phase concentration (kmol/kmol)
z	stage feed concentration (kmol/kmol)

S. de Oliveira Jr., *Exergy*, Green Energy and Technology,
DOI: 10.1007/978-1-4471-4165-5_5,

Greek Symbols

α	Parameter of Eq. 5.42
β	Parameter of Eq. 5.42
γ_i	Activity coefficient
δ	Percentage of irreversibility rate
ζ	$= \frac{1}{1-\frac{k-1}{\eta_{poli} k}}$
η_{poli}	Polytropic efficiency
θ	Carnot factor

Subscript

g	Gas or vapor phase
i	Chemical component
in	Inlet
j	Equilibrium stage
k	Unit
l	Liquid phase
out	Outlet
S	Steam

Superscript

_	average value
b	Boundary
e	Excess
0	Reference ($T = 298$ K, $P = 101.3$ kPa)

Abbreviations

C	Compressor
EV	Evaporator
HE	Heat exchanger, waste heat boiler
RX	Reactor
T	Tower

5.1 Introduction

The exergy analysis of chemical processes is one of the best examples to evidence the potential of application of this process assessment tool since it allows to evaluate the performance of energy conversion processes that happen in reactors, separators, and mixers in the same basis used to evaluate thermomechanical energy conversion processes. In the study described in this chapter, exergy analysis was applied to an acetaldehyde production plant in order to improve the use of energy inputs as well as to reduce the exergy destruction in the production processes. The analysis was performed by means of a thermodynamic model of the processes of the plant. The model was able to predict all relevant thermodynamic properties of the process streams. The vapor phase was considered ideal, while the nonidealities on liquid phase were corrected by Wilson equation. Based on the results of the analysis it was possible to propose a new configuration with lower overall exergy destruction rate and, as a consequence, lower specific energy consumption per kilogram of produced acetaldehyde. This production plant was chosen since it contains several energy conversion processes such as chemical reactions, heat transfer, and mass transfer and coupled heat and mass transfer (in the acetaldehyde absorption tower).

5.2 Acetaldehyde Production by Ethanol Partial Oxidation

Acetaldehyde, also called ethanal, is a colorless liquid with 294 K normal boiling point and requires pressurized storage systems. It is a chemical intermediary that can be used in the production of several substances, such as acetic acid, acetic esters, vinyl acetate, butiraldehyde (utilized to produce n-butanol), pyridine, and pentaeritritol. Johnson et al. [1] describe the acetaldehyde producion routes.

There are two main acetaldehyde production routes: the oxidation of ethylene in liquid phase and the partial oxidation of ethanol.

In the first route, a palladium and copper chloride catalyst is employed to promote the ethylene direct oxidation. In the second route, the oxidation occurs during the flow of a mixture of ethanol vapor and air through a bed of silver catalyst at temperatures between 800 and 900 K.

The production route studied in this chapter, shown in Fig. 5.1, is the second one. The chemical reactions that take place in this route are:

$$CH_3CH_2OH_{(g)} \rightarrow CH_3CHO_{(g)} + H_{2(g)}$$

$$H_{2(g)} + {}^{1}\!/_{2}O_{2(g)} \rightarrow H_2O_{(g)}$$

$$CH_3CH_2OH_{(g)} + 3O_{2(g)} \rightarrow 2CO_{2(g)} + 3H_2O_{(g)}$$

A hot stream of acetaldehyde, water, non-converted ethanol, and a mixture of gases (oxygen, nitrogen, hydrogen, and carbon dioxide) leaves reactor RX1 to the waste heat boiler HE2, where it is cooled to 443 K and produces low pressure steam (550 kPa).

After that, the stream follows to the acetaldehyde absorption tower T2 to be quenched with an alcoholic solution, which comes from the bottom of the distillation tower T3; acetaldehyde, water, and ethanol are condensed while gases rise up through the tower. The heat released during condensation is removed by the cooler HE3 and coils installed inside the column. The gases, saturated with water and ethanol, leave the top of tower T2 and go to the next tower T4, to be quenched

Fig. 5.1 Acetaldehyde production plant [2]

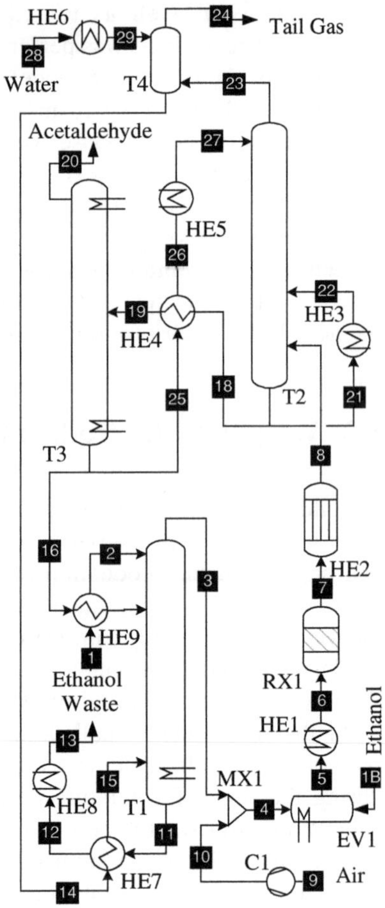

with chilled water. After that, the gases free from ethanol are released to the atmosphere at 288 K.

The liquid, a mixture of ethanol, water and acetaldehyde, leaves the column, is heated up by the pre-heater HE4, and follows to the distillation tower T3, which operates at 345 kPa. The pure acetaldehyde is removed at the top at 327 K, while the ethanol and water stream at bottom splits into two parts: one part goes to tower T2, and the other one is fed into tower T1.

The function of tower T1 is to strip the ethanol from water and recycle it to the process at azeotropic concentration; the water leaves the tower by the bottom as a waste. New raw ethanol is fed to top of the tower to cool it; the vapor of this tower is a mixture of water and ethanol, which feeds the evaporator EV1. In the evaporator EV1, operating at 340 K, the ethanol is mixed with air, compressed previously by the compressor C1.

Before feeding the reactor RX1, the ethanol–air mixture leaves the evaporator EV1 and goes to the preheater HE1 to be heated above dew point (393 K).

5.3 Thermodynamic Model

5.3.1 Introduction

The model developed by Vieira [3] takes into account the seven main compounds: ethanol, water, acetaldehyde, oxygen, nitrogen, hydrogen, and carbon dioxide; other chemical components are presented in a much lower concentration. The concentrations of the other components are so small that they can be neglected without any impact in the quality of the results predicted by the model.

The thermodynamic model hypotheses [3] are:

- Steady-state conditions for all processes;
- Ideal vapor phase;
- Raoult's law corrected by Wilson's equation;
- Gas solubility depends only on temperature;
- Control volumes are perfectly insulated;
- No temperature or concentration gradient inside control volumes (for equilibrium stages);
- No short-circuit or carry-over in the equilibrium stages.

The highest pressure occurs in acetaldehyde distillation tower T1 and it is equal to 345 kPa. As this pressure may be considered moderate, it is possible to assume ideal gas behavior for the vapor phase. However, this fact is not true for the liquid phase, in which the nonideality must be corrected. The Wilson equation was chosen to evaluate the liquid phase activity coefficients.

Henry's law estimated the dissolved gases behavior. The reactor RX1 was considered as a black box in which reactions, presented in 5.2, occur with given conversion rates for ethanol and oxygen, and acetaldehyde efficiency [3]. The air compression power was evaluated considering a polytropic compression of an ideal gas [3].

The exergy of substances were determined using the reference environment proposed by Szargut et al. [4].

In the model equations, i identifies the chemical components (acetaldehyde = 1, ethanol = 2, water = 3, oxygen = 4, nitrogen = 5, hydrogen = 6 and carbon dioxide = 7), j is the equilibrium stage number, and k names the unit the stage belongs to.

5.3.2 Process Modeling

5.3.2.1 Reactor Model

As written before, the reactor was modeled as a black box with given conversion yields which describe the molar variation of reactants and products for design

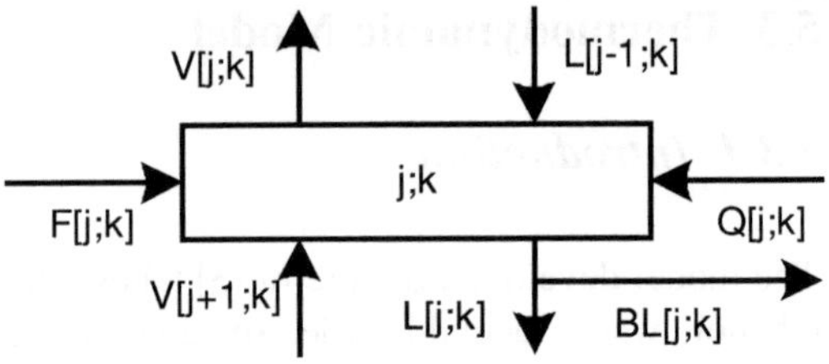

Fig. 5.2 Equilibrium stage

operating conditions: feed temperature of 400 K, feed pressure of 135 kPa, and an ethanol air ratio of 0.88 kmol/kmol.

For the inlet conditions indicated before, the considered conversion yields were:

- ethanol conversion = 57.85 %
- oxygen conversion = 75.95 %
- acetaldehyde conversion = 98.42 %

5.3.2.2 Equilibrium Stage Model

The separation steps are considered as equilibrium stages; the following sketch in the Fig. 5.2 represents one equilibrium stage:

The streams which leave the equilibrium stage $V_{i;j}$ and $L_{i;j}$ are the molar flow rate of the i component in the vapor and liquid phase respectively which are in thermodynamic equilibrium, and obey the following equations:

$$y_{i,j,k} - x_{i,j,k}K_{i,j,k} = 0 \tag{5.1}$$

$$\sum_i x_{i,j,k} - 1 = 0 \tag{5.2}$$

$$\sum_i y_{i,j,k} - 1 = 0 \tag{5.3}$$

$$\begin{aligned} & x_{i,j-1,k}L_{j-1,k} + y_{i,j+1,k}V_{j+1,k} + z_{i,j,k}F_{j,k} - x_{i,j,k}(L_{j,k} + BL_{j,k}) \\ & \quad - y_{i,j,k}V_{j,k} = 0 \end{aligned} \tag{5.4}$$

$$\begin{aligned} & h_{l_{j-1,k}}L_{j-1,k} + h_{g_{j+1,k}}V_{j+1,k} + h_{l_{j,k}}F_{j,k} - h_{l_{j,k}}(L_{j,k} + BL_{j,k}) \\ & \quad - h_{g_{j,k}}V_{j,k} + Q_{j,k} = 0 \end{aligned} \tag{5.5}$$

The Eq. 5.1 models the equilibrium liquid–vapor, while Eqs. 5.2–5.4 are the mass balances and Eq. 5.5 is the energy balance. To solve the previous set of equations it is necessary to define all inlet streams $L_{j-1;k}$, $V_{j+1;k}$ and $F_{j;k}$, their concentrations $x_{i;j-1;k}$, $y_{i;j+1;k}$ and $z_{i;j;k}$, the stage pressure $p_{j;k}$, the heat duty $Q_{j;k}$ and the ratio $B_{Lj;k/Lj;k}$. The solution will be the liquid $x_{i;j;k}$ and vapor $y_{i;j;k}$ concentration profile through the column, the liquid $L_{j;k}$, and the vapor $V_{j;k}$ molar rates, as well as the stage temperature $T_{j;k}$.

The exergy destroyed, or irreversibility rate $I_{j,k}$ is evaluated by the exergy balance:

$$\begin{aligned} &b_{l_{j-1,k}} L_{j-1,k} + b_{g_{j+1,k}} V_{j+1,k} + b_{l_{j,k}} F_{j,k} \\ &\quad - b_{l_{j,k}} (L_{j,k} + BL_{j,k}) - b_{g_{j,k}} V_{j,k} + Q_{j,k}\theta_{j,k} = I_{j,k} \end{aligned} \tag{5.6}$$

where $\theta_{j,\,k}$ is the Carnot factor, which is defined as follows:

$$\theta_{j;\,k} = 1 - \frac{T_0}{T^{\mathrm{b}}_{j,\,k}} \tag{5.7}$$

and $T^{\mathrm{b}}_{j,k}$ is the boundary temperature.

It is important to remind that temperature $T^{\mathrm{b}}_{j,k}$ is not equal to $T_{j,k}$. The last one is the temperature inside the stage. For example, if a given stage has an equilibrium temperature of 348 K and receives a heat transfer from a condensing steam at 428 K, then $T^{\mathrm{b}} = 428\,K$.

5.3.2.3 Equilibrium Multi-Stage Model

The columns T1, T2, and T3 are considered as an array of equilibrium stages. The set of equations is similar, and the only difference is the number of equations. The index j represents the stage number inside a unit k.

5.3.2.4 Heat exchangers

The energy balances for the fluids that receive and transfer heat (Eqs. 5.8 and 5.9) and the heat transfer equation (Eq. 5.10) for the heat exchangers without phase change, are:

$$h_{l_{1,k}} L_{1,k} - h_{l_{2,k}} L_{2,k} + Q_{j,k} = 0 \tag{5.8}$$

$$h_{l_{3,k}} L_{3,k} - h_{l_{4,k}} L_{4,k} - Q_{j,k} = 0 \tag{5.9}$$

$$Q_{j,k} - (UA)_{j,k} \tfrac{1}{2}[(T_3 - T_1) + (T_4 - T_2)] = 0 \tag{5.10}$$

where $(\mathrm{UA})_{j,k}$ is the product of the overall heat transfer coefficient by the heat transfer area of the heat exchanger. The irreversibility rate is calculated by:

$$b_{l_{1,k}} L_{1,k} + b_{l_{3,k}} L_{3,k} - b_{l_{2,k}} L_{2,k} - b_{l_{4,k}} L_{4,k} = I_{j,k} \tag{5.11}$$

For evaporators and reboilers the equations are simpler, since the heating medium is always low-pressure steam at T_{s}:

$$h_{l_{1,k}} L_{1,k} - h_{l_{2,k}} L_{2,k} + Q_{j,k} = 0 \tag{5.12}$$

$$Q_{j,k} - (UA)_{j,k}{}^{1}\!/\!{}_{2}[(T_S - T_1) + (T_S - T_2)] = 0 \tag{5.13}$$

$$b_{l_{1,k}} L_{1,k} - b_{l_{2,k}} L_{2,k} + Q_{j,k}\theta_{j,k} = I_{j,k} \tag{5.14}$$

with,

$$\theta_{j;k} = 1 - \frac{T_0}{T_S} \tag{5.15}$$

In case the heat exchanger is a cooler, the heat transfer temperature is not constant but varies between T_1 and T_2. In this case the Carnot factor should be an average value, estimated by:

$$\begin{aligned}\theta_{j;k} &= \frac{\int_{T_1}^{T_2} Q_{j;k}\theta_{j;k}\mathrm{d}T}{(T_2-T_1)Q_{j;k}} = \frac{\int_{T_1}^{T_2}(T-T_1)\left(1-\frac{T_0}{T}\right).\mathrm{d}T}{(T_2-T_1)(T_2-T_1)} \\ &= \frac{\left[\frac{T_2^2-T_1^2}{2}-(T_2-T_1)(T_1+T_0)+T_1T_0\ln\left(\frac{T_2}{T_1}\right)\right]}{(T_2-T_1)^2}\end{aligned}$$

5.3.2.5 Air Blower

The power required by the air blower is calculated assuming that a polytrophic process takes the air, which behaves as an ideal gas, from atmospheric pressure P_1 up to P_2.

The energy balance for an adiabatic control volume with the air blower gives (1 is inlet and 2 is outlet):

$$h_{g_{1,k}} V_{1,k} - h_{g_{2,k}} V_{2,k} - W_{j,k} = 0 \tag{5.16}$$

where,

$$h_{g_{2,k}} = \bar{c}_P T_1 \left(\left(\frac{P_2}{P_1}\right)^{\frac{\zeta-1}{\zeta}} - 1\right) \tag{5.17}$$

with,

$$\zeta = \frac{1}{1 - \frac{k-1}{\eta_{\mathrm{poli}} k}} \tag{5.18}$$

and

$$\eta_{\mathrm{poli}} = 0.88 \tag{5.19}$$

$$k = \frac{\bar{c}_P}{\bar{c}_V} \tag{5.20}$$

$$\bar{c}_P = \sum_i y_i \bar{c}_{P_i}(\bar{T}) \tag{5.21}$$

$$\bar{T} = \frac{T_2 + T_1}{2} \tag{5.22}$$

5.3.3 Thermodynamic Properties

The specific enthalpy and entropy of the vapor phase, h_g and s_g, are determined by Eqs. 5.23–5.27 [3].

$$h_g = \sum_i y_i h_{i,g} \tag{5.23}$$

$$h_{i,\,g} = h^0_{i,\,g} + \int_{T_0}^{T} c_{p_{i,\,g}} \mathrm{d}T \tag{5.24}$$

$$c_{p_{i,\,g}} = a_i + b_i T + c_i T^2 + d_i T^3 \tag{5.25}$$

$$s_g = \sum_i y_i s_{i,g} - R \sum y_i \ln(y_i) \tag{5.26}$$

$$s_{i,g} = s^0_{i,g} + \int_{T_0}^{T} \frac{c_{p_{i,g}}}{T} \mathrm{d}T - R \int_{P_0}^{P} \frac{\mathrm{d}P}{P} \tag{5.27}$$

The specific exergy of the vapor phase is given by Eq. 5.28.

$$b_g = \sum_i y_i b^0_{i,g} + RT_0 \sum_i y_i \ln(y_i) + \left(h_g - h^0_g\right) - T_0\left(s_g - s^0_g\right) \tag{5.28}$$

For the liquid phase, treated as nonideal, the determination of the specific enthalpy, entropy, and exergy are given by Eqs. 5.29–5.37 [3, 5].

$$h_l = \sum_i x_i h_{i,l} + h^e \tag{5.29}$$

$$h_{i,l} = h^0_{i,l} + \int_{T_0}^{T} c_{p_{i,l}} \mathrm{d}T \tag{5.30}$$

$$c_{p_{i,l}} = a_i + b_i T + c_i T^2 + d_i T^3 + e_i T^4 \tag{5.31}$$

$$h^{\mathrm{e}} = -RT^2\left(\frac{\partial g^{\mathrm{e}}/RT}{\partial T}\right)_{x,P} \tag{5.32}$$

$$g^e = RT\sum_i x_i \ln(\gamma_i) \tag{5.33}$$

$$s_l = \sum_i x_i s_{i,l} - R\sum_i x_i \ln(x_i\gamma_i) + s^{\mathrm{e}} \tag{5.34}$$

$$s_{i,1} = s_{i,1}^0 + \int_{T_0}^{T} \frac{c_{p_{i,l}}}{T}\mathrm{d}T \tag{5.35}$$

$$s^{\mathrm{e}} = \frac{h^{\mathrm{e}} - g^{\mathrm{e}}}{T} \tag{5.36}$$

$$b_1 = \sum_i x_i b_{\mathrm{i},1}^0 + RT_0\sum_i x_i \ln(x_i\,\gamma_i) + \left(h_1 - h_1^0\right) - T_0\left(s_1 - s_1^0\right) \tag{5.37}$$

The liquid vapor equilibrium is done by means of the partition coefficient K_i which represents the ratio of the vapor and liquid mol fractions of component i:

$$K_i = \frac{y_i}{x_i} = \frac{\gamma_i P_i}{P} \tag{5.38}$$

The vapor pressure is given by the Antoine equation. For every component, there is a particular set of constants a_i, b_i, and c_i as shown in Eq. 5.39 [8].

$$P_i = \mathrm{e}^{a_i - b_i(T + c_i)} \tag{5.39}$$

For components like acetaldehyde, O_2, N_2, and H_2 that are presented at temperatures below their critical temperatures, Eq. 5.38 is no more valid. To model the behavior of these substances it was used the empiric Eq. 5.40, where the coefficients are given by Lide [6]:

$$x_i = e^{[a_i + b_i 100/T + c_i \ln(100/T)]} \tag{5.40}$$

The activity coefficient γ_i is evaluated by the Wilson equation, being α and β determined for each pair according to Stichlmair (1998).

$$\ln(\gamma_i) = 1 - \ln\left(\sum_{k=1,n} x_k A_{i,k}\right) - \sum_{l=1,n} \frac{x_l A_{l,i}}{\sum_{j=1,n} x_j A_{l,j}} \tag{5.41}$$

$$A_{i,j} = \beta_{i,j}\mathrm{e}^{-\alpha_{i,j}RT} \tag{5.42}$$

5.4 Exergy Analysis of the Original Plant

5.4.1 Overall Analysis of the Plant

The set of nonlinear equations of the described model can be solved by using, for instance, the software EES® [7].

Table 5.1 summarizes the results obtained with the simulation of the original plant: mass flow rates (N), enthalpy flow rates (H); heat transfer rates (Q), exergy flow rates (B); exergy rates associated to heat transfers ($Q\theta$); irreversibility rates (I) and the percentage of the irreversibility rate (δ).

Table 5.1 shows that the main irreversibility rate, 458 kW and 42.6 % of the total irreversibility rate of the process, occurs in the ethanol oxidation reactor (RX1); these results were already expected since oxidation reactions on gaseous phase at high temperature are far away from the equilibrium. The reactor, together with the acetaldehyde distillation tower T3, responsible for 247 kW of exergy destruction, the waste heat boiler HE2, responsible for 118 kW, and the absorption tower T2, responsible for 105 kW of exergy destruction, accounts for more than 86 % of the total irreversibility rate of the process.

The simulation results were compared to the plant processes data. The data were collected during 7 days of stable run and their average values are shown in Table 5.2, as well their standard deviation. The stream number refers to the tags shown in black in Fig. 5.1. The stream STEAM T1 is the steam rate for the T1 reboiler, STEAM T3 refers to the steam rate for the T3 reboiler, while RFLX is the T3 internal reflux. The calculated mass flow rate results show a maximum deviation from the measured data of –14 %. The calculated temperature results show a maximum deviation from the measured data of 4.4 %.

In order to evaluate the performance of the separation and absorption processes, a detailed analysis of the distillation and absorption towers is discussed as follows.

5.4.2 Acetaldehyde Distillation

The objective of this process is to separate the acetaldehyde from water and the non-reacted ethanol. Acetaldehyde is removed at the top of the tower at a concentration higher than about 99.5 % molar basis, while at the bottom water and ethanol are withdrawn at a rate close to 2:1 molar basis. The tower has 40 theoretical strays and operates with 345 kPa at the bottom and 320 kPa in the last tray. The first theoretical stray represents the condenser and the last one (number 40) corresponds to the reboiler. In this way, the temperature increases from the top to the bottom of the tower. The feed of the tower is made in stray 15.

The concentration profiles and specific exergy through the acetaldehyde distillation tower are shown in Figs. 5.3 and 5.4:

Table 5.1 Original configuration mass, energy, and exergy balances [2]

Unit	N_{inlet} (kmol/s)	N_{outlet} (kmol/s)	$\sum(H)_{inlet}$ (kW)	$\sum(H)_{outlet}$ (kW)	$\sum Q$ (kW)	$\sum(B)_{inlet}$ (kW)	$\sum(B)_{outlet}$ (kW)	$\sum Q\theta$ (kW)	I (kW)	δ (% kW)
MX1	0.0218	0.0218	–2,150	–2,150	0	9,703	9,669	0	34	3.2
T1	0.0314	0.0314	–8,701	–8,340	360	9,872	9,947	114	39	3.6
EV1	0.0237	0.0237	–2,687	–2,611	76	11,934	11,934	19	19	1.8
C1	0.0129	0.0129	–99	–83	16	2	13	16	4	0.4
HE1	0.0237	0.0237	–2,611	–2,543	68	11,934	11,947	21	7	0.7
RX1	0.0237	0.0270	–2,543	–2,544	0	11,951	11,493	0	458	42.6
T2	0.4778	0.4779	–127,588	–127,604	–18	247,663	247,557	–1	105	9.7
HE3	0.4015	0.4015	–109,948	–110,535	–587	212,737	212,728	–9	0	0.0
T4	0.0310	0.0310	–5,456	–5,456	0	1,384	1,380	0	4	0.4
T3	0.0632	0.0632	–17,060	–16,831	224	33,538	33,595	304	247	22.9
HE2	0.0270	0.0270	–2,544	–3,094	–550	11,493	11,208	–167	118	11.0
HE7	0.0225	0.0225	–6,272	–6,365	–93	257	242	–12	3	0.3
HE8	0.0225	0.0225	–6,365	–6,413	–48	242	239	0	2	0.2
HE9	0.0087	0.0087	–2,368	–2,386	–18	4,201	4,198	–1	3	0.3
HE4	0.0632	0.0632	–17,310	–17,070	239	33,492	33,515	44	21	1.9
HE5	0.0500	0.0500	–13,873	–14,051	–178	24,149	24,136	–2	12	1.1
HE6	0.0178	0.0178	–5,093	–5,109	–16	56	56	0	0	0.0
Overall	0.0366	0.0395	–332,670	–333,187	–523	624,607	623,855	324	1.076	100

Table 5.2 Process plant data versus model results [2]

Stream		Measured		Model result	Deviation (%)
		Average	Standard deviation		
1	(kg/h)	600	1	595	−0.8
1B	(kg/h)	288	89	293	1.7
20	(kg/h)	725	206	755	4.1
9	(kg/h)	1,320	47	1,324	0.3
19	(kg/h)	7,684	294	6,608	−14.0
27	(kg/h)	5,051	167	5,019	−0.6
21	(kg/h)	40,007	164	41,975	4.9
18 + 21	(kg/h)	47,692	344	48,584	1.9
29	(kg/h)	1,155	83	1,155	0.0
RFLX	(kg/h)	2,430	129	2,639	8.6
16 + 25	(kg/h)	5,863	230	5,890	0.5
16	(kg/h)	811	87	871	7.4
STEAM T1	(kg/h)	695	27	649	−6.6
STEAM T3	(kg/h)	1,361	105	1,337	−1.8
7	(K)	841	3.9	817.0	−2.9
10	(K)	338	2.4	339.6	0.5
5	(K)	340	0.1	341.6	0.5
6	(K)	402	0.3	402.0	0.0
29	(K)	286	2.7	286.0	0.0
24	(K)	291	6.7	297.0	2.1
22	(K)	305	2.3	299.6	−1.8
19	(K)	351	0.0	353.0	0.7
21	(K)	332	1.6	315.0	−5.1
14	(K)	301	7.1	314.0	4.4
20	(K)	326	1	326.5	0.1
25	(K)	385	0.9	392.0	1.9
3	(K)	356	1.2	362.3	1.9
11	(K)	379	1.7	384.9	1.5

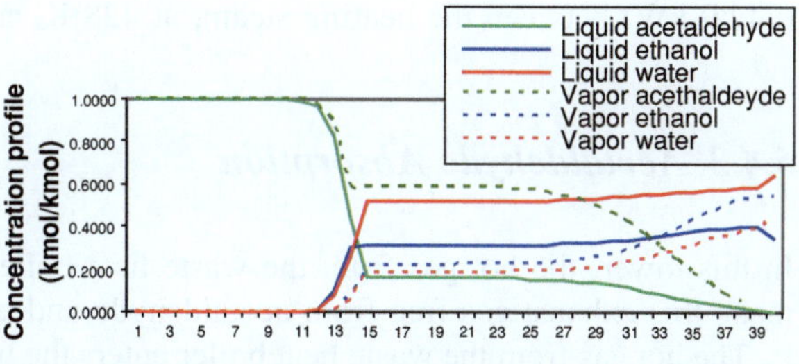

Fig. 5.3 Concentration profiles through the acetaldehyde distillation tower in the original configuration [3]

As shown in Fig. 5.4, the values of the specific exergy are higher as closer to the bottom of the tower the separation stray is. This behavior is due to the fact that the tower concentrates a pure product at its top and as the concentration increases, the chemical exergy also increases as it is several times higher than the physical

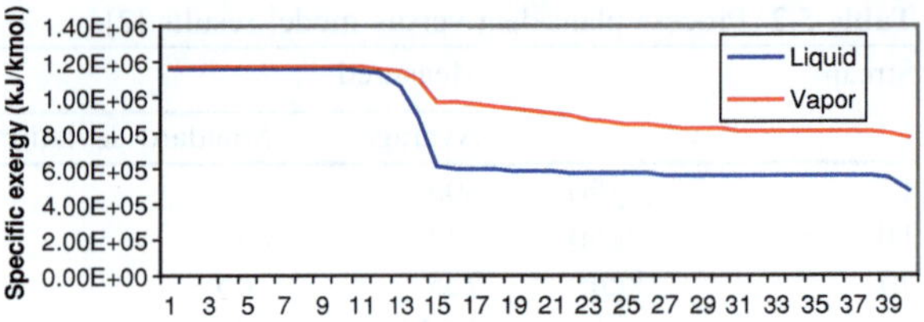

Fig. 5.4 Specific exergy through the acetaldehyde distillation tower in the original configuration [3]

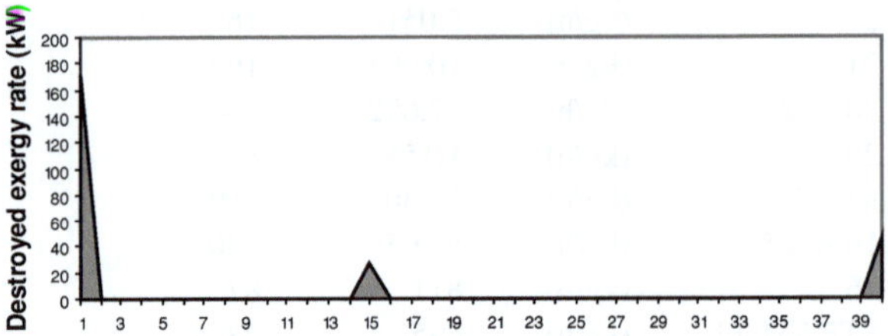

Fig. 5.5 Destroyed exergy rate through the acetaldehyde distillation tower in the original configuration [3]

exergy (that is reduced with lower temperature values). This behavior can also be observed in Fig. 5.3.

Figure 5.3 also points out that only a part of the equilibrium strays is effectively used to carryout the separation process because after stray 10 the acetaldehyde almost reaches 100 % concentration.

The exergy destruction rate through the tower is presented in Fig. 5.5 as, each stray is considered an equilibrium step, the irreversibilities happen in the condenser, reboiler, and around the feed stray. The condenser irreversibility is due to the heat transfer rate, 518.7 kW, between the condensing fluid and the refrigerant at a temperature difference of 20 K. The exergy destruction at the feed stray happens because of the difference between the feed concentration (7.58 % acetaldehyde, 32.35 % ethanol and 60.02 water) and the stray concentration (16.14 % acetaldehyde, 31.47 % ethanol and 52.39 % water), as well as the temperature difference between the feed (353 K) and stray (372 K) temperatures. At the reboiler the exergy destruction rate, 46.9 kW, is also due to the heat transfer rate (742.9 kW) between the heating steam, at 428 K, and the stray at 392 K.

5.4.3 Acetaldehyde Absorption

In this tower, the hot gas from the waste heat boiler is cooled and condensed to make this exhaust gas free from acetaldehyde and ethanol.

The hot gas from the waste heat boiler enters the bottom of the absorption tower and is cooled by an alcoholic solution. The heat transferred during the condensation of acetaldehyde, ethanol, and water is removed in heat exchanger 3 (see Fig. 5.1), where the bottom tower liquid is sent and is subsequently dispatched to a higher part of the tower. Heat is also transferred from the tower by means of water coils installed in each tray. Figure 5.6 shows that the main heat transfer rates occur

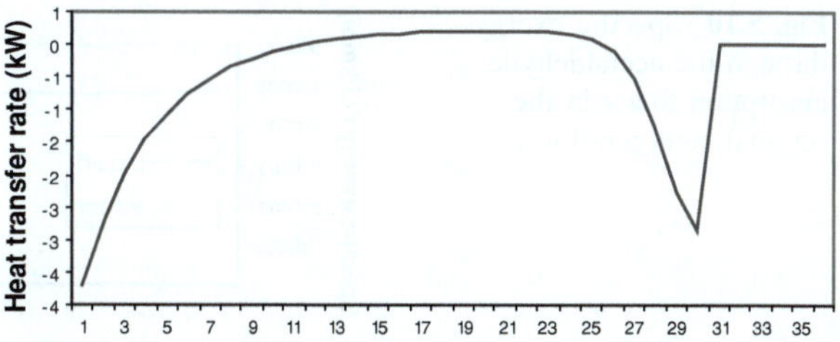

Fig. 5.6 Heat transfer rate in the acetaldehyde absorption tower in the original configuration [3]

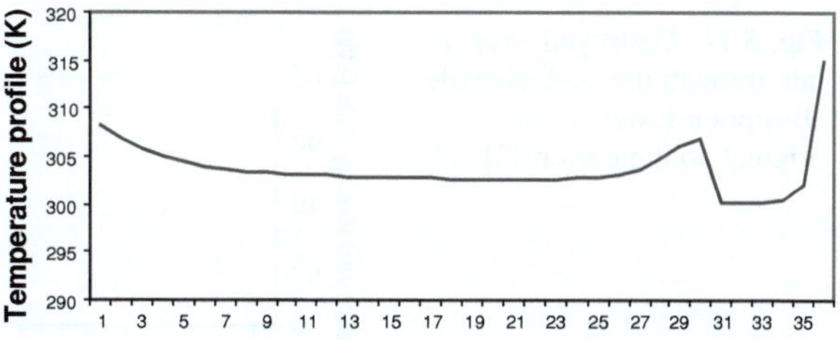

Fig. 5.7 Temperature profile through the acetaldehyde absorption tower in the original configuration [3]

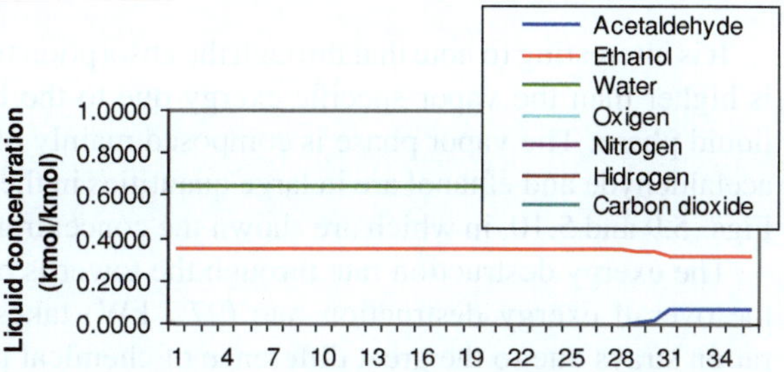

Fig. 5.8 Liquid concentration profile through the acetaldehyde absorption tower in the original configuration [3]

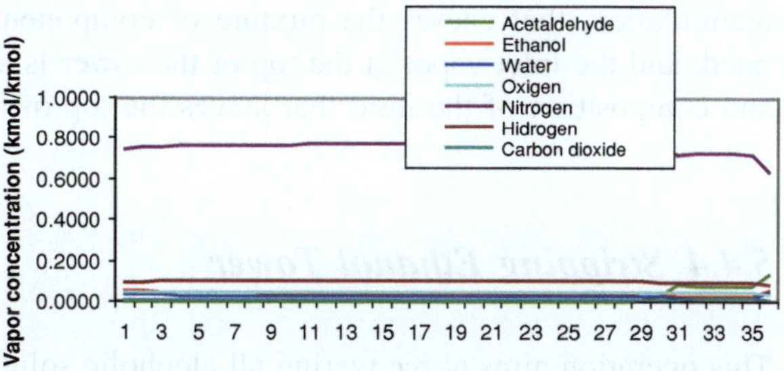

Fig. 5.9 Vapor concentration profile through the acetaldehyde absorption tower in the original configuration [3]

at the bottom and in the higher part of the tower due to the feed of a hot product at the top of the tower.

As part of the bottom liquid is sent to stray 31, the temperature profile through the tower has the behavior presented in Fig. 5.7. The small temperature increase observed in the higher part of the tower is due to the feed of the alcoholic solution at a higher temperature than the gases inside the tower.

Figures 5.8 and 5.9 present the liquid and vapor profile through the tower and Fig. 5.10 shows the profile of specific exergies through the absorption tower.

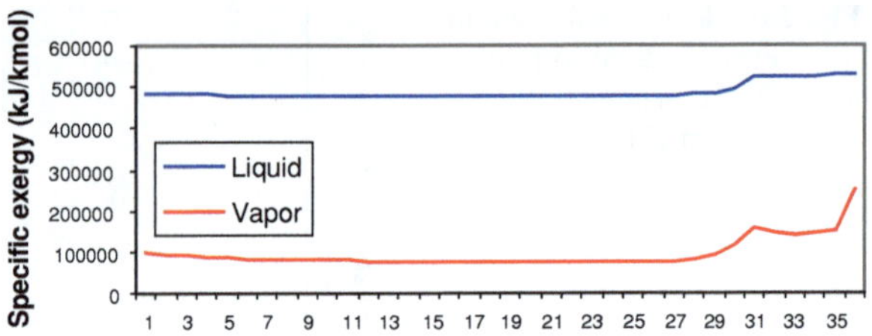

Fig. 5.10 Specific exergy through the acetaldehyde absorption tower in the original configuration [3]

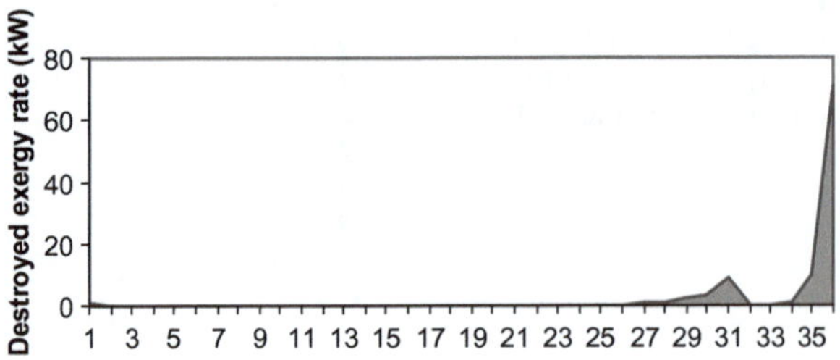

Fig. 5.11 Destroyed exergy rate through the acetaldehyde absorption tower in the original configuration [3]

It is interesting to note that through the absorption tower the liquid-specific exergy is higher than the vapor-specific exergy due to the higher chemical exergy of the liquid phase. The vapor phase is composed mainly of nitrogen and hydrogen while acetaldehyde and ethanol are in large quantities in the liquid phase, as can be seen in Figs. 5.9 and 5.10, in which are shown the concentration profiles of the two phases.

The exergy destruction rate through the tower is presented in Fig. 5.11. Most of the overall exergy destruction rate (97.7 kW) takes place in the first nine separation strays due to the great difference of chemical potential between the inlet and outlet fluid in every separation step. These differences occur due to the lateral recirculation, that causes the mixture of components that were previously separated, and the inlet vapor at the top of the tower is at a very different temperature and composition of the fluid that leaves the top of the tower.

5.4.4 Stripping Ethanol Tower

This operation aims at recovering all alcoholic solutions flows of the plant. These flows, with different concentrations of ethanol and water feed the distillation tower, and water without ethanol leaves the bottom of the tower, while at the top leaves a mixture of around 90 % ethanol, in molar basis.

The temperature profile in the ethanol stripping tower is shown in Fig. 5.12.

The strong temperature variations are caused by the liquid feed in the stages in the strays where these changes occur. Figures 5.13 and 5.14 present the concentration and specific exergy profiles through the ethanol stripping tower.

The exergy destruction through the ethanol stripping tower is shown in Fig. 5.15.

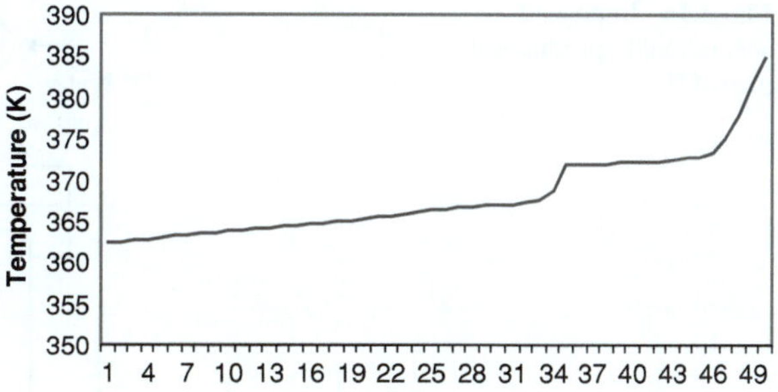

Fig. 5.12 Temperature profile in the ethanol stripping tower in the original configuration [3]

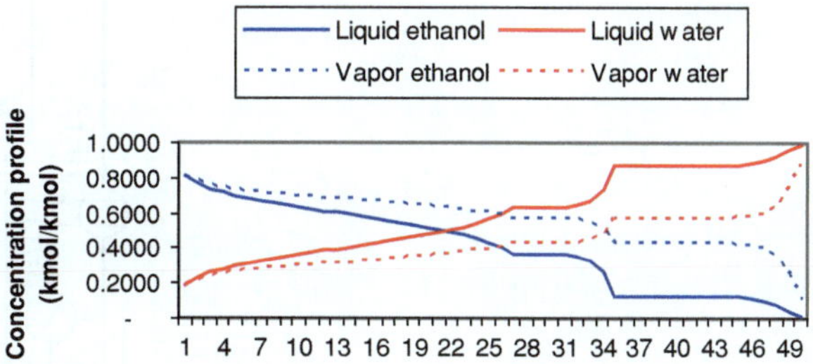

Fig. 5.13 Concentration profile through ethanol stripping tower in the original configuration [3]

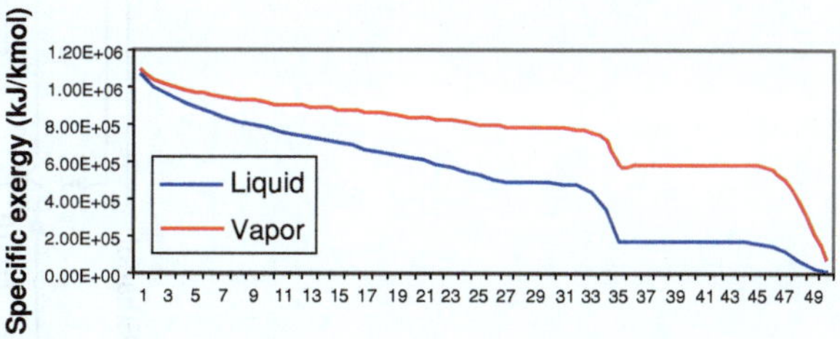

Fig. 5.14 Specific exergy through the ethanol stripping tower in the original configuration [3]

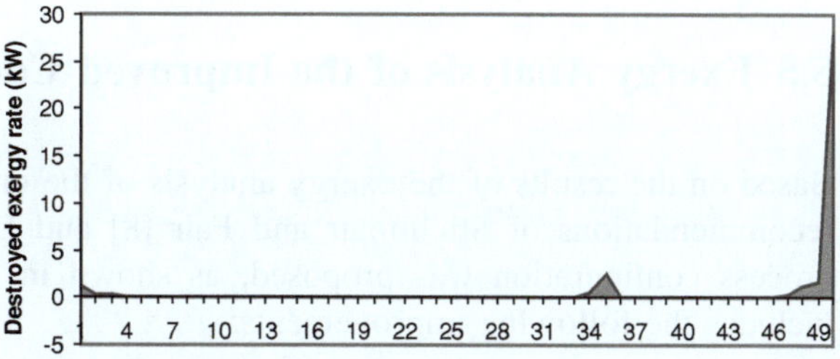

Fig. 5.15 Destroyed exergy rate through the stripping ethanol tower in the original configuration [3]

Here, the irreversibility is also caused also by the difference of chemical potential of the inlet and outlet separation stage solutions. These differences are higher in the feed region and in the base of the tower. In the reboiler besides the irreversibilities due to the chemical potential differences, there is still the exergy destruction associated to the heat transfer process.

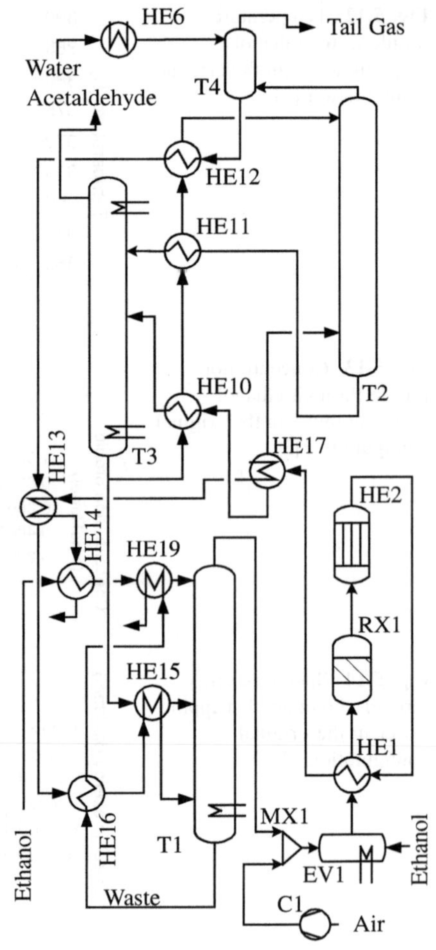

Fig. 5.16 Improved acetaldehyde production plant [2]

5.5 Exergy Analysis of the Improved Configuration

Based on the results of the exergy analysis of the original plant, as well as in the recommendations of Stichlmair and Fair [8] and Doldersum [9], an alternative process configuration was proposed, as shown in Fig. 5.16. This configuration includes the following improvements:

- Optimization of the heat exchanger network;
- Feeding the distillation tower T3 in two different points;
- Decreasing the recycle between the distillation tower T3 and the absorption tower T2;
- Choosing the best feed tray for tower T3;
- Installing a condenser between the waste heat boiler and the absorption tower T2;

- Preheating the reactor by using the hot gases from the waste heat boiler;
- Decreasing the amount of quench water of tower T4;
- Decreasing the reflux ratio of the distillation tower T3.

Tables 5.3 and 5.4 show the simulation results for the proposed configuration operating at the same feeding conditions of the original plant. In Table 5.3, it can be verified that the total irreversibility rate is lowered to 925 kW.

Data of Table 5.3 put in evidence that the improved configuration reduces the overall exergy destruction rate in 151 kW, or 14 % reduction, and that the most important differences occur in two components: the acetaldehyde distillation tower, with a reduction of 54 kW and the acetaldehyde absorption tower, with a reduction of 76 kW of the original irreversibility rate. The remainder reduction in the exergy destruction rate, 21 kW, is due mainly to the new heat exchangers network that interconnects the distillation and absorption towers.

The suppression of the intermediate recirculation in the absorption tower, possible with the inclusion of the condenser HE17, reduced the irreversibility caused by the mixture of streams with different concentrations. This effect can be observed in Fig. 5.17 in the region of the 31th stray. Another reduction in the exergy destruction inside the absorption tower is due to the reduction in the inlet temperature as well as the acetaldehyde concentration, as shown in Fig. 5.17 in the region of the 36th stray. The exergy destruction rate caused by the inclusion of the new condenser (6 kW in HE17) is largely compensated by the reduction in the reduction of the exergy destruction rate inside the absorption tower.

There was also a significant gain in the efficiency of the acetaldehyde distillation tower, caused by the reduction of the feeding rate of the tower (from 0.0500 to 0.0300 kmol/s) and reflux rate (from 0.0167 to 0.0117 kmol/s) and also by segregating the feeding flows.

The reduction in the feeding rate is consequence of the inclusion of the condenser HE17, since it allowed the reduction of the recirculation of the alcoholic solution from the bottom of the distillation tower to the top of absorption tower (from 0.0500 to 0.0300 kmol/s, or 40 % of the original flow rate). Another consequence of including the condenser HE17 is the segregation of the feeding flows of the distillation tower (outlet flows of HE10 and HE11 as shown in Fig. 5.16). In this way, the richer acetaldehyde flow (exit of HE11) is fed in a higher position of the tower, while the poorer flow (exit of HE10) is fed nine strays below. Figure 5.18 presents the exergy destruction rate in the distillation tower for the original and improved configuration.

Another important improvement in the new configuration is the reduction of the reflux rate in the distillation tower, because, as shown in Fig. 5.4, 25 % of the separation stages of the original configuration do not have any effective separation function. By reducing the reflux rate the concentration profile is elongate (see Fig. 5.19) which provides the utilization of all separation steps with the consequent energy reduction in the reboiler, from 743 to 543 kW or, 26.9 %.

According to the better thermal integration in the new plant, the heating demand (steam demand) dropped to 840 kW, or 28 % of the original value

Table 5.3 Mass, energy, and exergy balances improved acetaldehyde production plant [2]

Unit	N_{inlet} (kmol/s)	N_{outlet} (kmol/s)	$\sum (H)_{\text{inlet}}$ (kW)	$\sum (H)_{\text{outlet}}$ (kW)	$\sum Q$ (kW)	$\sum (B)_{\text{inlet}}$ (kW)	$\sum (B)_{\text{outlet}}$ (kW)	$\sum Q\theta$ (kW)	I (kW)	δ (%kW)
MX1	0.0202	0.0202	–1,784	–1,785	0	8,187	8,157	0	31	3.3
T1	0.0439	0.0439	–12,216	–11,920	297	8,951	9,009	90	33	3.6
EV1	0.0222	0.0222	–2,322	–2,243	79	10,422	10,440	38	20	2.1
C1	0.0129	0.0129	–99	–86	12	2	13	12	1	0.2
HE1	0.0222	0.0222	–2,243	–2,178	65	10,440	10,452	18	5	0.6
RX1	0.0222	0.0248	–2,178	–2,178	0	10,452	10,016	–1	436	47.4
T2	0.0450	0.0450	–9,108	–9,163	–55	18,493	18,464	0	29	3.2
T4	0.0452	0.0452	–9,663	–9,663	0	1,397	1,392	0	5	0.5
T3	0.0423	0.0423	–11,152	–11,034	117	24,766	24,806	232	193	21.0
HE2	0.0248	0.0248	–2,178	–2,779	–601	10,016	9,690	–133	143	15.6
HE17	0.0248	0.0248	–2,844	–3,320	–476	9,672	9,618	–49	6	0.6
HE18	0.0098	0.0098	–2,627	–2,627	0	5,620	5,614	0	6	0.6
HE11	0.0322	0.0322	–8,788	–8,642	145	17,168	17,181	18	5	0.5
H10	0.0098	0.0098	–2,627	–2,572	55	5,614	5,625	17	6	0.7
HE12	0.0300	0.0300	–8,391	–8,415	–25	14,490	14,489	–1	0	0.0
HE13	0.0336	0.0336	–9,535	–9,461	74	1,205	1,213	11	3	0.3
HE16	0.0336	0.0336	–9,461	–9,405	56	1,213	1,222	11	2	0.2
HE15	0.0336	0.0336	–9,405	–9,387	18	1,222	1,226	3	0	0.0
HE14	0.1300	0.1300	–36,694	–36,720	–25	439	435	–2	2	0.2
HE19	0.0366	0.0366	–10,278	–10,283	–5	812	811	–1	0	0.0
HE6	0.0324	0.0324	–9,263	–9,288	–24	101	102	1	0	0.0
Overall	0.7071	0.7098	–153,593	–153,862	–268	160,581	159,872	215	925	100.0

Table 5.4 Process thermal demand [2]

	Thermal demand (kW)		
	Original	Improved	Reduction (%)
Reboiler (T = 1)	360.4	297.0	–18
Reboiler (T = 3)	742.9	543.0	–27
Heater (HE = 1)	68.0	–	100
Overall	1,171.3	840.0	–28
	Specific energy consumption (kJ/kg)		
Overall	5,622	4,032	–28

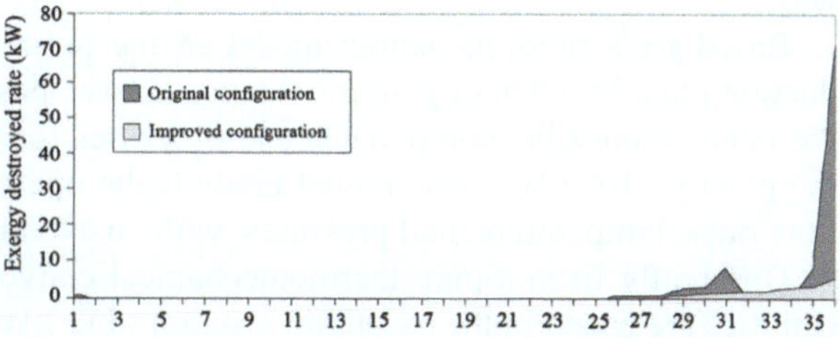

Fig. 5.17 Exergy destruction rate comparison in the absorption tower [3]

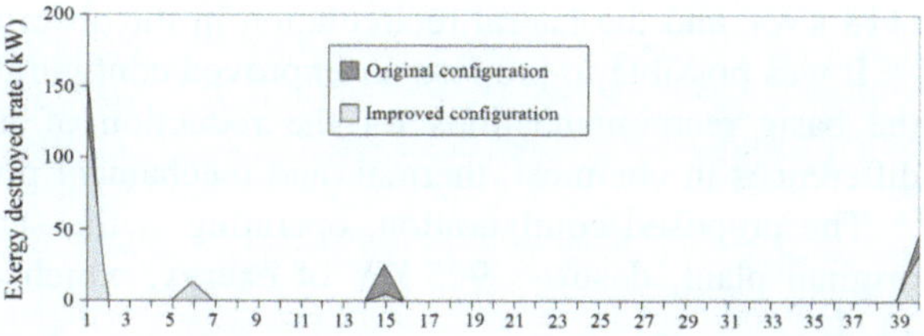

Fig. 5.18 Exergy destruction rate comparison in the distillation tower [3]

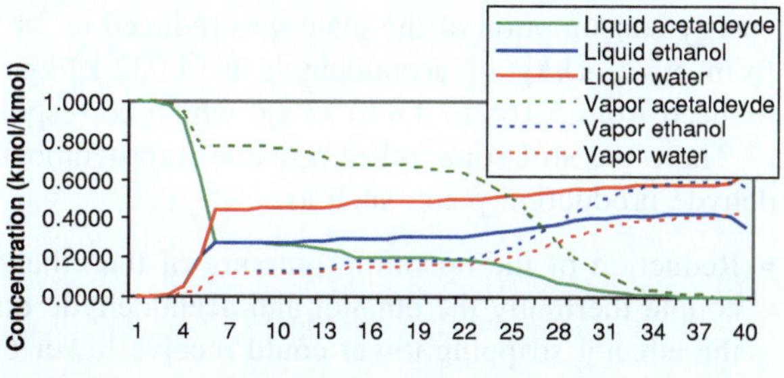

Fig. 5.19 New concentration profiles in the distillation tower [3]

(1,171 kW). The steam demand is originally needed in the reboilers of the ethanol stripping tower and acetaldehyde distillation tower, as well as in the preheater HE1. In the new configuration, the preheater HE1 utilizes the outlet gases of the waste heat boiler, instead of steam.

Table 5.4 summarizes the results described before and presents the specific energy consumption per kg of produced acetaldehyde.

Considering the exergy destruction as the comparison criteria, one has a reduction from 1,076 to 925 kW, or 14 %, and specific exergy destruction from 5,165 kJ/kg, for the original configuration, to 4,440 kJ/kg, for the improved one.

5.6 Concluding Remarks

This chapter described how the evaluation and improvement of chemical processes by means of the use of the exergy analysis can be done in a very straightforward way.

Based on a thermodynamic model of the processes of an acetaldehyde production plant by ethanol partial oxidation, it was possible to quantify and identify the main responsible components and processes for the exergy destruction rate in the plant (1,076 kW). This model predicts the operating parameters values, mass flow rates, temperature, and pressures, with an average deviation of 0.67 ± 3.5 %.

Differently from a pure thermomechanical conversion plant, the main irreversibilities are found in the oxidation reaction (458 kW), the high internal reflux rate in the acetaldehyde tower (247 kW), the heat transfer in the waste heat boiler (118 kW), and the lateral recirculation in the absorption tower (105 kW).

It was possible to propose an improved configuration of the plant by following the basic recommendations for the reduction of irreversibilities: minimize the differences in chemical, thermal, and mechanical potentials.

The proposed configuration, operating at the same feeding conditions of the original plant, destroys 925 kW of exergy, which correspond to a reduction of 14 %.

As a consequence of the reduction in the exergy destruction rate, the specific energy consumption of the plant was reduced in 28 % compared the original one, from 5,622 kJ/kg of acetaldehyde to 4,032 kJ/kg. The exergy destruction was reduced from 5,165 to 4,440 kJ/kg, which corresponds to 14 % of reduction.

There are still some other possible improvements to be applied to the acetaldehyde production plant, such as:

- Reduction of the operating pressure of the ethanol stripping tower in order to couple thermally the ethanol and acetaldehyde distillation towers. In doing so, the ethanol stripping tower could receive lower exergy flows.
- Use of heat pumps to generate low-pressure steam by using the thermal wastes of the plant.
- The hydrogen available in the exhaust gases of the acetaldehyde absorption tower could be used for combustion purposes.

References

1. Johnson WK, Fink U, Sakuma Y (1998) Acetaldehyde. In: CEH product review, SRI international
2. Vieira U Jr, Oliveira S Jr (2005) Exergy analysis of an acetaldehyde production plant. In: Proceedings of the 18th International conference on efficiency, costs, optimization, simulation and environmental impact of energy systems, Trondheim
3. Vieira U Jr (2004) Exergy analysis of an acetaldehyde production plant. Master dissertation, Polytechnic School of the University of Sao Paulo, Sao Paulo (in Portuguese)
4. Szargut J, David RM, Steward F (1988) Exergy analysis of thermal, chemical, and metallurgical processes. Hemisphere Publishing, New York
5. Perry RH, Green DW (eds) (1984) Perry's Chemical Engineers' Handbook. 6th. Mcgraw-Hill, London
6. Lide DR (ed) (2004) CRC Handbook of chemistry and physics. 85th ed. CRC Press, Boca Raton
7. Klein SA (2011) Engineering equation solver—EES, F-chart software, www.fChart.com
8. Stichlmair J, Fair JR (1998) Distillation: principles and practice. Wiley-Liss, New York
9. Doldrsum A (1998) Exergy analysis proves viability of process modification. Energ Convers Manage 38:1781–1789

Chapter 6
Exergy Analysis and Parametric Improvement of the Combined Production of Sugar, Ethanol, and Electricity

Greek Symbols

λ	Renewability exergy index

Subscripts

cogeneration	Cogeneration system
mill	Related to the mill

Abbreviations

BIGCC	Biomass integrated gasification combined cycle
BPST	Back pressure steam turbine
CEST	Condensing-extraction steam turbine
HRSG	Heat recovery steam generator
LHV	Lower heating value
SuSC	Supercritical steam cycle
Trad	Tradition mill

6.1 Introduction

Sugarcane culture was the first agriculture activity developed in Brazil, after the arrival of the Portuguese in 1500. This Brazilian agroindustry has evolved from a typical single product industry (sugar) to a polygeneration plant (sugar, ethanol, and electricity) nowadays [1]. In the future, other products might be obtained considering different energy conversion routes (cellulosic ethanol, chemicals) and/or further processing other by-products such as stillage (biodigestion, concentration) and trash to generate more electricity and improve the recycle of nutrients in the crop. Hence, the sugarcane industry is a very suitable platform for the development of biorefineries.

S. de Oliveira Jr., *Exergy*, Green Energy and Technology,
DOI: 10.1007/978-1-4471-4165-5_6, © Springer-Verlag London 2013

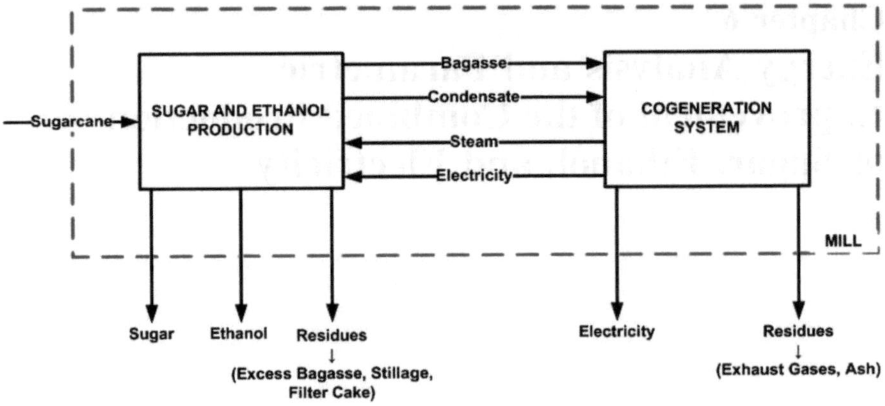

Fig. 6.1 Scheme of a sugar and ethanol mill [44]

The amount of sugarcane processed in Brazil, in the harvest 2010/2011, was 624.991 million tons, producing 38.7 million tons of sugar (46.2 % of the processed sugarcane) and 27.7 million m^3 of ethanol (53.8 % of the processed sugarcane). The total area of production corresponds to 8.0 million hectares, near 15 % of the total land available for agriculture [45].

Figure 6.1 represents a sugar and ethanol mill composed of two control volumes: one that contains the sugar and ethanol production processes, and other with the utilities plant. This scheme indicates that the mill can be a sustainable industrial plant because of the use of sugarcane as raw material and sugarcane bagasse as the fuel of the utilities plant. The analysis of the ensemble of processes that take place in a sugar and ethanol mill evidences that it can be considered as a converter of solar energy, used in the formation of the sugarcane, into sugar, ethanol, and also electricity.

In Brazil, the contribution of sugarcane and other biomasses in the electricity matrix is still marginal, approaching 5 %. In 2009, power generation using bagasse represented 75 % of the biomass-based generation, or 4.6 GW. Most of this power is consumed in mill to fulfill the electric requirements of the process, with 25 % of total sold to the grid [2].

Commonly, bagasse-fired boilers raise steam to 300 °C and 21 bar that is used in backpressure turbines, responsible for the electromechanical demands of the mill. Backpressure steam (2.5 bar) is used to fulfill the thermal requirements of the process, and its condensate is returned to the boiler. Normally, the electromechanical energy produced is for internal use only. However, some mills already use steam with higher parameters (42–67 bar), generating an excess of electricity that is sold to the grid. Also, there is a tendency in the sector to replace old boilers by new ones with greater capacity (80–120 bar, for instance). These systems are based on backpressure (BPST) and condensing-extraction steam turbines (CEST). Evaluation regarding the potential power generation using these conventional systems have been made by different authors [3–11].

In the future, some authors suggest that biomass integrated gasification combined cycles (BIGCC) is the best alternative to cogeneration plants in sugarcane mills. These systems might attain 35–40 % efficiency for the conversion of power [12–14]. Supercritical steam cycles (SuSC) might also attain these efficiency values, what makes them an alternative to gasification-based systems [15]. As for SuSC applied to sugarcane mills, Pellegrini et al. [16] compared these with BIGCC systems.

Hence, this chapter presents an approach to the problem of exergy optimization of cogeneration systems in sugarcane mills, considering the impact of different cogeneration configurations on the exergy-based cost of sugar, ethanol, and electricity, as well some economic features of the use of exergy as a cost measure in sugarcane mills to assess an economic feasibility analysis of the different cogeneration alternatives.

Furthermore, as in the last years a great discussion is being made regarding the renewability of different biomass-derived fuels, this chapter also describes a new perspective to the renewability of the energy conversion processes inside the mill, based on the use of the renewability exergy indicator (λ) described in Chap. 2.

6.2 Energy Conversion in the Production of Sugar, Ethanol, and Electricity

The sugarcane industrial stage processing can be separated into 5 different control volumes, as shown in Fig. 6.2: extraction system, juice treatment, sugar production, ethanol production, and cogeneration system. As shown in Fig. 6.2, there is a large interaction among different processes, what means that changes in one of them will influence the performance of the others and also the production costs of the products of the mill.

Table 6.1 shows the quantities of sugar and ethanol that can be produced per ton of sugarcane according to the operating strategy chosen by the mill.

A brief description of each process is given below, based on different works [17–19]:

- **Extraction System**: sugarcane is composed mainly by fiber and juice (a sugar–water solution), in which sucrose is dissolved. Thus, the aim of this process is to recover as much juice as possible, but also to produce a final bagasse in suitable conditions for fast burning in the boilers. It is a great consumer of water, in order to facilitate the recovery of sugars. There are two types of extraction systems: milling and diffusion. Both systems require previous cane preparation, using knives and shredders that operate with direct drive steam turbines or electric motors. Regarding the performance of these systems two points deserve special attention:
  - Energy requirements: mills require more mechanical energy than diffusers (15 kWh/t of cane against 10 kWh/t of cane), even though diffusers demand low pressure steam to heat the juice during the extraction. The impact of using

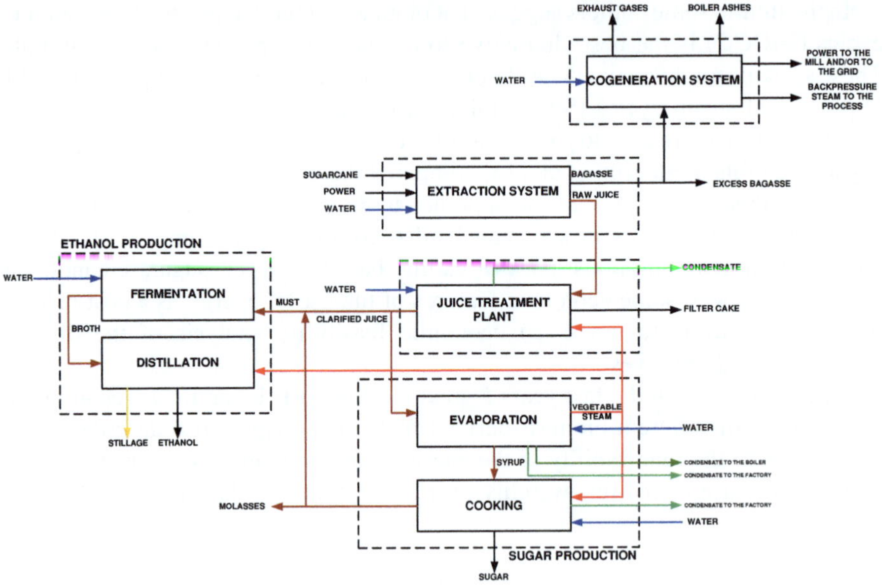

Fig. 6.2 Sugar and ethanol production processes [28]

Table 6.1 Sugar and ethanol quantities produced for 1 ton of sugarcane [44]

Production strategy	Sugar (kg)	Ethanol (L)
Only sugar	120	7 (residual)
Sugar and ethanol (50/50)	60	42
Only ethanol	–	85

one system or another in the production costs of sugar, ethanol, and electricity has been evaluated in Ensinas et al. [7, 8].

– Extraction efficiency: diffusers are able to extract 99 % of the total sugars in the sugarcane, while mills are limited to 97 %. Nevertheless, the moisture content of bagasse from diffusers is also higher, what lowers the boiler efficiency. Moreover, the use of mills allows the production of a higher purity juice (higher quality of the sugar) due to the possibility of juice extraction in the first part of the crushing part.

• **Juice Treatment**: raw juice from the extraction system is treated in order to remove sugar impurities, using chemicals, as to improve the quality of the final products. There are some very few differences between the treatment of the juice to sugar and to ethanol productions, mainly regarding the addition of chemicals [17]. Raw juice from the extraction system is mixed with filtered juice (leading to mixed juice) and then heated up to 105 °C, using vegetable steam from the multiple-effect evaporator. During the heating process some chemicals are added in order to adjust the pH of the juice and agglutinate suspended solids, different heat exchangers are employed in the process. After the flash tank, the juice is left in a decanter to rest, where suspended solids are collected at the bottom of the decanter as mud and sent to filters, while clarified juice is pumped to the sugar and ethanol productions. In the filters, part of the juice is recovered from the mud, and suspended matter with salts formed and fine bagasse is extracted as filter cake. It is worth to highlight that the higher the thermal integration between the heat exchanger network and the multiple-effect extraction is, the lower will be the steam consumption in the process. This aspect has been studied by different authors [7, 8, 12, 20–22]. A lower steam consumption in the process results in a higher potential for power generation.

- **Sugar Production**: clarified juice obtained in the treatment plant undergoes a concentration process by removing the water contained in it. The first stage of concentration is carried out in a multiple-effect evaporator. This equipment is responsible for the concentration of juice into syrup, and the production of vegetable steam (evaporated water from the juice in the different effects of the evaporator) used for heating purposes in other parts of the process (treatment plant, cooking, and distillery). Commonly, Robert-type 5 effect multiple-evaporators are employed in Brazilian sugarcane mills, although falling-film evaporators are being used in new or retrofit projects. Steam extractions to other processes are made at the first and second effect, occasionally at the third effect. One advantage of falling-film evaporators (specially, plate type) is the possibility of decreasing temperature difference among the stages (decrease the pressure difference) [22]. It is well-known in the sugar literature [18] that extracting steam for the last stages of the evaporator lead to a reduction in the amount of backpressure steam needed in the first step. However, due to the low temperature of the last stages, the use of the steam generated in these stages is very limited and even impossible. Thus, with the use of falling-film evaporators, it would be possible to augment the pressure of the last stages (consequently, increasing the temperature of the steam generated), allowing more steam to be extracted from the last stages to be used as heating source in other processes [20]. Due to the high viscosity of the syrup leaving the multiple-effect, it is no longer possible to concentrate it in normal evaporators. Thus, it is used equipment called pans, which operate under vacuum conditions in a discontinuous way (heat requirements for this process is fulfilled by vegetable steam from the multiple-effect evaporator). Evaporation of the water creates a mixture of crystal coated in a sugary solution, which is called the cooked paste. The cooked paste moves to the centrifuging sector and is discharged into the centrifuges. The centrifugal force separates sucrose crystals from the sugar solution. The process is completed by washing the sugar with water and steam while it is still inside the basket. The removed sugar solution returns to the cookers for recovery of the remaining dissolved sugar, until it is more exhausted. From this point, the sugar solution is called end syrup or molasses and is sent to make ethanol. Sugar extracted by the centrifuges has high moisture level, being sent to drying before it is packed.
- **Ethanol Production**: The Melle–Boinot fermentation process (cell-recycle batch fermentation) is most commonly used in ethanol distilleries in Brazil. Its

main characteristic is the recovery of yeasts through fermented wine centrifuging. Part of clarified juice is mixed with molasses from the sugar production (the mixture is called must), and sent to the fermentation vats. Inside the vats, sugars are transformed into ethanol. During the reaction, there is a strong release of carbon dioxide, the solution gets hotter and some secondary products are formed such as: superior alcohols, glycerol, aldehydes, etc. Cooling water is used to maintain the solution inside the vats at 32 °C in order to not jeopardize the fermentation kinetic, and to avoid the formation of these by-products in excess. The mixture leaving the vats, called fermented broth, is sent to the centrifuges to recover the yeast. Recovered yeast concentrate, called yeast milk, returns to the tanks for treatment. The light centrifuging phase, or "deyeasted" broth, is sent to distillation columns. Ethanol in the broth is recovered by distillation, which uses the different boiling points of the various volatile substances to separate them. The operation is performed using seven columns spread through four sets (superimposed columns): distillation, rectification, dehydration, dehydration agent recovery. From the distillation and rectification sets hydrated alcohol is obtained. This alcohol can be further dehydrated using a dehydration agent (cyclo-hexane, mono ethylene glycol, molecular sives), producing anhydrous alcohol. The hydrated alcohol, the end product from the purification processes (distillation) and rectification, is a binary alcohol–water mixture reaching a level of 96°GL. The anhydrous alcohol, the end product from the dehydration processes, is a binary alcohol-water mixture reaching a level of 99.7°GL. The main by-product of the ethanol production is stillage, composed by water (96 % wt.), mineral solids, and small level of ethanol (0.02°GL). Heat demand of the distillation process is supplied by backpressure steam from the turbines and/or vegetable steam. Over the last years, the steam consumption in the distillation columns has been continuously decreased due to improvements in the columns as well as to a better thermal integration of the process, considering the use of regenerative heat exchangers and, most recently, the multiple-effect distillation concept [23]. Other options to reduce the steam consumption are underdevelopment, such as extractive fermentation and the use of membranes to concentrate ethanol–water solutions (pervaporation).

- **Cogeneration System**: steam-based systems (Rankine cycle-based) are employed in all Brazilian mills. Bagasse generated in the extraction system is sent to the cogeneration plant to raise steam to be used in backpressure turbines. This equipment is responsible for the fulfillment of the electro-mechanical demands of the mill, while backpressure steam is used to satisfy the thermal requirements of the process, and its condensate is returned to the boiler. Normally, the electro-mechanical energy produced is for internal use only. However, some mills already use steam with higher parameters (42–67 bar), generating an excess of electricity that is sold to the grid. Around 5 % of the bagasse is stored to be used during start-up after production stops. The amount of excess bagasse in sugarcane mills is directly related to the steam consumption in the process and the existence or not of condensing turbines (in this case, bagasse is burned and the steam generated is used to generate more electricity for the grid). Currently, mills have excess bagasse varying from 5 to 30 % of the total production. For a long time, bagasse was considered as a problem waste, and cogeneration plants and steam demands in the process were designed as so to eliminate all bagasse. Nowadays, the possibility of selling bagasse to third parties and the generation of electricity to the grid has turned bagasse into a valuable by-product of the process.

6.3 Modeling Approach for Sugar and Ethanol Production Processes

In order to investigate the performance of the different energy conversion processes inside the mill, a global model of the co-production of sugar, alcohol, and electricity was developed to simulate the steady state operation of sugarcane mills.

The model allows the evaluation of different configurations of the cogeneration system, as well as of diverse heat exchanger networks inside the mill, considering different levels of integration among the processes. It is composed of mass, energy, and exergy balances, heat and mass transfer equations and equations for determination of thermodynamic properties for sucrose–water and ethanol–water solutions. Also, relations to determine the liquid–vapor equilibrium of water-sucrose in the multiple-effect evaporator and vacuum pans were implemented.

Thermodynamic properties for sucrose–water solutions were calculated as suggested in Nebra and Parra [24], while for ethanol–water solutions, the correlations cited in Modesto et al. [25] were used. As for bagasse, the lower heating value (LHV) was calculated by the correlation given in Channiwala and Parikh [26], and its exergy was determined considering the methodology developed in Szargut et al. [27] for solid fuels, based on its LHV and composition. Finally, the exergy of sugarcane was set equal to weighted sum of the exergy of the juice (sucrose–water content) and the exergy of the fiber. Values for the exergy of different streams in a sugarcane mill are given as reference in Table 6.2.

The model assesses all major processes involved in the production of sugar and ethanol (extraction system, juice treatment plant—pH correction and heating, evaporation, cooking, fermentation, and distillation), as well as the reuse of part of the condensate generated by the use of vegetable steam (extracted steam from the multiple-effect evaporator) as imbibition in the extraction system and dilution water in the ethanol process. The input parameters for the models were taken from a real sugarcane mill in a field research performed during harvests of 2005 and 2006 [28].

In what follows, a brief description of the characteristics of the model is provided, considering the five control volumes shown in Fig. 6.2.

- The extraction system is evaluated through mass balances of sucrose and fiber, considering as inputs of the process sugarcane and imbibition water, and as output raw juice and bagasse. It is assumed that bagasse leaves the extraction

Table 6.2 Specific exergy values for different streams in a sugarcane mill [44]

Stream	Temperature (°C)	Pressure (bar)	Solid content (wet basis) (%)	Exergy (kJ/kg)
Sugarcane	25	1.0	28.5[a]	5273
Bagasse	25	1.0	50.0[b]	9654
Raw juice	35	1.0	15.5	2751
Clarified juice	115	2.0	16.3	2915
Syrup	62	0.2	60.0	10546
Sugar	60	0.2	99.9	17485
Molasses	60	0.2	76.7	13474
Must	44	1.0	16.0	2847
Hydrated alcohol (ethanol)	25	1.0	–	27217
Stillage	89	1.0	–	84
Process steam	140	2.5	–	676

[a] 12.5 % fibers; 16.0 % sucrose, and other solids
[b] 50 % moisture content

with 50 % moisture content, 96 % of the sucrose coming from the sugarcane is recovered in the raw juice (recovery ratio), and the amount of imbibition water is set to 2 kg of water/kg of fiber in the sugarcane [18]. For the mechanical energy requirements, it was considered a consumption of 15 kWh/tc. These values were also checked with the parameters used in a real sugarcane mill in Brazil.

- The main processes in the juice treatment plant are: heating of juice, flash before decanter, filtered juice recovery from the decanter, and addition of chemical products. Energy balances were used to calculate the amount of vegetable steam required in the heat exchangers, considering 2 % of losses. The amount of filtered juice produced is set to 0.2 kg of filtered juice/kg of mixed juice, and it was considered that for each ton of cane, 4 kg of filter cake is produced. These numbers were taken from a real sugarcane mill, and they are in accordance with the literature [18]. The addition of chemicals was modeled as a simple dilution with water, and the proportion is 15 kg/tc. Flash tank was evaluated based on the assumption that the vapor generated is in equilibrium with the juice according to the Modified Raoult's Law.
- The production of sugar is calculated using data for solids and sucrose concentration of the syrup leaving the multiple-effect evaporator and the different flows of the cooking process (vacuum pans). Thus, the mass flow rates of the streams were calculated through mass balances, considering that the vapor and the sucrose–water solution are in equilibrium at the outlet of any equipment, as in the flash tank. The steam consumption in these processes is determined by energy balances. The pressures of each multiple-effect evaporator stage (1.80, 1.34, 0.94, 0.56, and 0.20 bar) and vacuum pans (0.20 bar) were also taken from the real sugarcane mill [28].
- The amount of ethanol produced is based on the stoichiometric conversion of sugars into ethanol (0.511 L of ethanol/kg of sugars), considering 89 % efficiency for the fermentation process (conversion of sugars into ethanol) and 99 %

for the distillation process (separation of ethanol from the ethanol–water solution). Furthermore, steam consumption for the distillation process is evaluated at 3.5 kg of low pressure steam (2.5 bar)/L of ethanol for a conventional mill, and 1.6 kg of low pressure steam/L of ethanol for a multi-pressure distillation [29].

In addition, exergy-based cost balances were developed to evaluate the production costs of sugar, ethanol, and electricity. No capital costs were considered. Thus, based on the Exergetic Cost Theory [30], the exergy-based cost of all input exergy streams were set to unity. The criterion used to distribute costs among different products in a given control volume was that each product has the same importance [31] because of the high energetic interaction of the five analyzed control volumes. Thus, their exergy-based costs were set equal (equality criterion). With cost balances it is possible to show the cost formation process inside the mill and evaluate the impact of improving the mill exergy efficiency on the cost formation process. Also, it was assigned cost zero to the excess bagasse not used in the cogeneration plant, as well as to the stillage generated in the ethanol production.

Besides, nowadays most sugarcane producers are paid based on the sugar content in the sugarcane, the participation of sugarcane in the sugar, and ethanol costs and market prices for these products [1]. Therefore, cost formation methodologies used nowadays are based solely on mass balances and conversion factor for the production of sugar and ethanol. According to this methodology, bagasse has no cost, and whenever electricity is sold to the grid, the sugarcane producer does not participate on this additional revenue. This asymmetry leads to a question related to the attribution of cost to the bagasse.

The use of exergy-based costs allow the attribution of cost to the bagasse, giving a new perspective in relation to the profitability of the products of the mill, and the possibility of a new payment system for the sugarcane. In this way, thermoeconomic cost (US$/kJ) balances were developed based on the exergy-based (kJ/kJ) cost balances. Again, no capital costs were considered. The aim is to evaluate the thermoeconomic cost formation of the products, which will be used to measure the profitability of each of them.

6.4 Exergy Analysis of a Traditional Sugarcane Mill

Nowadays, the efforts related to the optimization of the sugar and ethanol yield are centered in developing sugarcane varieties that have a higher sucrose yield per cultivated area, and in reducing the sucrose losses during the processes inside the mill using modern control systems and better production techniques. Also, regarding the development of different varieties, optimization techniques are being used to choose sugarcane varieties that would benefit sugarcane production and industrial systems, by reducing crop residue and increasing final energy production [32]. One should keep in mind that a higher content of sucrose means less fiber (less bagasse will be available). Related to energy conversion processes inside the mills,

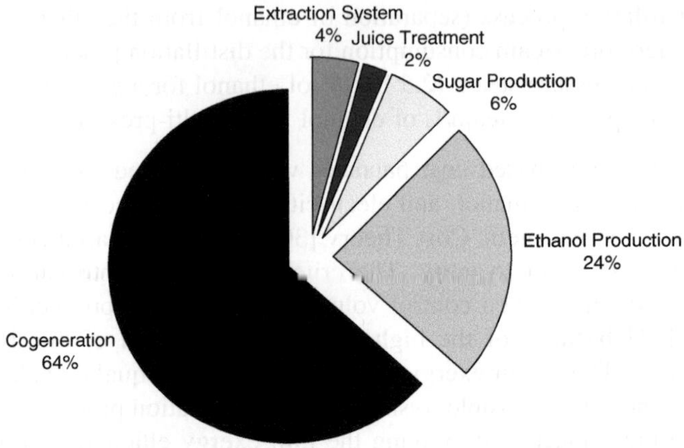

Fig. 6.3 Distribution of irreversibilities in a conventional mill [28]

two interrelated approaches may be cited: reduction of steam consumption, and the improvement of the cogeneration system, allowing higher generation of electricity.

In order to investigate the performance of the different energy conversion processes, a global model of the co-production of sugar, alcohol, and electricity was developed [10]. The model allows the evaluation of different configurations of the cogeneration system, as well as, of the heat exchanger network inside the mill. It was implemented in the engineering equation solver (EES®) [33] and simulated considering a steady state operation, with approximately 50 % of the cane crushed to sugar production and 50 % to hydrated alcohol production.

As a starting point, an exergy analysis was developed for a traditional sugarcane mill, producing sugar and ethanol, with no excess electricity generation and a process steam consumption of 490 kg/tc (kg per ton of sugarcane). The results show that the exergy destruction inside the mill is 729 kWh/tc (as a reference, sugarcane exergy is 5273 kJ/kg or 1465 kWh/tc). Furthermore, excess bagasse and stillage account for 128 and 15 kWh/tc, respectively. Both may be considered as exergy losses. Thus, the total irreversibilities in the mill are 872 kWh/tc.

Figure 6.3 shows the distribution of the exergy destruction among the processes, considering data from a typical Brazilian mill.

The cogeneration system is responsible for almost 65 % of the exergy destruction inside the mill. The ethanol production comes in second place, followed by the sugar production. Hence, modifications in the cogeneration system would have a higher impact in the overall efficiency than modifications on other sub-systems alone. However, due to the high level of interaction between each sub-system (Fig. 6.2), other modifications should be evaluated. In addition, the rational use of excess bagasse to generate electricity and/or cellulosic ethanol will improve the exergy performance of the mill.

Some comments regarding the main sources of irreversibilities in each control volume depicted in Fig. 6.3 can be drawn.

6.4.1 Extraction System

The objective of this process is to separate juice from the fiber of the sugarcane, producing raw juice and bagasse. In this study, a traditional mill has been considered; hence the separation is mechanically driven. In traditional sugarcane mills, simple stage backpressure steam turbines are employed, using medium pressure steam. The total exergy destruction in this system is 30.1 kWh/tc of which: 60 % is related to the separation process itself and 40 % in the turbine. To improve the exergy performance of the separation process, a better recovery ratio should be sought, even though mills are limited to 97 % of sucrose recovery, while diffusers might attain 99 %. Thus, for the sugarcane mill studied, the replacement of the steam turbine by electric motors is a good opportunity to increase the exergy performance of the process, since the electricity required to drive the motors could be generated in more efficient turbines.

6.4.2 Juice Treatment

The exergy destruction in processes involved in the juice treatment accounts for 15.8 kWh/tc, where 50 % is related to heat transfer and 40 % to the decanter and filter, due to the production of filter cake, treated here as a residue. Very little may be done to reduce the amount of filter cake produced. As for the heat transfer processes, the temperature differences between the juice and the condensing steam is the main source of entropy generation. In the sugarcane mill evaluated here, the thermal demand has been supplied by steam from the first effect of the multiple-effect evaporator (118 °C), resulting in high temperature differences to heat juice coming from the extraction system (35 °C). The following procedures could be implemented to minimize the exergy destruction in the juice heating process:

- The use of regenerative heaters, recovering the thermal exergy from the condensates of the multiple-effect evaporator;
- The use of steam extracted from other effects of the multiple-effect evaporator.

6.4.3 Sugar Production

In the sugar production control volume, vacuum pans are responsible for 46 % of the exergy destruction, while the multiple-effect evaporator accounts for 31 %, and the barometric condenser for 17 %. The amount of exergy destruction is 44.2 kWh/tc.

The main entropy generation sources in the evaporator and vacuum pans are the separation and heat transfer processes. Taking the multiple-effect evaporator, the first effects present lower temperature differences than the last ones; however, the former requires more exergy to evaporate 1 kg of water. Thus, there must be a balance between the efficiency of the effect and the exergy requirement to evaporate the water.

This balance might be attained by a more uniform distribution of the exergy destructions among the effects, considering the extraction of steam from different effects and a pressure distribution among the processes, which minimize the temperature differences in the effect. In the present study, a five effect evaporator has been evaluated, in which the first effect accounts for over 50 % of the total exergy destruction, followed by the fifth (24 %), fourth (12 %), third (8 %), and second (6 %).

As for the vacuum pans, the use of steam from another effect (second or third) could decrease the temperature difference of the process. Also, syrup leaving the multiple-effect evaporator could enter the cooking process with higher solid content, thus reducing the thermal demand of this process, which presents lower exergy efficiency when compared to the evaporator.

6.4.4 Ethanol Production

The total exergy destruction in the ethanol production is 176.9 kWh/tc, while stillage produced in the distillation represents an exergy loss of 14.9 kWh/tc. The main sources of entropy generation are:

- Chemical reaction in the fermentation vat;
- Heat dissipation due to the exothermic nature of fermentation reactions;
- Heat and mass transfer in the distillation columns.

All in all, the following concepts could be applied to improve the exergy performance of the ethanol production:

- Increase the ethanol concentration at the inlet of the distillation, thus reducing the thermal requirements:
  - Better control of the temperature inside the vats [34];
  - Extractive fermentation [35].
- Decrease the steam consumption in the distillation columns:
  - Better thermal integration;
  - Multiple-effect distillation schemes [23, 29];
  - Use of pervaporation systems [36].

A review of alternatives to integrate different fermentation and distillation technologies to improve the technical and economical performance of the ethanol production may be found in Cordona and Sanchez [37] and Haelssig et al. [38].

6.4.5 Cogeneration System

The exergy destruction inside the cogeneration system is 460.1 kWh/tc of which 97 % takes place in the boiler. In order to improve the exergy efficiency of this equipment, higher temperatures and pressures should be considered as well as better combustion systems. Furthermore, the use of exhaust gases to dry bagasse also increases the efficiency. Actually, bagasse drying may be accomplished by different technologies, each of them with an associated improvement in the exergy performance of the mill [39]. Some mills still make use of pressure reduction valves to complement the steam sent to the process, whenever the electromechanical demands are fulfilled. In the present study, 10 % of the process steam comes from a pressure reduction station. As for the turbines, most of them use medium pressure steam (21 bar and 300 °C) with isentropic efficiencies in the range of 55 % (single stage) to 65 % (multiple stages), hence their exergy performance is poor. Modern turbines for sugarcane mills available in Brazil may achieve 80–85 % isentropic efficiency, with temperatures up to 540 °C and pressures up to 120 bar (this operating condition may cause fouling in the superheater due to the potassium content in the bagasse). Finally, excess bagasse imposes a very large exergy loss to the mill if it is not used.

On the whole, there are many possibilities for a better performance of subsystems of a cogeneration facility in a sugarcane mill: higher values of steam parameters; improvements in the furnace, reducing the excess air needed; better arrangements of heat transfer areas, decreasing the temperature of exhaust gases; replacement of the reduction valve by an ejector, reducing the steam demand in the multiple-effect evaporator, or by a turbine to generate electricity to the grid. Furthermore, more advanced cogeneration systems could be used such as: SuSC and/or biomass gasification combined cycles (BIGCC). However, the use of these configurations requires modifications in the heat exchanger network in order to reduce the backpressure steam consumption.

6.5 Improving the Combined Production of Sugar, Ethanol, and Electricity

Due to the impact of the performance of the cogeneration plant in the overall performance of the mill, some improvements in the configuration and operating conditions of these plants are described and compared afterward. These alternatives are:

- Better thermal integration of the heat exchange processes [7, 20, 40];
- Different configurations for the cogeneration plant [28].
  - BPST;
  - CEST;
  - SuSC;
  - BIGCC.

BPST systems are the most common configuration in Brazilian mills, composed of bagasse-fired boilers and backpressure steam turbines, as shown in Fig. 6.4. These systems are only capable of generating excess electricity to be sold during

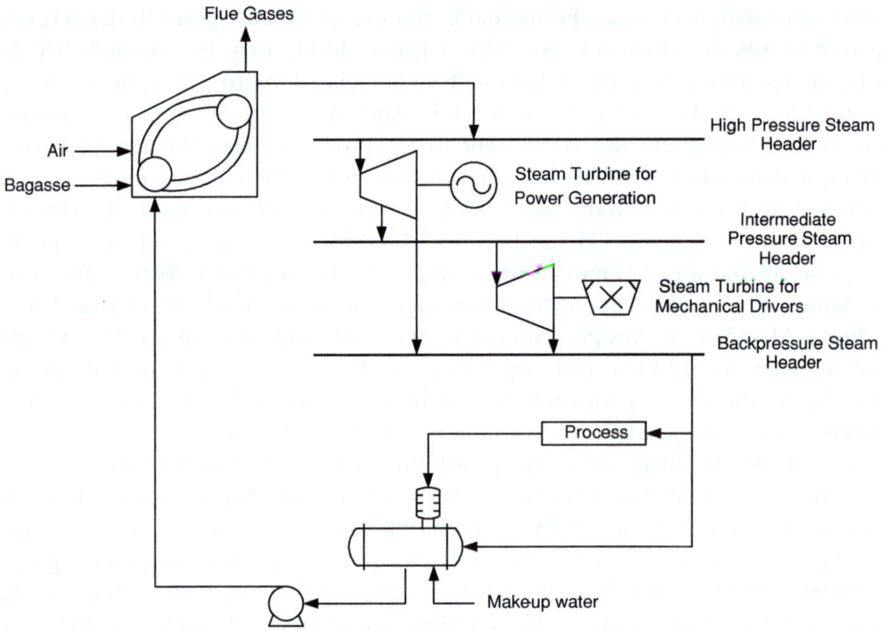

Fig. 6.4 System with backpressure steam turbines [10]

harvest season. Almost all available bagasse is consumed. In order to sell electricity during the whole year, CEST systems may be employed as an option for BPST (Fig. 6.5). The condensing-extraction turbine supplies steam for the process and the excess is sent to the condenser. In these systems, a reduction in the steam consumption by the process is necessary (from 490 to 391 kg/tc [10], considering a better integration of the heat exchanger network with the extractions from the multiple-effect evaporator, and also a decrease in the steam consumption in the distillery to 2.8 kg/tc, commonly used in new projects. Hence bagasse can be stored during harvest season to be used during off-season. For both, BPST and CEST systems, different steam generation temperatures and pressures were used: 42 bar/400°C, 42 bar/450°C, 67 bar/480°C, 67 bar/515°C, 80 bar/520°C, 100 bar/520°C, and 120 bar/540°C. Also, the reduction of steam consumption was evaluated separately from the use of condensing turbines in order to understand its impact on sugar, ethanol, and electricity production.

Table 6.3 summarizes some technical parameters used in the simulation of BPST and CEST systems.

The SuSC configuration, presented in Figs. 6.6 and 6.7, is in fact an improvement of the CEST systems, using very high steam parameters and regenerative heat exchangers to preheat boiler feedwater. These modifications improve the steam cycle efficiency. Table 6.4 presents values for specific mass flowrate, temperature, and pressure for each flow indicated in Fig. 6.6. The SuSC

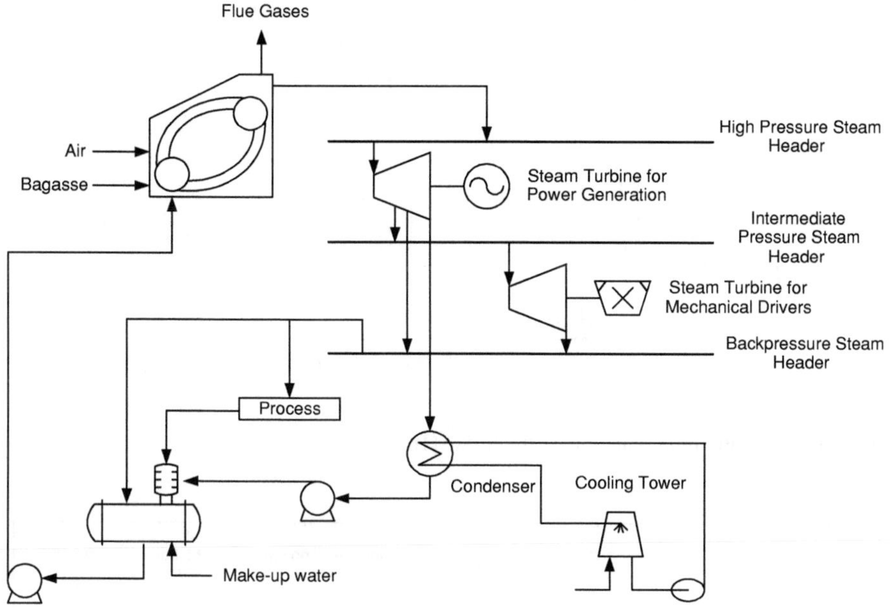

Fig. 6.5 System with backpressure and condensing steam turbines [10]

Table 6.3 Technical parameters of BPST and CEST systems [44]

Boiler efficiency (%, LHV basis)[a]	85
Isentropic efficiency of turbogenerators (%)	80–82
Electric generator efficiency (%)	95
Mechanical drive turbine isentropic efficiency (%)	55
Pumps isentropic efficiency (%)	70

Boiler feed water temperature −115 °C
[a] Excess air – 35 %, boiler flue gases temperature −167 °C

is capable of generating an excess electricity of 142 kWh/tc, reducing the exergy destruction inside the mill by nearly 12 %.

Considering the possibility of using gas turbines to generate electricity, biomass gasification based-systems are one possibility. Sugarcane bagasse and trash are dried and sent to a gasifier, in which a low calorific value gas (produced gas) is obtained. This gas is used as fuel in a gas turbine, and the exhaust gases are sent to a heat recovery steam generator (HRSG) that produces superheated steam to a condensing-extraction turbine. This study evaluated two configurations for BIGCC technology: low pressure air-blown and high pressure air-blown. For the low pressure air-blown configuration (BIGCC I and II), illustrated in Fig. 6.8, the off-design operation strategy adopted was de-rating the firing temperature in the turbine. For the high pressure air-blown gasifier (BIGCC III), presented in Fig. 6.9, the compressor blast-off was used, with the air extracted being sent to the gasifier [16]. For these

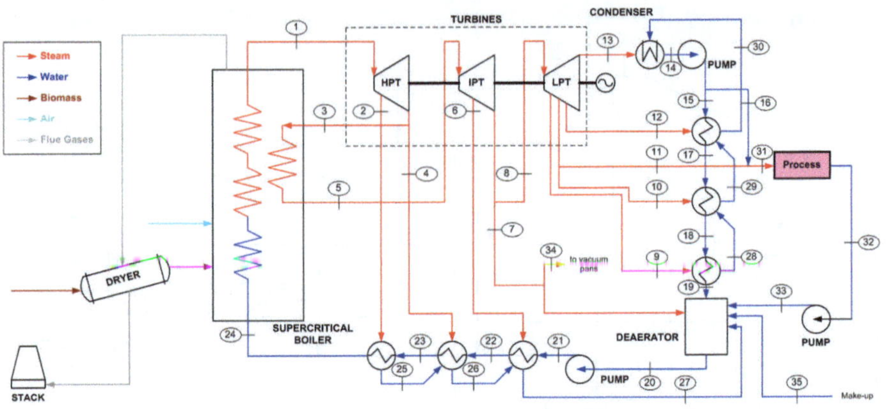

Fig. 6.6 Schematic representation of the SuSC [16]

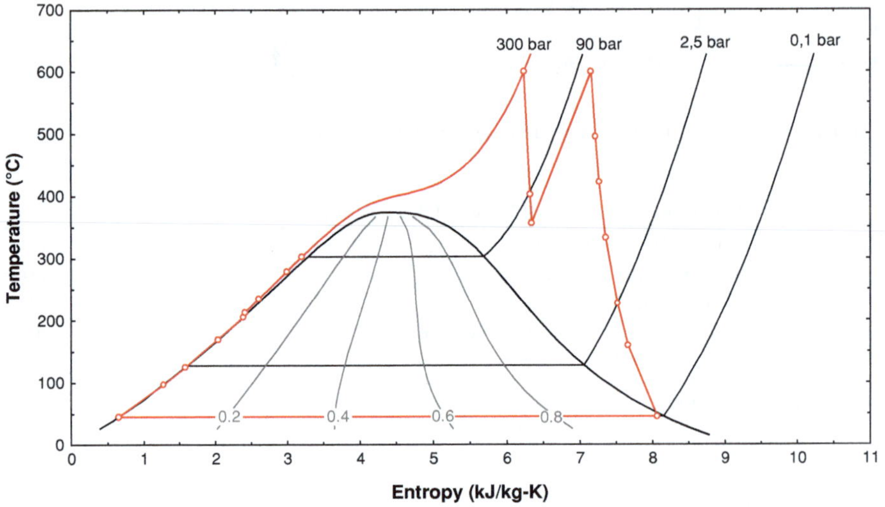

Fig. 6.7 T–s diagram for the SuSC system [16]

configurations, steam consumption in the process was decreased to 277 kg/tc, through a total integration of the multiple-effect evaporator with the heat-exchanger network as well as the use of multiple-effect distillation systems [20].

The gasifier gas composition was obtained by means of a chemical equilibrium model developed by Pellegrini and Oliveira Jr. [10]. The model allows the evaluation of the composition of the produced gas, under different pressures and temperatures, as well as for different compositions of the biomass, as well as its LHV. Table 6.5 presents the composition of the produced gas generated in both low pressure and high pressure air-blown gasifiers. For both cases, it was

Table 6.4 Calculated values for specific mass flowrate, temperature, and pressure for the SuSC configuration [16]

Tag	Mass flowrate (kg/tc)	Temperature (°C)	Pressure (bar)
1	668	600.0	300.0
2	45	402.5	87.1
3	551	356.5	62.2
4	72	356.5	62.2
5	551	600.0	62.2
6	17	495.4	30.6
7	52	422.7	17.6
8	477	422.7	17.6
9	16	333.1	8.0
10	9	227.7	2.5
11	260	227.7	2.5
12	17	159.9	1.0
13	176	45.8	0.1
14	217	45.8	0.1
15	198	46.0	17.6
16	18	46.0	17.6
17	198	97.6	17.6
18	198	125.4	17.6
19	198	169.4	17.6
20	668	206.0	17.6
21	668	213.2	300.0
22	668	235.0	300.0
23	668	278.9	300.0
24	668	303.0	300.0
25	45	301.0	87.1
26	116	277.9	62.2
27	133	235.0	30.6
28	16	170.4	8.0
29	24	127.4	2.5
30	41	99.6	1.0
31	278	140.0	2.5
32	264	120.0	2.5
33	264	120.3	17.6
34	7	422.7	17.6
35	21	70.0	17.6

considered an equivalence ratio of 0.3 and that all bagasse is converted into produced gas (no charcoal production).

The steam generation at 40 bar and 400 °C, in the atmospheric configuration I, is due to limitations to superheat steam at the HRSG because of the low temperature of the gas turbine exhaust gases. Thus, in atmospheric configuration II, only saturated steam is produced in the HRSG and its superheating takes place in the regenerative heat exchanger of the gases cleaning system. This low exhaust

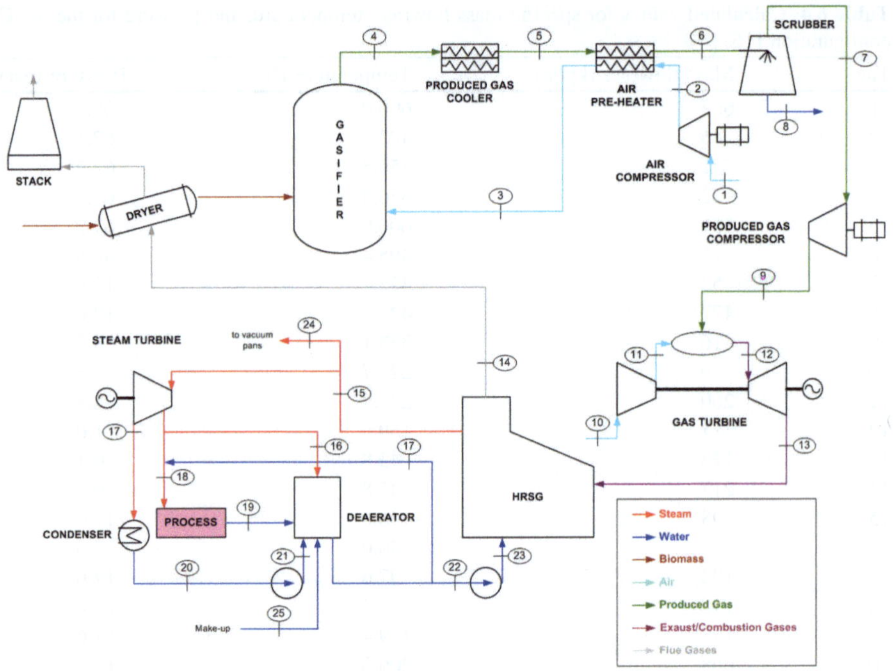

Fig. 6.8 Schematic representation of the atmospheric BIGCC [16]

gases of the gas turbine temperature is a consequence of the operation strategy adopted to simulate the turbine (de-rating). For the gas turbine simulation fired with produced gas, two approaches were considered [41, 42]:

- De-rating for the low-pressure air-blown gasification (turbine inlet temperature (TIT) decrease, with the air mass flowrate constant in the compressor);
- Compressor blast-off for the high pressure air-blown gasification (same TIT as the design point, extracting the excess air from the compressor before the combustion chamber).

The design conditions were defined based on ALSTOM GT-11 operating under ISO conditions, and these parameters were used in the off-design simulation.

The calculated values for specific mass flowrate, temperature, and pressure for each flow of every BIGCC configuration are shown in Tables 6.6, 6.7 and 6.8.

Tables 6.9 and 6.10 show the main technical characteristics of the evaluated combined cycle-based gasification systems.

For the SuSC and BIGCC studied configurations it was considered the use of bagasse driers. For the SuSC system, the exhaust gases from the boiler are used to dry bagasse up to 40 % moisture content. As for the BIGCC systems, it is possible to dry bagasse to 10 % moisture content, since the amount of exhaust gases from the HRSG is higher than that of the SuSC boiler.

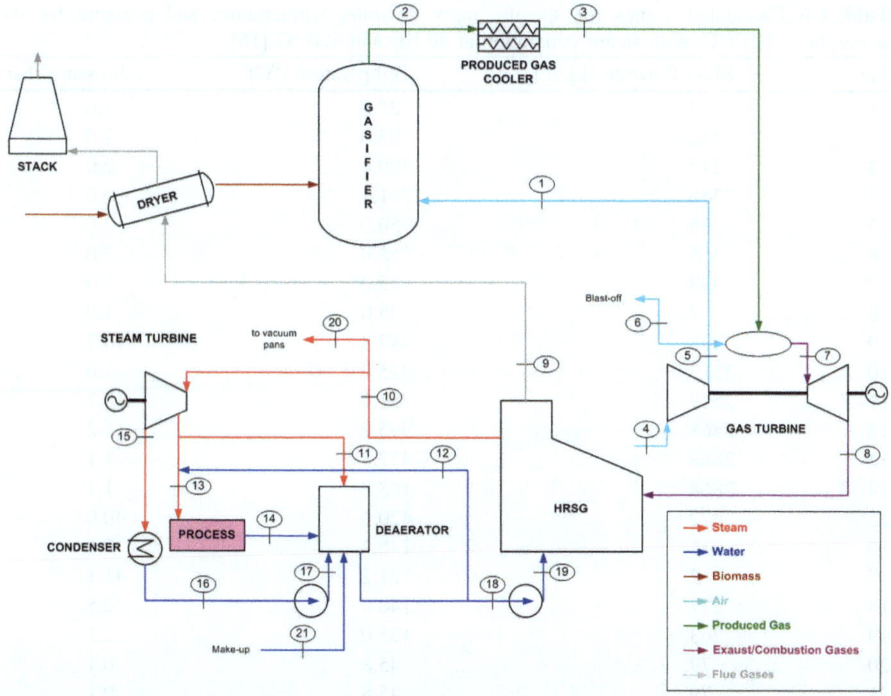

Fig. 6.9 Schematic representation of the pressurized BIGCC [16]

Table 6.5 Produced gas composition for the two configurations of gasifier (low and pressure air-blown) [16]

	Molar fraction (%)	
Component	Low pressure air-blown	High pressure air-blown
CH_4	0.2	0.8
CO	23.2	23.0
H_2	22.9	20.2
CO_2	10.3	9.5
H_2O	5.6	9.2
N_2	37.4	37.0
Ar	0.5	0.4
LHV (kJ/kg)	5137	5939

6.6 Exergy-Based Comparison of Alternatives

Figures 6.10 and 6.11 present the exergy-based cost for sugar, ethanol, and electricity for all configurations, as well as the amount of excess electricity generated in each case.

Table 6.6 Calculated values for specific mass flowrate, temperature, and pressure for the atmospheric BIGCC, with steam generation at 40 bar and 400 °C [16]

Tag	Mass flowrate (kg/tc)	Temperature (°C)	Pressure (bar)
1	212	25.0	1.0
2	212	103.6	2.0
3	212	300.0	2.0
4	355	761.4	2.0
5	355	350.0	2.0
6	355	258.0	2.0
7	348	35.0	1.0
8	7	35.0	1.0
9	348	447.8	15.7
10	2519	25.0	1.0
11	2519	451.9	15.7
12	2868	995.0	15.2
13	2868	457.8	1.1
14	2868	188.8	1.1
15	372	420.0	40.0
16	11	146.4	2.5
17	2	121.2	41.1
18	277	140.0	2.5
19	263	125.0	2.5
20	79	45.8	0.1
21	79	45.8	0.1
22	79	45.8	2.5
23	374	120.4	2.5
24	372	121.2	41.1
25	8	420.0	40.0
26	22	70.0	2.5

It is possible to show a decrease in the exergy-based cost of sugar and ethanol with the use of more efficient cogeneration systems. By more efficient systems, one should understand systems with higher output electricity generation.

The decrease of exergy-based cost of sugar and ethanol for BPST system is related to a reduction in the irreversibilities in the cogeneration system (low efficiency of the steam cycle and the amount of excess bagasse not used in the system). In this way higher temperatures and pressures for the steam produced in the boiler, which imply higher amounts of bagasse consumed in the cogeneration system, have a positive effect on the exergy-based cost of electricity and process steam. This in turn leads to smaller production costs for sugar and ethanol, although the exergy destruction in their production processes is the same as in the traditional mill.

Reduction of the steam consumption in the process leads to a decrease of the exergy destruction in the sugar and ethanol production processes with an increase of the exergy loss related to a greater amount of excess bagasse. Thus, the net effect of this alternative is higher sugar and ethanol exergy-based cost.

Table 6.7 Calculated values for specific mass flowrate, temperature, and pressure for the atmospheric BIGCC, with steam generation at 80 bar and 510 °C [16]

Tag	Mass flowrate (kg/tc)	Temperature (°C)	Pressure (bar)
1	212	25.0	1.0
2	212	103.6	2.0
3	212	261.2	2.0
4	355	751.0	2.0
5	355	311.2	2.0
6	355	237.2	2.0
7	349	35.0	1.0
8	6	35.0	1.0
9	349	447.2	15.7
10	2516	25.0	1.0
11	2516	451.9	15.7
12	2864	994.6	15.2
13	2864	457.6	1.1
14	2864	218.3	1.1
15	326	510.0	80.0
16	7	149.3	2.5
17	2	121.9	81.1
18	277	140.0	2.5
19	263	120.0	2.5
20	38	45.8	0.1
21	38	45.8	0.1
22	38	45.8	2.5
23	326	120.4	2.5
24	326	121.9	81.1
25	6	510.0	80.0
26	20	70.0	2.5

In order to use the excess bagasse generated, CEST systems are introduced. These systems present an average 6 % decrease in the exergy-based cost of sugar and ethanol, related to the reduction in the steam consumption in the process, and smaller exergy-based costs of electricity and process steam. The smaller costs of electricity and process steam are connected to an increase in the excess electricity of 56 % (average) and the use of all available bagasse in the cogeneration system. Hence, the unitary exergy-based cost for bagasse consumed is smaller.

From the analysis above, it is possible to consider that the assumption of assigning zero cost to the excess bagasse has a great impact on the exergy-based cost of process steam and electricity, and, as a consequence, on the sugar and ethanol costs. Such assumption tends to increase the cost of bagasse consumed in the boiler whenever there are greater amounts of excess. Thus, the analysis indicates that the rational use of bagasse must be sought in sugarcane mills in order to improve its performance. Also, exergy-based analysis could be used to compare alternative uses for the excess bagasse.

Table 6.8 Calculated values for specific mass flow rate, temperature, and pressure for the pressurized BIGCC, with steam generation at 80 bar and 510 °C [16]

Tag	Mass flowrate (kg/tc)	Temperature (°C)	Pressure (bar)
1	212	451.9	15.7
2	355	857.6	15.7
3	355	550.0	15.7
4	2251	25.0	1.0
5	2251	451.9	15.7
6	61	451.9	15.7
7	2333	1119.0	15.2
8	2333	541.6	1.1
9	2333	185.7	1.1
10	362	510.0	80.0
11	12	149.3	2.5
12	2	120.0	81.1
13	277	140.0	2.5
14	263	120.0	2.5
15	70	45.8	0.1
16	70	45.8	0.1
17	70	45.8	2.5
18	362	120.4	2.5
19	362	120.0	81.1
20	6	510.0	80.0
21	20	70.0	2.5

Table 6.9 Technical characteristics of the gasification systems [44]

Parameter	Atmospheric I	Atmospheric II	Pressurized
Equivalence ratio	0.3	0.3	0.3
Gasifier operating pressure (bar)	2	2	16
Air temperature at the gasifier inlet (°C)	300	261	452
Gases temperature at the outlet of the gasifier (°C)	761	751	858
Gases temperature after the cleaning system (°C)	35	35	550
Gas turbine combustor gases outlet temperature (°C)	995	995	1119
Gas turbine exhaust gases temperature (°C)	458	458	542
HRSG flue gases temperature (°C)	189	218	186
Steam pressure (bar)	40	80	80
Steam temperature (°C)	400	510	510

As far as the electrification of mechanical drivers are concerned, it represents a 2 % average reduction in the sugar and ethanol exergy-based costs, while generating around 7 kWh/tc of excess electricity.

Advanced cogeneration systems benefit from: (a) lower exergy destruction in the sugar and ethanol processes related to better thermal integration and (b) higher

Table 6.10 Technical parameters used to simulate the BIGCC systems [44]

HRSG pinch point (°C)	10
HRSG approach point (°C)	5
Compressors isentropic efficiency (%)	80
Isentropic efficiency of turbogenerators (%)	79–80
Electric generator efficiency (%)	95
Pumps isentropic efficiencies (%)	70

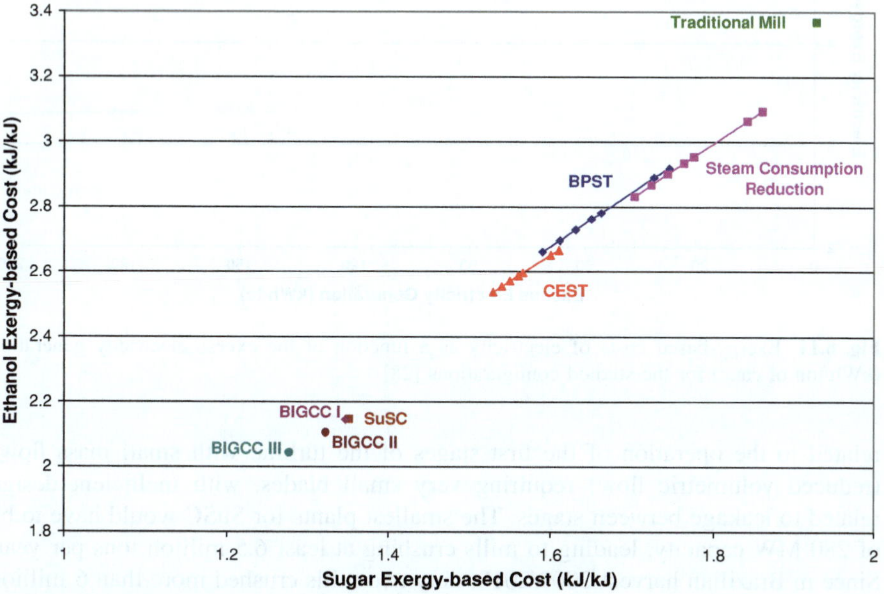

Fig. 6.10 Exergy-based costs of sugar and ethanol for the studied configurations [28]

bagasse-to-electricity efficiencies. Thus, these systems allow exergy-based cost for sugar and ethanol 15 % (average) smaller than costs from CEST systems.

SuSC and BIGCC's systems are able to produce three times more electricity when compared to current available options (BSPT and CEST). Yet, comparing the SuSC system with the BIGCC systems, the first generates 10 kWh/tc less electricity than atmospheric BIGCC systems, while the pressurized one is able to generate 30 % more electricity. The reason for this difference is that atmospheric systems need a produced gas compressor to inject the produced gas into the gas turbine combustor, requiring almost 50 kWh/tc. Thus, more advanced systems can increase the bagasse-to-electricity efficiency up to 36 %.

Although, SuSC are not competitive with pressurized BIGCC as far as electricity generation is concerned, the technology to implement SuSC plants in sugarcane mills seems to be closer to commercial scale than BIGCC. However, SuSC systems are not suitable for small installed capacities, due to problems

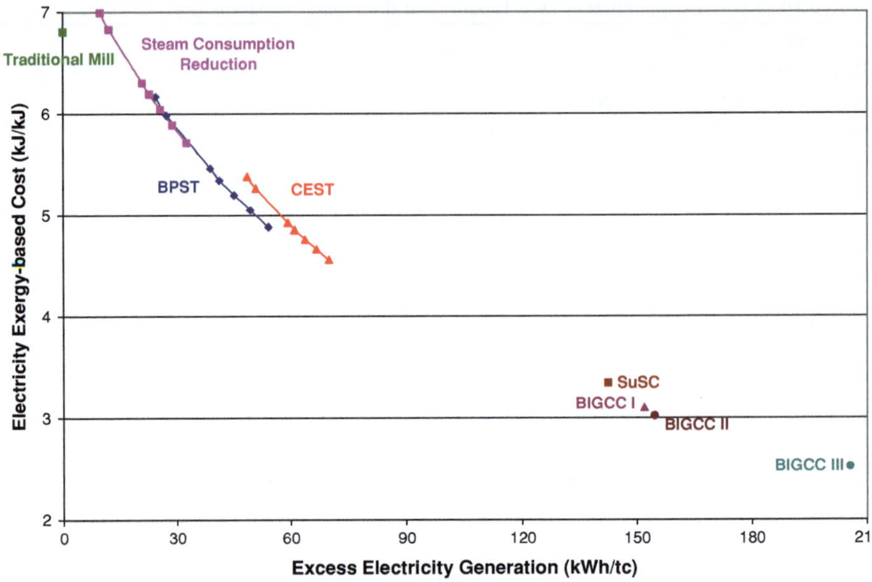

Fig. 6.11 Exergy-based costs of electricity as a function of the excess electricity generation (kWh/ton of cane) for the studied configurations [28]

related to the operation of the first stages of the turbine with small mass flows (reduced volumetric flow) requiring very small blades, with inefficient design related to leakage between stages. The smallest plants for SuSC would have to be of 280 MW capacity, leading to mills crushing at least 6.5 million tons per year. Since in Brazilian harvest 2009/2010, only two mills crushed more than 6 million tons, the realization of SuSC plants might be feasible for centralized plants to be constructed near a pool of mills, instead of inside a single mill.

If electricity generation during the whole year is considered, then SuSC and BIGCC configurations need a complementary fuel (sugarcane trash), as all bagasse would be consumed during the milling season. The BPST configuration is not suitable for electricity generation during the whole year, and the CEST configuration is capable of generating using only bagasse. Yet, problems related to the storage of bagasse and sugarcane trash should be addressed carefully. One possible solution is the drying of these biomasses and their processing into briquettes or pellets, which are more suitable to storage.

Regarding the determination of the production costs in the mill, exergy-based cost analysis allowed the identification of reductions in the production costs of sugar, ethanol, and electricity as a consequence of a better exergy performance of the energy conversion processes inside the mill, mainly in the cogeneration plant. Such decrease in turn leads to higher revenues for the mills.

The use of exergy as a criterion to assign cost in the mill gives a new perspective related to how those are formed. Then, the increase in the revenue is

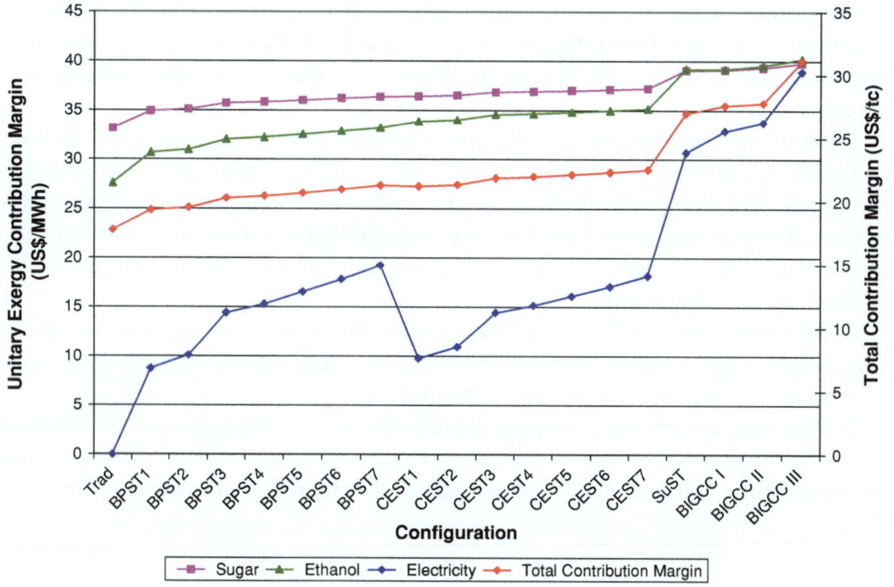

Fig. 6.12 Unitary exergy and total contribution margin [28]

associated to a difference of the total production cost for a base case and that of the new configuration. Furthermore, this procedure assigns costs to bagasse, and, hence, one is able to evaluate the weight of sugarcane on the electricity cost, and then it is possible to modify the way sugarcane producers are paid.

The attribution of costs based on exergy may be used as a managerial tool for production planning and controlling, evaluating the contribution margin (defined as the difference between the price of a product and its variable production cost) of each of the products. It may help in economic optimizations regarding the best mix of production as well as the choice for the varieties of sugarcane that should be harvested. Such analysis relates prices, costs, and volumes of production, indicating the participation of each product in the total result of the business [43]. It is important to notice that this analysis did not take into account fixed costs, for instance those related to the purchase of equipment.

As an example of such analysis, Fig. 6.12 presents the contribution margin for sugar, ethanol, and electricity for the different configurations (the values are given in US$/MWh exergy-basis to ease the comparisons). The prices used for the analysis were those being developed in Brazil for the 2008/09 sugarcane harvest (sugarcane—US$ 13.72/tc, sugar—US$ 234.00/t, ethanol—US$ 341.00/m^3, electricity—US$ 65.00/MWh). No other input costs, other than sugarcane, were considered. Yet, around 60–70 % of total input cost in sugarcane mill comes from the purchase of sugarcane.

It is possible to show that a better thermodynamic performance of the energy conversion processes results in better economic performance, increasing the contribution margin of the mill, since there is new revenue from the selling of electricity. Also, the more efficient is the bagasse-to-electricity efficiency, the more important this product is to the profit of the mill, due to its higher unitary contribution margin.

Figure 6.12 also indicates that for conventional systems the increase in the total contribution margin is slight, even though there is new revenue from the selling of electricity. On the other hand, with more advanced cogeneration systems, the impact of higher amounts of electricity to the grid is more representative. It is also interesting that for more efficient mills, the unitary contribution margin of the product tends to be similar.

A final comment on the use of the contribution margin analysis is that it is a very simple way to define the break-even point of any modification in the process as well as the introduction of new technologies.

In that way, as far as the economic feasibility of the configurations, a traditional economic/financial analysis is essential to evaluate if the thermodynamic benefits are translated into economic gains and provide proper returns to needed investments.

6.7 Renewability of the Combined Production of Sugar, Ethanol, and Electricity

The sustainability of the combined production of sugar, ethanol, and electricity can be assessed by evaluating the renewability exergy indexes (λ of the studied configurations).

Figure 6.13 shows the renewability exergy index (λ values for the mill as a whole and also for the cogeneration system alone for the different analyzed configurations).

As for the exergy-based costs of sugar and ethanol, the renewability exergy index has better values for more efficient configurations. Furthermore, only SuSC and BIGCC's systems have a value greater than 1, indicating that the processes may not be considered renewable from a Second Law of Thermodynamics point of view.

Comparing the cogeneration systems, it is interesting to see a better environmental performance of BPST systems related to CEST. This is a consequence of the inefficient operation of the latter as an electricity generation system. However, this result should not be understood as BPST are better than CEST systems, since the high steam consumption in the process BPST results in higher exergy destruction in the sugar and ethanol production processes, with worse environmental performance for the mill as a whole.

It is interesting to compare λ values of these cogeneration systems with those from conventional thermoelectric plants, as presented in Chap. 2, which varies from 0.18 to 0.43. Thus it may be argued that CEST systems have similar environmental performance of thermoelectric plants with efficiencies lower than 30 %,

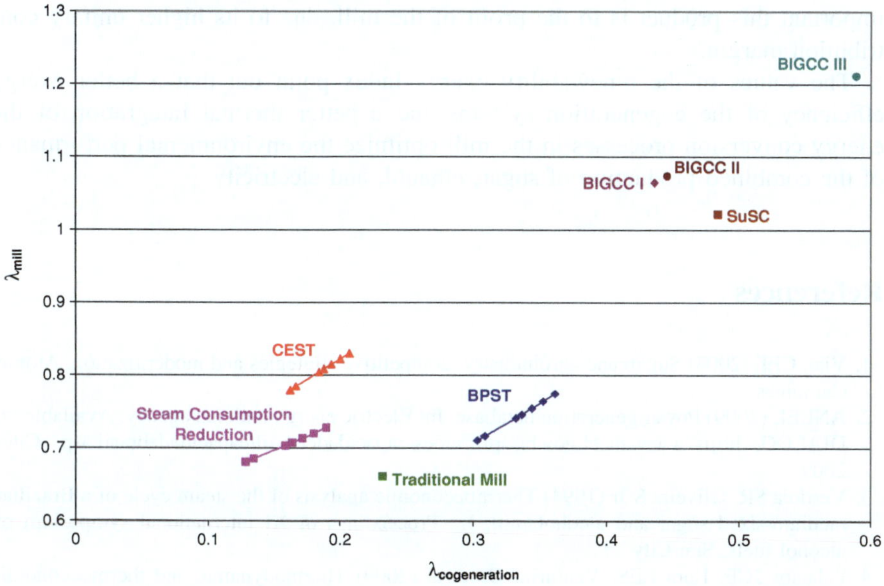

Fig. 6.13 Renewability exergy index for the mill and for the cogeneration systems [28]

while BPST are comparable to plants with 50 % efficiency. On the other hand, advanced cogeneration systems present better environmental performance than thermoelectric power plants.

All in all, the bagasse-to-electricity conversion efficiency must be optimized in order to obtain the best environmental performance of the mill as a whole.

6.8 Concluding Remarks

The results described in this chapter show that a better thermodynamic performance of the cogeneration system is related to a decrease in the exergy destruction of all energy conversion processes. Furthermore, a better performance is translated into a decrease in the exergy-based cost of sugar and ethanol.

Conventional cogeneration systems are able to generate up to 80 kWh/tc of excess electricity depending on the steam consumption in processes. Furthermore, advanced cogeneration systems allow up to 200 kWh/tc, which represents a potential of 111 TWh/year generation or 25 % of the Brazilian electricity consumption.

It is possible to show that a better thermodynamic performance of the energy conversion processes results in better economic performance, increasing the contribution margin of the mill, since there is new revenue from the selling of electricity. Also, the more efficient is the bagasse-to-electricity efficiency, the more important this product is to the profit of the mill, due to its higher unitary contribution margin.

The values of the renewability exergy index point out that a better exergy efficiency of the cogeneration systems and a better thermal integration of the energy conversion processes in the mill optimize the environmental performance of the combined production of sugar, ethanol, and electricity.

References

1. Vian CEF (2003) Sugarcane agroindustry: competitive strategies and modernization. Átomo, Campinas
2. ANEEL (2009) Power generation database. In: Electric energy national agency. Available via DIALOG. http://www.aneel.gov.br/aplicacoes/capacidadebrasil/capacidadebrasil.asp. Cited 2009
3. Vertiola SR, Oliveira S Jr (1994) Thermoeconomic analysis of the steam cycle of a Brazilian medium-sized sugar and alcohol mill. In: Proceedings of XI international symposium on alcohol fuels, Sun City
4. Palacios JCE, Lora EES, Venturini OJ et al (2009) Thermodynamic and thermoeconomic evaluation of different technologies and steam parameters for plant modernization alternatives in an autonomous distillery. In: Proceedings of the 22nd international conference on efficiency, costs, optimization, simulation and environmental impact of energy systems, Foz de Iguaçú
5. Barreda del Campo ER, Cerqueira SAAG, Nebra SA (1998) Thermoeconomic analysis of a cuban sugar cane mill. Energy Convers Manag 39:1773–1780
6. Prieto MGS, Nebra SA (2004) Thermodynamic and exergetic cost analysis of two steam based cogeneration schemes proposed for a Brazilian sugar mill. In: Proceedings of the 17th international conference on efficiency, costs, optimization, simulation and environmental impact of energy systems, Mexico City
7. Ensinas AV, Nebra SA, Lozano MA et al (2007) Analysis of process steam demand reduction and electricity generation in sugar and ethanol production from sugarcane. Energy 48:2978–2987
8. Ensinas AV, Modesto M, Nebra SA (2007) Analysis of different cane juice extraction systems for sugar and ethanol production: influences on electricity generation and final products exergetic costs. In: Proceedings of the 20th international conference on efficiency, costs, optimization, simulation and environmental impact of energy systems, Padova
9. Pellegrini LF, Oliveira S Jr (2007) Exergy analysis of sugarcane bagasse gasification. Energy 32:314–327
10. Pellegrini LF, Oliveira S Jr (2007) Exergy efficiency of the combined sugar, ethanol and electricity production and its dependence of the exergy optimization of the utilities plants. In: Proceedings of the 20th international conference on efficiency, costs, optimization, simulation and environmental impact of energy systems, Padova
11. Palacios-Bereche R, Pereira PA, Nebra SA et al (2009) Energetic evaluation of cogeneration systems in sugar cane plants in Brazil case studies. In: Proceedings of the 22nd international conference on efficiency, costs, optimization, simulation and environmental impact of energy systems, Foz de Iguaçú
12. Ogden JM, Hochgreb S, Hylton M (1990) Steam economy and cogeneration in cane sugar factories. Int Sugar J 92:131–140
13. Larson ED, Williams RH, Leal MRLV (2001) A review of biomass integrated-gasifier/gas turbine combined cycle technology and its application in sugarcane industries, with an analysis for Cuba. Energy Sust Dev 5:54–75

14. Walter A, Overend RP (1998) Analysis of BIG-GT cycles in the sugarcane industry. In: Proceedings of the 10th biomass European congress, Wurzburg
15. Beér JM (2007) High efficiency electric power generation: the environmental role. Energy 33:107–134
16. Pellegrini LF, Oliveira S Jr, Burbano JC (2010) Supercritical steam cycles and biomass integrated gasification combined cycles for sugarcane mills. Energy 35:1172–1180
17. Camargo AC (1990) Energy conservation in the sugar and ethanol industry. In: Camargo AC (ed) Handbook for energy conservation in the sugar and ethanol industry. Instituto de Pesquisas Tecnológicas, São Paulo (in Portuguese)
18. Hugot E (1986) Handbook of cane sugar engineering. Elsevier, Amsterdan
19. Payne JH (1989) Unit operations in the production of cane sugar. Nobel/STAB, São Paulo
20. Avram P, Stark T (2004) Integration of ethanol production with a sugar factory producing maximum cogeneration. Int Sugar J 106:126–137
21. Christodoulou P (1996) Energy economy optimization in the separation processes: optimizing the separation of sucrose/water and non-sugars. Int Sugar J 98:419–430
22. Morgenroth B (1996) Experiences with the falling film plate evaporator—latest state in the development of evaporation technology as well as concepts for energy efficient process schemes in the cane sugar industry. In: International sugar symposium, Luxor
23. Seemann F (2003) Energy reduction in distillation for bioethanol plants. Int Sugar J 105:420–423
24. Nebra SA, Parra MIF (2005) The exergy of sucrose–water solutions: proposal of a calculation method. In: Proceedings of the 18th international conference on efficiency, costs, optimization, simulation and environmental impact of energy systems, Trondheim
25. Modesto M, Nebra SA, Zemp RJ (2005) A proposal to calculate the exergy of non-ideal mixtures ethanol–water using properties of excess. In: Proceedings of the 14th European biomass conference and exhibition: biomass for energy industry and climate protection, Paris
26. Channiwala AS, Parikh PP (2002) A unified correlation for estimating HHV of solids, liquid and gaseous fuels. Fuel 81:1051–1063
27. Szargut J, David RM, Steward F (1988) Exergy analysis of thermal, chemical, and metallurgical processes. Hemisphere Publishing, New York
28. Pellegrini LF, Oliveira S Jr (2011) Combined production of sugar, ethanol and electricity: thermoeconomic and environmental analysis and optimization. Energy 36:3704–3715
29. Pellegrini LF, Modesto M, Nebra SA et al (2008) Modern concept for ethanol distilleries: maximization of the electricity surplus. In: Proceedings of the 12th Brazilian congress of thermal sciences and engineering, Belo Horizonte
30. Lozano MA, Valero A (1993) Theory of the exergetic cost. Energy 18:939–960
31. Gaggioli RA, Wepfer WJ (1980) Exergy economics: I. cost accounting applications, II. benefit-cost conservation. Energy 5:823–837
32. Florentino HO, Sartori MMP (2003) Game theory in sugarcane crop residue. Biomass Bioenerg 25:29–34
33. Klein SA (2011) Engineering equation solver—EES, F-chart software. www.fChart.com
34. Magazoni FC, Monteiro JB, Deucher R et al (2009) Cooling of ethanol fermentation process using absorption chillers. In: Proceedings of the 22nd international conference on efficiency, costs, optimization, simulation and environmental impact of energy systems, Foz de Iguaçú
35. Kargupta K, Datta S, Sanyal SK (1998) Analysis of the performance of a continuous membrane bioreactor with cell recycling during ethanol fermentation. Biochem Eng J 1:31–37
36. Sorin M, Ayotte-Sauvé E, Sadeghi F et al (2009) Thermodynamic equipartition and energy efficient membrane networks. In: Proceedings of the 22nd international conference on efficiency, costs, optimization, simulation and environmental impact of energy systems, Foz de Iguaçú
37. Cordona CA, Sanchez OJ (2007) Fuel ethanol production—process design trends and integration opportunities. Bioresource Technol 98:2415–2457
38. Haelssig JB, Tremblay AY, Thibault J (2008) Technical and economic considerations for various recovery schemes in ethanol production by fermentation. Ind Eng Chem Res 47:6185–6191

39. Morgenroth B, Batstone D (2005) Development and prospects for drying bagasse by steam. Int Sugar J 107:410–415
40. Rein P (2007) Cane sugar engineering. Verlag Dr. Albert Bartens, Berlin
41. Walter A (2000) Simulation of gas turbines operating in off-design condition. In: Proceedings of the 13th international conference on efficiency, costs, optimization, simulation and environmental impact of energy systems, Enschede
42. Zanetti AA, Pellegrini LF, Oliveira S Jr (2007) Thermoeconomic analysis of a BIGCC cogeneration system using natural gas and sugarcane bagasse as complementary fuels. In: The 20th international conference on efficiency, costs, optimization, simulation and environmental impact of energy systems, Padova
43. Hirschey M (2005) Fundamentals of managerial economics. South-Western College Publishing, Mason
44. Pellegrini LF (2009) Analysis and thermo-economic and environmental optimization applied to the combined production of sugar, alcohol and electricity. Ph.D. Thesis, Polytechnic School of the University of São Paulo, São Paulo
45. National Supply Company—CONAB (2011) Third assessment of the sugarcane harvesting. http://www.conab.gov.br/OlalaCMS/uploads/arquivos/11_01_06_09_14_50_boletim_cana_3o_lev_safra_2010_2011..pdf

Chapter 7
Exergy and Renewability Analysis of Liquid Biofuels Production Routes

Symbols

B	Exergy rate/flow rate (kW)
b	Specific exergy (kJ/kg, kJ/kmol)
H/C, O/C	Atomic ratio of the elements
FA	Fatty acid
FFA	Free fatty acid
FFB	Fresh fruit bunches
G	Glycerol
HHV	Higher heating value (kJ/kg)
LHV	Lower heating value (kJ/kg)
ME	Methyl ester
TG	Triglycerides

Greek Symbols

β	Parameter defined by Eq. 7.1
η	Efficiency
λ	Renewability exergy index

Subscripts

b	Exergy
ch	Chemical exergy
bio	Biomass
de	Deactivation
dest	Destroyed
global	Related to the whole plant/process
nr	Non-renewable
p	Product, useful effect
r	Raw material
u	Utilized/required
util	Utilities plant
w	Waste

S. de Oliveira Jr., *Exergy*, Green Energy and Technology,

DOI: 10.1007/978-1-4471-4165-5_7,

7.1 Introduction

Ethanol and biodiesel produced from different renewable feedstocks constitute the most widely used alternative fuels for internal combustion engines [1–4]. These biofuels can be used neat or blended with gasoline and conventional diesel fuel, respectively, and as a consequence their utilization allows decreasing fossil fuel consumption as well as increasing energy security. Additionally, since their carbon content has a vegetable origin, it has been argued that they have the potential to be CO_2-neutral [5]. However, concerns exist about feedstocks or raw materials, including the impact they may have on biodiversity and land use and competition with food crops [6].

A wide variety of raw materials has been used for ethanol production. Sugar containing crops such as sugarcane can be converted into high-grade ethanol through milling, fermentation, and distillation. Developments in bioprocesses are being made to allow the use of amilaceous and lignocellulosic materials to produce ethanol through hydrolysis, fermentation, and distillation. On the other hand, vegetable oils and animal fats have been the conventional raw materials for biodiesel production. These natural sources of triglycerides can be converted into fatty acid methyl esters or biodiesel by transesterification and esterification reactions.

An effective tool to analyze the production processes of biofuels from an integrated point of view is provided by exergy analysis since vehicles require the exergy of fuels to be converted in kinetic energy (or kinetic exergy) for transportation purposes. In addition, exergy allows comparisons, using the same physical basis, among all inflows and outflows, regardless if they are mass or energy streams [7, 8].

Exergy analysis has been used to evaluate biodiesel production from cooking oils [9]. Similar studies have been developed using palm oil as a raw material [10, 11]. The combined production of sugar, ethanol, and electricity, considering different configurations of the cogeneration plant, has been analyzed using exergy-based costs [12].

In this chapter, liquid biofuels production routes are analyzed and compared by using exergy analysis for evaluating the quality of the energy conversion processes, and to assess the renewability of such processes based on the consumption of fossil exergy and the irreversibilities associated with the production processes. This analysis is performed by means of a simulation tool implemented using the EES® software [13] and the data used in the analysis were taken from biofuels production pilot plants as well as industrial units. The developed simulation tool allows also identifying alternatives for process optimization.

The biofuels production routes are based on sugar and ethanol combined production from sugarcane (first generation ethanol), ethanol production from amilaceous and lignocellulosic material (second generation ethanol), and biodiesel production from African palm oil.

The comparative study includes all production stages, from the harvest to the final product (biofuel). For sugar and ethanol production from sugarcane the following steps were considered: growing and transport of sugarcane, milling, juice clarification,

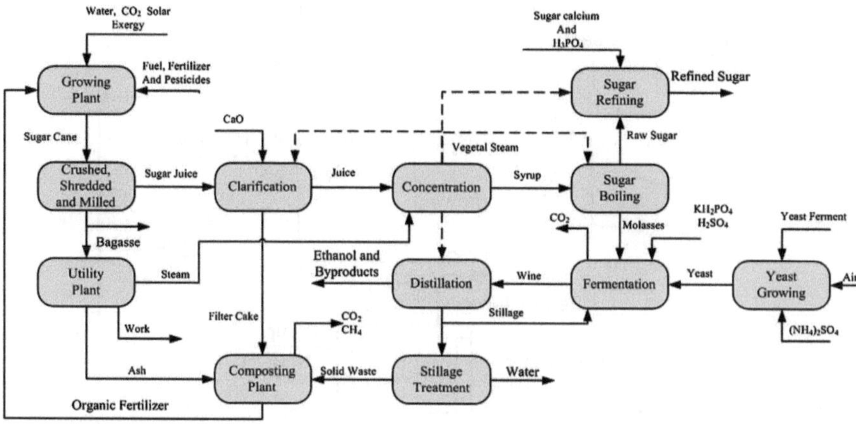

Fig. 7.1 Scheme of sugar and ethanol production process from sugarcane [14]

concentration, sugar boiling and refining, fermentation, distillation, and dehydration. Ethanol production from amilaceous and lignocellulosic material takes into account: growing and transport of biomass, pretreatment, hydrolyses, purification, fermentation, distillation, and dehydration. Biodiesel from palm oil considers: growing and transport of palm fruits, oil extraction plant, biodiesel production, and purification. Also, the utilities plant responsible for the generation of steam and electromechanical power required in the processes and residues treatment are evaluated.

7.2 Ethanol Production Process from Sugarcane

Sugar and ethanol production stages from sugarcane are shown in Fig. 7.1. The scheme is based on a specific plant located in Colombia. A total of 120 t/ha year of sugarcane are produced, and 60 t/ha year of residual biomass as leaves, and other lignocellulosic material are left on the land as protecting material.

Sugar and ethanol production can be separated into five control volumes, as presented in Chap. 6: extraction system, juice treatment, sugar production, ethanol production, and cogeneration plant.

In the ethanol production control volume takes place the fermentation process, whereby yeast, subjected to anaerobic conditions, modifies its metabolic route to convert sugars into ethanol as shown in the following chemical irreversible path [15]:

$$C_{12}H_{22}O_{11} + H_2O \rightarrow 2C_6H_{12}O_6 \rightarrow 4C_2H_5OH + 4CO_2$$

Sucrose in presence of enzymes absorbs water and splits into reducing sugars (glucose and fructose) which are finally converted into ethanol releasing CO_2. The theoretical fermentation reaction yield is 51 %; however, it is only possible to reach between 89 and 91 % of this theoretical conversion. Furthermore, during the

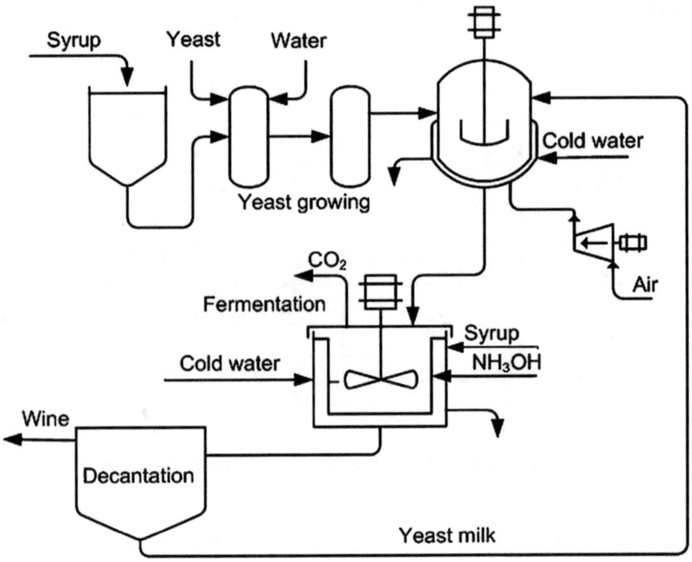

Fig. 7.2 Scheme of the sugar fermentation process

fermentation process other compounds are produced, such as: aldehydes, heavy alcohols, fatty acids, and residual biomass.

As shown in Fig. 7.2, the fermentation process is divided in two parts: yeast growing and syrup fermentation. Yeast growing requires an initial syrup supply and a constant oxygenation to guarantee aerobic conditions. During the fermentation process about 2 % of the molasses is used for yeast growing in aerobic conditions.

Additionally, agitation and refrigeration are required to maintain a constant temperature in the reactor (33 °C).

Syrup fermentation is accomplished under anaerobic conditions with constant agitation and maintaining temperature between 32 and 35 °C. After fermentation, the mixture is decanted, and the separated wine is send to distillation while the yeast milk is returned to the process. The nature of the syrup which is fed to the fermentation process depends on the raw material. It is composed of glucose and water when comes from the hydrolysis of starch, or sucrose and water in the case of sugarcane.

Ethanol at 96 % w/w is produced in the distillation process. Normally two distillation columns are used and some by-products as aldehydes and heavy alcohols are recovered. Stillages (water together with other by-products) are separated, and then about 70 % of this liquid mixture is sent again to the fermentation process for increasing the process efficiency. Finally, the stillages are carried to the stillage treatment plant where the water is treated and the solids are separated and sent to the composting plant, where they are mixed with ashes and the filter cake to obtain an organic fertilizer. At the end of the process the product is dehydrated using molecular sieves to produce anhydrous ethanol at 99.8 % w/w.

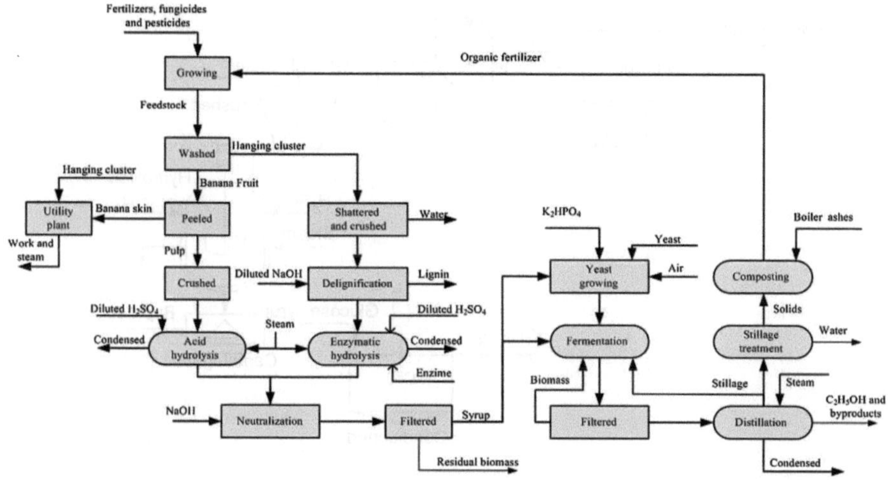

Fig. 7.3 Scheme of the ethanol production process from using starch and cellulose as feedstock [14]

7.3 Ethanol Production from Amilaceous and Lignocelullosic Biomass

Developments in bioprocesses allowed the use of amilaceous and lignocellulosic materials to produce ethanol through hydrolysis, fermentation, and distillation [2, 16, 17].

In Colombia, banana fruit surplus production amounts to 850,000 t/year and it is generated 1,150,000 t/year of associated residual biomass [18]. This material is considered biomass waste. In some farms it is treated in composting plant, but generally there is not adequate practice for its use originating environmental problems.

Looking for solving this problem it has been proposed to use the banana fruit surplus and residual biomass to produce ethanol. Four production routes were analyzed according to the biomass used as feedstock: banana pulp, banana fruit, hanging cluster, and banana skin.

The stages of ethanol production from banana fruit and its biomass residuals are present in Fig. 7.3. A total of 13.4 t/ha of dry biomass is produced, but only residual banana fruit and the clusters support are used as feedstock to produce ethanol. Two producing routes for hydrolysis reaction are studied:

1. The banana fruit is peeled and the banana pulp is subjected to acid hydrolysis, taking advantage of amilaceous material, while the banana skin is used in boilers as fuel.
2. The clusters support is subjected to enzymatic hydrolysis, taking advantage of lignocellulosic material.

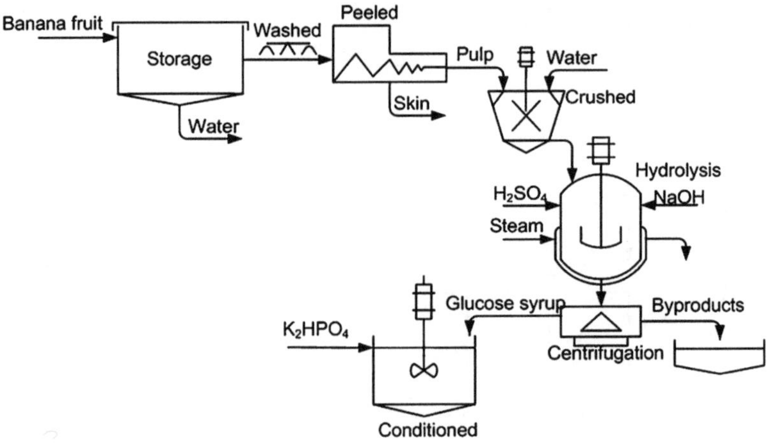

Fig. 7.4 Scheme of the starch acid hydrolysis process

Hydrolysis is a chemical or biochemical process which allows the production of reducing sugars from starch and lignocellulose. It is an indispensable and intermediate step in ethanol production, since microorganisms that promote fermentation are not able to directly metabolize the original raw materials.

Hydrolysis can be carried out in two ways: acid (chemical route) or enzymatic (biochemical route) [19–21]. Because of its low cost and availability, sulfuric acid (H_2SO_4) is most often used in acid hydrolysis. Hydrochloric (HCl) and nitrous (HNO_2) acids are alternatively used. Enzymes commonly used in enzymatic hydrolysis are α-amylase and cellulases [22, 23]. In general, cellulose is converted into glucose, and hemicellulose into pentose and hexose [24, 25]. Hydrolysis can be represented by the following reaction [25]:

$$(C_6H_{10}O_5)n + nH_2O \xrightarrow{\text{acid or enzymatic medium}} n\,C_6H_{12}O_6$$

Figure 7.4 shows the different production steps, from raw material reception to sugar syrup production, that biomass has to undergo in order to transform its starch content into sugars by acid hydrolysis. When banana fruit is used as a raw material, it is possible to process the entire fruit or only its pulp. In the former case, the fruit is chopped and crushed. In the later case banana fruit has to be peeled. Then, the feedstock is ground and water is added to it until acquiring a proper consistency for the reaction. This is a critical step since it implies heat and mechanical work consumption.

In the acid hydrolysis reaction, diluted sulfuric acid is added and the mixture is stirred and heated by steam during 10 h at 100 °C. After that time interval the syrup obtained is neutralized using NaOH which forms Na_2SO_4. Then, the mixture is filtered by centrifugation, and the syrup and residues are separated. The syrup is conditioned for fermentation with proteins and minerals as K_2HPO_4.

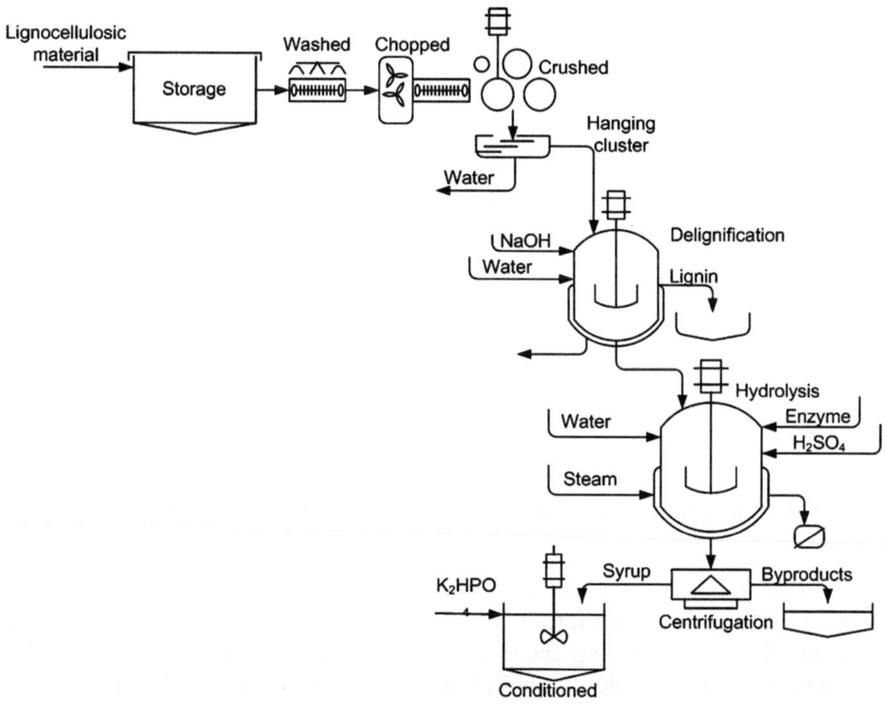

Fig. 7.5 Scheme of the lignocellulosic material enzymatic hydrolysis process

During the acid hydrolysis diluted H_2SO_4 is used for reducing the pH of the mixture which is shaken and heated by steam until 100 °C. After 6 h, about 95 % of the starch chains are transformed into glucose [18].

Figure 7.5 shows the different stages that biomass has to undergo in order to transform its cellulosic material content into sugars by enzymatic hydrolysis. The lignocellulosic material is shattered and crushed before passing through a delignification process which is carried out at ambient temperature using NaOH to increase the pH. Then, the material is hydrolyzed by adding sulfuric acid and the enzyme for 5 h at 50 °C. The lignin is a by-product that can be sold as agglutinative agent or as feedstock for the food animal industry. Finally, the mixture is also neutralized and filtered before fermentation.

During enzymatic hydrolysis organic enzyme is used as an agent for obtaining glucose and diluted H_2SO_4 is employed for reducing the pH of the mixture which is shaken and heated until 50 °C. After 10 h of treatment, approximately 70 % of cellulosic material is transformed into glucose.

The mixture is also neutralized and filtered before being prepared for fermentation. The syrup can be marketed as sweetener or used as raw material to produce ethanol.

The fermentation, distillation, and dehydration processes for ethanol production are carried out in a similar way as ethanol is obtained from sugarcane. Electromechanical energy and steam are generated in the utilities plant using banana skin and another hanging cluster as fuel, with similar parameters of sugarcane mills.

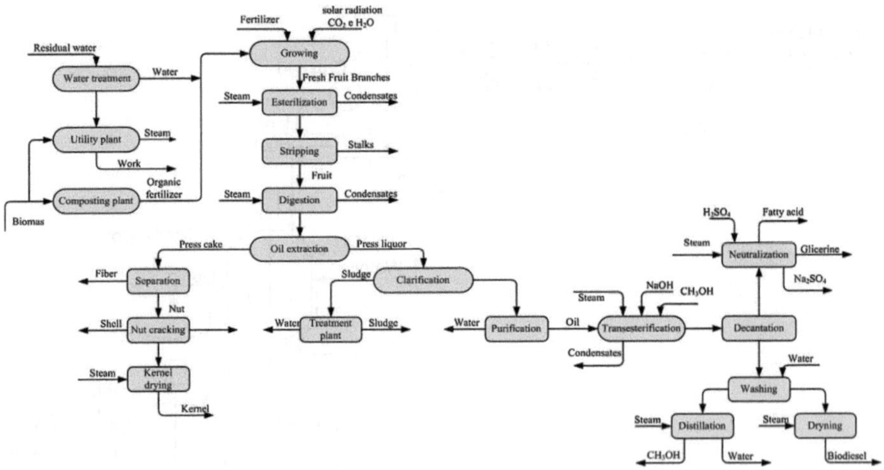

Fig. 7.6 Scheme of biodiesel production process from African palm oil [14]

Residual biomass generated during hydrolysis process is sent to the cogeneration plant to produce steam to be used in backpressure turbines. This equipment is responsible for supplying the electromechanical demands of the plant. The cogeneration plant is made of a boiler that generates steam at 22 bar and 333 °C, that is expanded in backpressure turbine till 2.6 bar with isentropic efficiency of 70 %. The electromechanical energy produced is for internal use only. A boiler working at 10 bar is used to fulfill the thermal requirements of the process and the condensate returns to the boilers.

7.4 Biodiesel Production Process

Biodiesel production stages are shown in Fig. 7.6. Biomass production is around 30–36 t/ha, with a 75 % of fresh fruit bunches (FFB). The biodiesel production involves three control volumes: palm oil milling, biodiesel production plant, and utilities plant. Palm oil milling involves the following steps:

- Fruit Reception: in order to obtain good-quality palm oil, it is essential that the damage to the fruit be minimal and therefore the handling of the fruit bunches (FFB) from the field to the sterilizers must be carried out with high care.
- Sterilization: it is carried out by placing the sterilizer at a steam pressure of 2.6 bar during approximately 60 min. The objectives of sterilization are: prevention of further rises in the free fatty acid (FFA) of the oil due to enzymatic reaction; facilitation of mechanical stripping; preparation of the pericarp for subsequent processing, and preconditioning of the nuts to minimize kernel breakage.

- Stripping: its objective is the separation of the sterilized fruit from the bunch stalks.
- Digestion: its objectives are to reheat the sterilized fruits, to loose the pericarp from the nuts, and to break the oil cells before passing to the oil extraction unit. The best digestion conditions are obtained by mixing the fruits at a temperature between 95 and 100 °C for approximately 20 min. Heating is done from direct steam injection.
- Oil Extraction: oil extraction is generally carried out using continuous screw presses comprising a perforated horizontal cage in which two screws or worms run. There are two products from the press: a mixture of oil, water, and solids; and a press cake containing fibers and nuts.
- Clarification: the crude oil from the press has an average composition of 66 % oil, 24 % water, and 10 % non-oily solids (NOS). The crude oil is screened to remove fibrous materials and then pumped to a continuous settling tank where it is separated into two parts: oil and sludge.
- Nut and Fiber Separation: when the oil is extracted from the digested fruit, a cake of nuts and fiber is produced and fed, via a conveyor, to a vertical column having an upward airflow at a velocity of 6 m/s. At this velocity all the fiber is moved upward or held in suspension, and the nuts drop to the bottom of the column. The fiber is led to a cyclone for use as a boiler fuel while the nuts pass to a rotating polishing drum installed at the bottom of the column.
- Nut and kernel treatment: this treatment covers four distinct operations: nut conditioning, nut cracking, kernel-shell separation, and kernel drying.

Figure 7.7 shows a scheme of a biodiesel or methyl ester (ME) production plant. The data used in this study came from a pilot plant designed to test several raw materials with a capacity to process 1 t of oil per day.

The first step is the mixing of methanol with the selected catalyst (NaOH). For refined palm oil, composed mainly by triglycerides (TG), a 6:1 molar ratio of methanol to oil (100 % excess alcohol) and a 0.6 % by weigh of NaOH were used. In the case of crude palm oils having free fatty acid (FFA) contents in the range 3–5 % by weight, it is necessary to increase the alcohol excess (12:1 molar ratio) and to use an additional quantity of catalyst required to neutralize the FFA:

$$\text{FFA} + \text{NaOH} \rightarrow \text{Soapstock} + H_2O$$

The second and main step is the transesterification reaction:

$$\text{TG} + 3CH_3OH \Leftrightarrow 3\text{ME} + \text{G}$$

The alcohol–catalyst mixture is combined with palm oil in the reactor and agitated for 1 h at 60 °C. Once the reaction is completed, the reactor content is separated in two phases, one rich in ME and the other in glycerol (G). The separation step can be promoted by gravity using a settling vessel and/or by centrifugation. The lighter ME-rich phase can also contain catalyst and free glycerol traces; variable concentrations of bonded glycerol, monoglicerydes, and diglycerides (depending on the reaction yield); soaps (proportional to the oil FFA

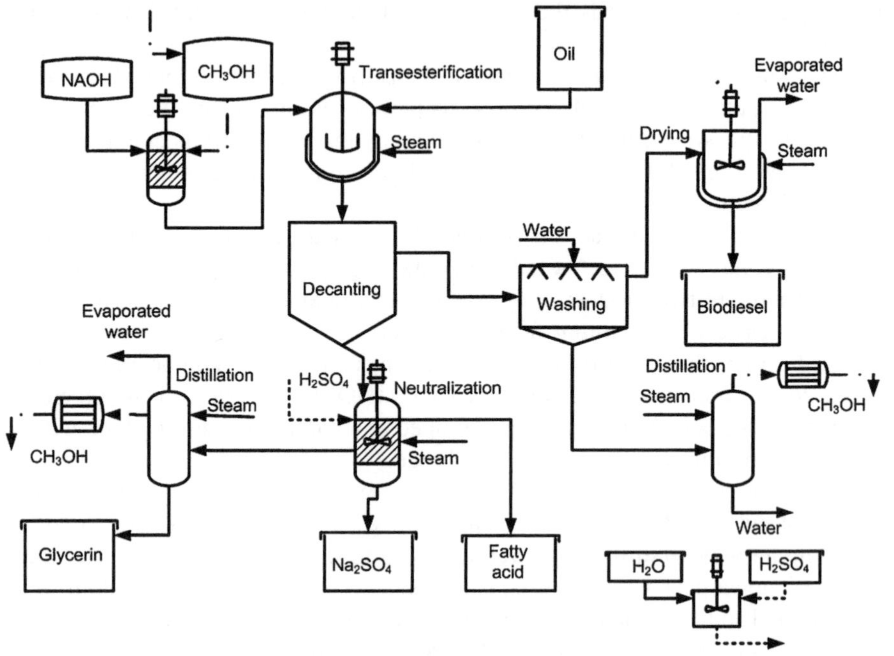

Fig. 7.7 Scheme of the biodiesel production process

content); and a substantial amount of the excess methanol. On the other hand, the denser rich glycerol phase contains most of the catalyst used and soap formed, the rest of the excess methanol, and any water formed in the occurring secondary reactions.

After separation from the denser phase, the ME-rich phase is washed gently with fresh water. In this step, it is necessary to guarantee a close contact between water and the washed phase in order to remove almost all the methanol present. This removal is favored by the chemical affinity between water and methanol. The water also removes soaps formed by dissolution. Following the washing step, any remaining water is removed from the ME phase by a vacuum flash process or a normal distillation. Once dried, the biodiesel can be sent to storage. On the other hand, the used water must be treated in order to be reused in the process or to be disposed adequately and specially for recovering the methanol.

The denser phase is only about 50 % glycerol and so it has little value and disposal may be difficult. Also, the methanol content requires the glycerol to be treated as hazardous waste. The glycerol refining step begins with the addition of a diluted acid, such as phosphoric or sulphuric one, to split the soaps into FFA and salts. The added acid also neutralizes the catalyst present. This neutralization step requires heating and mixing. The FFA is not soluble in the glycerol and will rise to the top where it can be removed. The salt precipitates out and can be filtered and dried. The methanol and water in the glycerol are removed by evaporation.

7.5 Modeling Approach and Simulation of Biofuels Production Processes

The developed model aims at describing the steady state operation of the four biofuels production routes. It is composed of mass, energy and exergy balances, heat and mass transfer equations, and thermodynamic properties correlations.

For sugar and ethanol production processes derived from sugarcane, the model developed by Pellegrini and Oliveira [26] was used. This model has already been used to evaluate different configurations of cogeneration systems in Brazilian sugarcane mills [27]. Thermodynamic properties of sucrose-water solutions were calculated according to the correlations given in [28]. Exergy of ethanol–water solutions were taken from Modesto and Nebra [29].

For organic compounds whose elemental compositions are known, b_{ch} is calculated as a function of the lower heating value (LHV) and the elementary composition [30] as given by Eq. 7.1

$$b_{ch} = \beta \, \text{LHV} \tag{7.1}$$

For a solid material having exclusively atoms of carbon, hydrogen, and oxygen, the value of β can be calculated according to Eq. 7.2:

$$\beta = \frac{\left(1.0438 + 0.1882\frac{\text{H}}{\text{C}} - 0.2509\left(1 + 0.7256\frac{\text{H}}{\text{C}}\right)\right)}{\left(1 - 0.3035\frac{\text{O}}{\text{C}}\right)} \tag{7.2}$$

The elemental composition of different kinds of biomass (palm oil fiber, sugarcane bagasse, banana fruit, banana skin, hanging cluster of banana brunch), higher and lower heating values (HHV and LHV), necessary to develop the exergy analysis, were obtained by experimental analysis carried out at the Thermal Laboratory in National University of Colombia, and they were analytically corroborated using expressions proposed in literature [31, 32].

The composition of palm oil, biodiesel, and kernel oil were obtained by chromatographic analysis and its properties were calculated using the Joback method of contribution groups [33].

The thermodynamic properties and chemical exergy of other substances like: NaOH, H_2SO_4, Na_2SO_4, CaO, CH_3OH, and KH_2PO_4, were obtained from different bibliographic sources [30, 34–36].

Technical parameters needed for sugar and ethanol production from sugarcane were taken from a Colombian sugarcane mill with a milling capacity of 3 million tons per year. The extraction oil plant was modeled using technical parameters of a Colombian palm oil extraction plant and the technical parameters for biodiesel production were obtained from a biodiesel pilot plant built at the National University of Colombia. The technical parameters for ethanol from amilaceous and lignicellulosic materials were obtained from analyses carried out at the Bioprocesses Laboratory of thc National University of Colombia, based on the design of a pilot plant with capacity of processing 4,000 kg of material per day.

This model was implemented and simulated in EES® software [13] by using its data base of thermodynamic properties for H_2O, CH_3OH, C_2H_5OH, and ideal gases such as CO_2, H_2O, O_2, CO, N_2, CH_4.

7.6 Exergy Evaluation of Biofuels Production Processes

For the sake of evaluating the exergy performance of biofuels production routes, the energy conversion processes that take place during each one of the routes can be classified in two types, in order to identify products and exergy consumptions:

- Separation of substances: as sugarcane milling, sugarcane juice concentration and boiling, sugar refining, palm oil extraction, and ethanol distillation. In these processes water steam and mechanical work are used to separate substances. For example, in sugarcane milling, mechanical work is used to separate sugarcane into bagasse and juice. In concentration operations, steam and mechanical work is used to concentrate the sugar juice, getting the vegetable steam and hot water used in other plant processes.
- Chemical reactions: as hydrolysis of banana pulp and hanging cluster, sugar fermentation, oil transesterefication, and combustion reaction in boilers. For example, in the hydrolysis process the syrup is the product, the biomass the raw materials, and steam and work are required to drive the process. In a utilities plant steam and work are the products and the biomass used is the fuel.

With these distinctions, the exergy evaluation of a given biofuel production processes is carried out considering the products obtained (B_p) or the useful exergy and the exergy required in the production processes (B_u), as shown by Eq. 7.3:

$$\eta_b = \frac{B_p}{B_u} \tag{7.3}$$

The results of the exergy efficiency of the biofuels production processes are shown in Fig. 7.8

The results shown in Fig. 7.8 can be explained by the following reasons:

- The hydrolysis of banana pulp exhibits a better exergy efficiency (57.4 %) than the hydrolysis of the hanging cluster (20.3 %) due especially for the higher content of amilaceous material in banana pulp (80.2 %) with relation to cellulosic material in hanging cluster (40.9 %) [38, 39].
- The fermentation process exhibits similar performance for all studied raw materials because the process conditions are similar for the three cases. Nevertheless, the exergy efficiency is higher when sugarcane is used since the sugar molecule for fermentation is sucrose (the syrup sent to fermentation in the case of the hydrolysis of the amilaceous and lignocellulosic materials is glucose).
- The lower exergy efficiencies correspond to the utilities plant due to the high exergy destroyed in the boilers. This can be explained by two well-known factors:

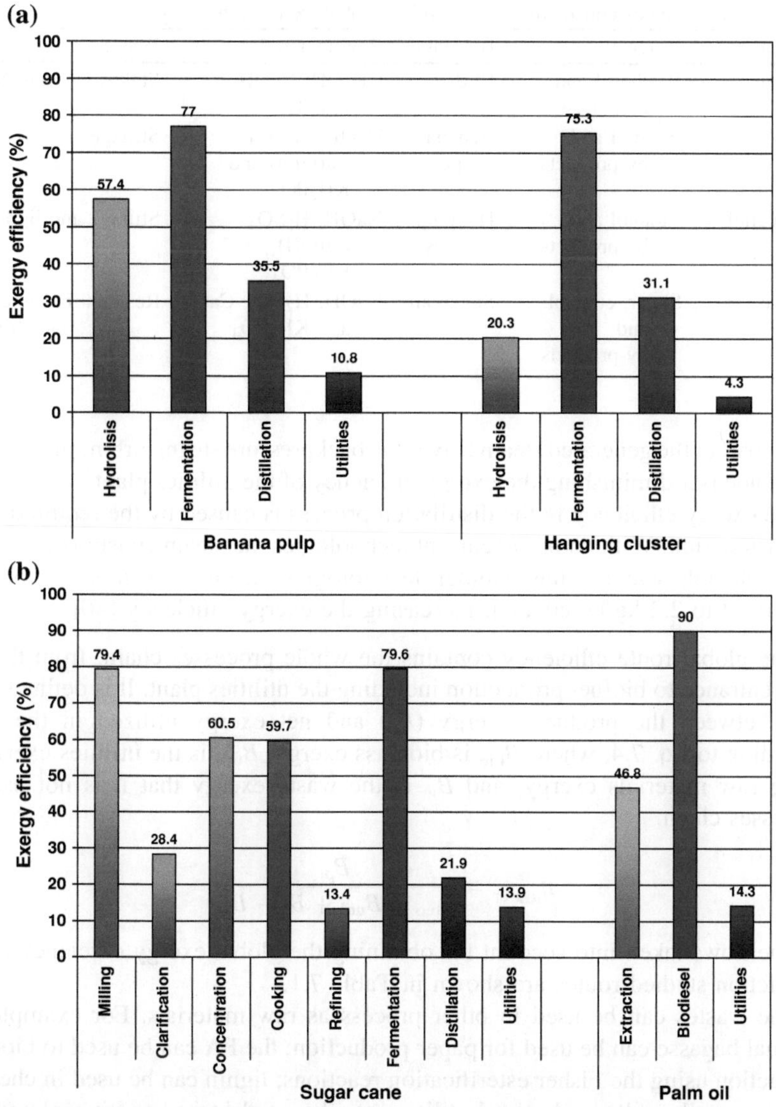

Fig. 7.8 Exergy efficiency of biofuels production processes. **a** Amilaceous material in banana pulp and lignocellulosic material in hanging cluster of banana to produce ethanol. **b** Sucrose in sugarcane to combined production of sugar and ethanol, and palm oil to produce biodiesel [37]

high exergy destruction in combustion reaction and high temperature difference between the combustion gases and generated steam. The worst case is for ethanol production using banana hanging clusters, because the amount of steam used for mechanical work consumed in the plant is higher than the amount of steam needed in the production chain processes (hydrolysis and fermentation), which implies

Table 7.1 Substances considered for obtaining global exergy efficiency

Substances	Products	Biomass	Raw material	Waste
Palm oil	Biodiesel and kernel	FFB	CH_3OH, NaOH e H_2SO_4	Stalks, FA, and Na_2SO_4
Starch	Ethanol and by-products	Banana pulp	NaOH, H_2SO_4, $Ca(OH)_2$ and KH_2PO_4	Stillage
Lignocellulose	Ethanol and by-products	Hanging cluster	NaOH, H_2SO_4, $Ca(OH)_2$, and KH_2PO_4	Stillage and lignin
Sucrose	Sugar, ethanol, and by-products	Sugarcane	NaOH, H_2SO_4, CaO, and KH_2PO_4	Residual bagasse, filter cake, and stillage

that part of the generated steam leaves the backpressure steam turbine and is sent to a condenser diminishing the exergy efficiency of the utilities plant.

- The exergy efficiency in the distillation process is caused by the required steam consumption. With common current technologies the steam consumed is 3.8 kg/kg ethanol, while using another technologies steam consumption could be reduced to 2.2 kg/kg ethanol, increasing the exergy efficiency [40].

The global route efficiency contains the whole processes chain, from the biomass entrance to biofuel production including the utilities plant. It is defined as the ratio between the products exergy (B_p) and net exergy utilized in the route, according to Eq. 7.4, where B_{bio} is biomass exergy, B_{util} is the utilities exergy, B_r is the raw materials exergy, and B_w is the waste exergy that it is not used on processes chain.

$$\eta_{\text{b,global}} = \frac{B_p}{B_{\text{bio}} + B_{\text{util}} + B_r - B_w} \tag{7.4}$$

The flows taken into account for obtaining the global exergy efficiency for the production studied routes are shown in Table 7.1.

The wastes can be used in other process as raw materials. For example: the residual bagasse can be used for paper production; the FA can be used to biodiesel production using the Fisher esterification reactions; lignin can be used in chemical industry; stalks, filter cake, and stillage can be used in composting plants. The results for global exergy efficiency are presented in Table 7.2.

The exergy efficiency of biodiesel production was obtained by considering only biodiesel as the product. If glycerin is also taken as a product, the exergy efficiency increases to 96.3 %.

When African palm oil is used a high exergy efficiency is obtained due to various factors [41]:

- The lower exergy destroyed in oil extraction (345.3 MJ/t-FFB) and in the biodiesel production plant (221.1 MJ/t-FFB) in comparison with other processes:

Table 7.2 Global exergy efficiency for biofuel production routes

Biomass	$\eta_{B,Global}$ (%)
FFB of palm oil	74.7
Sugarcane	45.5
Banana pulp (starch)	35.1
Banana hanging cluster (lignocellulose)	12.2

in banana pulp hydrolysis exergy it is destroyed 895.6 MJ/t-pulp and in syrup fermentation 743.5 MJ/t-pulp.

- The high chemical exergy in products (10,172 MJ/t-FFB) in comparison with other processes such as ethanol production from banana pulp (2,458 MJ/t-pulp).
- The high exergy in residual biomass that can be used in other process: the stalks obtained in palm oil extraction has 2,079 MJ/t-FFB in comparison with exergy in residual bagasse (552 MJ/t-sugarcane).

When sugarcane is used to produce sugar and ethanol the best behavior in comparison with the other biomass studied (amilaceous and lignicelullosic material) is obtained. This result can be explained by the high chemical exergy in products (2,487 MJ/t-sugarcane) in comparison with 1,774 MJ/t-banana when banana pulp is used, the energetic integration of the production plant as well as the residual bagasse that can be used in another process [41].

The result obtained for global exergy efficiency in sugar and ethanol production can be compared with the value of 43.5 % obtained for sugar and ethanol plant working at similar conditions in Brazil [40].

When lignocellulosic material is used, the worst result is obtained due to various factors: the low efficiency in the hydrolysis process, the high mechanical work consumed in stirrers leading to the use of additional biomass in the utilities plant.

7.7 Renewability Analysis of Liquid Biofuels Production Routes

The evaluation of the renewability of energy conversion processes related to the conversion of biomass into biofuels is done by calculating the renewability exergy index introduced in Chap. 2. These calculations take into account the exergy of the products and the exergy required and destroyed during the phases of growing, feedstock transport, processing plant, and residues treatment. It is also considered that the non-renewable exergy required in equipment and plant construction is negligible when compared to the other exergy considered values [42–44].

The mass, volume, and exergy performance parameters, as well as the steam and work specific consumption for analyzed biofuels production routes are summarized in Table 7.3. When hanging cluster is used, the results obtained are the worst ones. This result is the combination of various factors: the moisture content of raw material is high (94 %), the cellulosic content is low (41 % dry basis),

Table 7.3 Mass and exergy performance and steam and mechanical work consumed on biofuels production processes

Biomass		Field production (t raw material/ha)	Volume/mass performance	Exergy performance MJ/ha	Steam kg/t raw material	Work MJ/t raw material
Banana (starch)		12.97	4,495 L/ha	1.042×10^5	324.7	103.4
Banana pulp (starch)		9.51	3,696 L/ha	8.570×10^4	457.5	140.0
Hanging cluster (lignocellulose)		0.37	46 L/ha	1.061×10^3	36.1	35.9
Palm oil		25.00	5,743 L/ha	2.023×10^5	611.7	93.7
Sugarcane	Ethanol	120.00	1,673 L/ha	2.977×10^5	449.3	103.8
	Sugar		14,740 kg/ha			

the hydrolysis efficiency is also low (55 %), and low productivity per hectare [25]. Currently, this production route is energetic unfeasible since the use of steam and work is greater than the energy content in the ethanol produced. The results for the routes that use banana fruit and banana pulp shows that the best is the one that employs the starch in banana pulp for ethanol production and utilizes the banana skin as the boiler fuel. If only sugarcane is used to produce ethanol, it can be obtained 10,800 L/ha [45] showing that this is the higher mass performance, explained for the best biomass production (120 t/ha).

When sugarcane is used to produce sugar and ethanol, it is obtained the highest exergy performance, due to the high mass production and sugar specific exergy. If sugarcane is used only to produce ethanol, the exergy performance result is 2.589×10^5 MJ/ha.

For the calculation of the renewability exergy index for each biofuel production, some considerations were made, and the values of the terms involved in Eq. 2.98 are shown in Table 7.4.

- Raw materials used in growing are: $(NH_3)_2PO_4$, KCl, NH_3SO_4, B, Zn, Mg, and $C_{19}H_{39}NO$, diesel oil and gasoline consumed in growing and biomass transport, are also evaluated.

Fossil fuel consumption is considered as deactivation exergy in composting plant. In ethanol production from banana waste, the solid stillage and boiler ashes are mixed; in ethanol from sugarcane solid stillage, filter cake, residual bagasse, and boiler ashes are mixed and in biodiesel production stalks, fiber fruit shell nut, sludge, Na_2SO_4, and boiler ashes are mixed, all to produce an organic fertilizer, decreasing the use of mineral fertilizers.

- CH_4 produced in the composting plant is burned, without any use. It could be utilized to generate electricity.
- The biomass as leaves and other lignocellulosic materials are left on the field as protecting material, and they are not considered as wastes consuming deactivation exergy.

Table 7.4 Values for the terms involved in λ calculation for biofuels production (kJ/kg-biomass)

	Terms	Banana Pulp	Banana	Hanging cluster	Sugarcane	Palm oil
B_{nr}	B_p	2,458	1,774	191	2,487	10,172
	Growing	123	123	123	76	282
	Process	59	96	12	11	509
	B_{de}	12	13	2	18	173
B_W	Utilities plant	352	349	119	252	281
	Composting plant	474	502	58	468	3,408
	Fermentation	36	26	3	49	–
	Others	–	–	–	–	685
	B_{dest}	4,938	4,828	1,207	2,626	3,134

- The CO_2 emissions in fermentation process, composting plant, and combustion gases in boilers are considered as wastes and are evaluated with their exergy value.
- As a first approach, the contaminated water by fertilizer and pesticides is not taken into account, due to the lack of information for water treatment.

In each one of the production routes, the following considerations were made:

or

Ethanol from sugarcane

- The products considered are: sugar, ethanol and as by-products, aldehydes, and heavy alcohols.
- As raw materials in process plant were considered: NaOH, H_2SO_4, CaO, and KH_2PO_4.
- Sugarcane bagasse is used as boiler fuel.

Ethanol from amilaceous and lignocellulosic material

- The products considered are ethanol and as by-products, aldehydes, and heavy alcohols.
- As raw materials in process plant were considered: NaOH, H_2SO_4, and KH_2PO_4.
- The residual biomass obtained in hydrolysis is used as boiler fuel and lignin extracted in enzymatic hydrolysis is considered as waste.

Biodiesel

- The products considered are: biodiesel and kernel oil.
- As raw materials in process plant were considered: CH_3OH, NaOH e H_2SO_4
- As waste are considered the FA, glicerine.
- The fiber from palm oil fruit is used as boiler fuel.

It can be observed in Table 7.4, that there are three quantities that have higher values: exergy in products, gaseous waste emissions, especially emissions in composting plant, and the exergy destroyed. These are the most important terms that affect the renewability exergy index results.

Table 7.5 Renewability exergy index and global exergy [37]

Biomass	ηb Global	λ
Banana pulp	35.1	0.41
Banana fruit	24.5	0.30
Hanging cluster	12.2	0.13
Sugarcane	45.7	0.72
Palm oil	74.7	1.20

The non-renewable exergy in raw material used in growing of biomass and biofuels plant production have similar values except for biodiesel production, where CH_3OH is used.

The exergy of waste produced in the utilities plant is composed of the boiler gases, the composting plant produces the CH_4 that could be used for electric generation, the fermentation waste is the CO_2 produced, and other terms considered the glycerin and FA to biodiesel production. These substances can be raw materials for other production process but currently are only wastes.

Although exergy is destroyed in all production processes, between 75 and 83 % is generated in utilities plant. This term represents the inefficiencies on process production, and is the higher term considered in λ results.

The results for λ and global exergy efficiency are showed in Table 7.5 for all studied production process.

The results show a direct relationship between the global exergy efficiency and the renewability exergy index due to the strong influence of the destroyed exergy and the exergy of the emissions of the processes in the value of the index.

The results obtained when sugarcane is used to sugar and ethanol production are similar to those obtained for Brazilian plant production of 43.5 % for global efficiency and 0.66 for Renewability Indicator [12]. This is the best result for ethanol production showing that sugarcane is the best raw material because of the high quantity of products that are obtained and less exergy is destroyed in the processes chain. Nevertheless, the index can present a better result when the exergy of emissions leaving the control volume are used instead of being destroyed. For instance, if the release of CH_4 produced in composting plant is used to as fuel of power generation plant with exergy efficiency of 30 %, λ can reach 0.90.

Furthermore, as presented in Chap. 6, Pellegrini et al. [12] showed that it is possible to increase both the global efficiency and the renewability exergy index of the mill if there is a better exergy performance of the cogeneration plant, allowing the generation of excess electricity to be sold to the grid by using supercritical steam cycle and BIGCC systems. With these systems it is possible to attain λ values higher than one (Fig. 6.13).

When lignocellulosic material from hanging cluster is used, the lowest λ values are obtained due to the lower efficiency in enzymatic hydrolysis and the higher consumption of steam and mechanical work.

When amilaceous material in banana fruit is used, the best results are obtained for banana pulp because the acid hydrolysis efficiency and amilaceous content in biomass are higher, but it is necessary to improve the process looking for better results.

Table 7.6 Global efficiency and renewability exergy index for improved ethanol production processes [46]

Biomass	$\eta_{b\ Global}$ (%)	λ
Banana pulp	46.5	0.68
Banana fruit	42.3	0.45
Banana skin	17.1	0.23
Hanging cluster	22.3	0.34

Examples of possible improvements in the ethanol production from lignocellulosic and amilaceous materials are:

- The acid hydrolysis time could be reduced to 3 h at 50 °C, or 6 h for enzymatic hydrolysis (half the current time). In this case, work and steam requirements will be diminished.
- The water used for diluting the biomass for hydrolysis can be reduced by a factor of two. If this improvement could be done, it will reduce H_2SO_4, NaOH, power, and steam consumptions.
- The steam consumed in distillation can be reduced from 3.7 up to 2.4 kg steam/kg ethanol.
- The conventional boiler used to generate steam for electromechanical power generation could be changed for a boiler producing steam at 120 bar and 510 °C.

If such process modifications were made, the λ and $\eta_{b\ Global}$ values shift to those shown in Table 7.6.

Although the results are better than the previous ones, the ethanol production processes from lignocellulosic and amilaceous material are still non-renewable according to the renewability exergy index.

Among all simulated processes, the best results are obtained when palm oil is used to produce biodiesel, due to the higher exergy in the products and less exergy destroyed in oil extraction and biodiesel production. This is the unique process that can be considered as renewable because the result of the renewability indicator is higher than one. Although, ethanol from sugarcane presents a higher yield of exergy per hectare.

7.8 Concluding Remarks

The exergy evaluation of the biofuels production routes shows that chemical reactions, such as hydrolysis, fermentation, and combustion, are the main causes of exergy destruction and this is main factor affecting the renewability of biofuel production processes.

The global exergy efficiency of palm oil used to produce biodiesel is high due to the characteristics of the transesterification chemical reaction, low exergy destroyed in oil extraction, and the biodiesel high chemical exergy in relation to the exergy consumed in its production.

Sugarcane exhibits the better global exergy efficiency results for ethanol production in comparison with banana pulp or banana hanging cluster.

New researches are necessary for improving the results obtained when starch in banana fruit or lignocellulosic material from banana production is used for ethanol production.

When the waste produced in a given process is treated or used as raw materials it improves the values of λ. For the studied processes, it is shown that it is important to use the CH_4 produced in the composting plant as a fuel to generate power.

The renewability evaluation using the λ indicator is a function of the control volume adopted. Only when the control volume considers all process chain involved in a production route, the effective renewability of process will be assessed.

The results obtained for λ show that although biomass is used as raw material only when palm oil is used to produce biodiesel the process can be considered renewable, due to the irreversibilities that take place along the production routes. Therefore, it is necessary to improve the global exergy efficiency to obtain better results.

References

1. Agarwal AK (2007) Biofuels (alcohols and biodiesel) applications as fuels for internal combustion engines. Prog Energ Combust 33:233–271
2. Demirbas MF (2009) Biorefineries for biofuel upgrading: a critical review. Appl Energ 86:S151–S161
3. Carraretto C, Macor A, Mirandola A et al (2004) Biodiesel as alternative fuel: experimental analysis and energetic evaluations. Energy 29:2195–2211
4. Hsieh WD, Chen RH, Wu TL et al (2002) Engine performance and pollutant emission of an SI engine using ethanol–gasoline blended fuels. Atmos Environ 36:403–410
5. Zidansek A, Blinc A, Jeglic A et al (2009) Climate changes, biofuels and the sustainable future. Int J Hydrogen Energ 34:6980–6983
6. Naik SN, Goud VV, Rout PK et al (2010) Production of first and second generation biofuels: a comprehensive review. Renew Sust Energ Rev 14:578–597
7. Ayres RU (1998) Eco-thermodynamics: economics and the second law. Ecol Econ 26:189–209
8. Rosen MA (2002) Can exergy help us understand and address environmental concerns? Int J Exergy 2:214–217
9. Talens L, Villalba G, Gabarrell X (2007) Exergy analysis applied to biodiesel production. Resour Conserv Recy 51:397–407
10. Velásquez-Arredondo HI, Benjumea P, Oliveira S Jr (2007) Exergy and environmental analysis of the palm oil biodiesel production process. In: Proceedings of the 20th International conference on efficiency, costs, optimization, simulation and environmental impact of energy systems, Padova
11. Velásquez-Arredondo HI, Benjumea P, Oliveira S Jr (2007) Exergy analysis of palm oil biodiesel production by base catalyzed methanolysis. In: Proceeding of the 19th international congress of mechanical engineering, Brasilia
12. Pellegrini LF, Oliveira S Jr (2011) Combined production of sugar, ethanol and electricity: thermoeconomic and environmental analysis and optimization. Energy 36:3704–3715

13. Klein SA (2011) Engineering equation solver—EES, F-Chart software, www.fChart.com
14. Velásquez-Arredondo HI, Ruiz Colorado AA, Oliveira S Jr (2010) Ethanol production from banana fruit and its lignocellulosic residues. Energy 35:3081–3087
15. Camargo CA (coord.) (1990) Energy conservation in sugar and alcohol. Instituto de Pesquisas Tecnológicas, São Paulo (in Portuguese)
16. Clark JH, Deswarte FEI, Farmer TJ (2009) The integration of green chemistry into future biorefineries. Biofuels Bioprod Bioref 3:72–90
17. Lange J (2007) Lignocellulose conversion: an introduction to chemistry, process and economics. Biofuels Bioprod Bioref 1:39–48
18. Bohórquez C, Herrera S (2005) Determinación de las mejores condiciones de hidrólisis del banano verde de rechazo. Facultad de Minas. Universidad Nacional de Colombia
19. Spano LA, Medeiros J, Mandels L (1976) Enzymatic hydrolysis of cellulosic wastes to glucose. Resour Recovery Conserv 1:279–294
20. Wyk Van JPH (1999) Hydrolysis of pretreated paper materials by different concentrations of cellulase from penicillium funiculosum. Bioresour Technol 69:269–273
21. Movagharnejad K, Sohrabi MA (2003) Model for the rate of enzymatic hydrolysis of some cellulosic waste materials in heterogeneous solid–liquid systems. Biochem Eng J 14:1–8
22. Jennylynd A, Byong H (1997) Glucoamylases: microbial sources, industrial applications and molecular biology-review. J Food Biochem 21:1–52
23. Cao Y, Tan H (2002) Effects of cellulase on the modification of cellulose. Carbohydrate Research 337:1291–1296
24. Mohamed AF, Hossam M, Ahmed ED (1983) Effect of peracetic acid, sodium hydroxide and phosphoric acid on cellulosic materials as a pretreatment for enzymatic hydrolysis. Enzyme Microb Technol 5:421–424
25. Nouri M (1991) Catálisis ácida vs. hidrólisis enzimática en la industria almidonera. Alimentación Equipos y Tecnología 1991:141–145
26. Pellegrini LF, Oliveira S Jr (2007) Exergy efficiency of the combined sugar, ethanol and electricity production and its dependence of the exergy optimization of the utilities plants. In: Proceedings of the 20th international conference on efficiency, costs, optimization, simulation and environmental impact of energy systems, Padova
27. Velásquez HI, Pellegrini LF, Oliveira S (2008) Ethanol and sugar production process from sugar cane: renewability evaluation. Proceedings of the 12th brazilian congress of thermal sciences and engineering, Belo Horizonte. v. p. (CD-ROM)
28. Nebra SA, Fernández-Parra MI (2005) The exergy of sucrose-water solution: proposal of a calculation method. In: Proceedings of the 18th international conference on efficiency, costs, optimization, simulation and environmental impact of energy systems, Trondheim
29. Modesto M, Nebra SA (2005) A proposal to calculate the exergy of non ideal mixtures ethanol-water using properties of excess. In: Proceedings of 14th European biomass conference, Paris
30. Szargut J, David RM, Steward F (1988) Exergy analysis of thermal, chemical, and metallurgical processes. Hemisphere Publishing, New York
31. Hugot E (1986) Handbook of sugarcane engineering, 3rd edn. Elsevier Science Publishers, New York
32. Channiwala SA, Parikh PP (2002) A unified correlation for estimating HHV of solid, liquid and gaseous fuels, fuel 81:1051–1063
33. Reid RC, Prausnitz JM, Poling BE (2000) The properties of gases & liquids. 5th edn. McGraw-Hill
34. Smith J, Van Ness HC, Abbott MM (2003) Introduction to chemical engineering. McGraw-Hill, México, D.F (in Spanish)
35. Ball DW (2004) Physical chemistry. 1st edn. Thomson, México (in Spanish)
36. Moran MJ, Shapiro HN (2006) Fundamentals of engineering thermodynamics. Ed. Jhon Wiley & Song, New York

37. Velásquez-Arredondo HI, Oliveira S Jr, Benjumea P (2009) Exergy analysis of biofuels production routes. In: Proceedings of 20th international congress of mechanical engineering, Gramado
38. Hoyos LM, Pérez YM (2005) Pretratamiento de banano de rechazo de la zona de urabá para la obtención de un jarabe azucarado. Facultad de Minas. Universidad Nacional de Colombia
39. MontesVN, Torrez CL (2004) Hodrólisis del banano verde de rechazo. Facultad de Minas. Universidad Nacional de Colombia
40. Pellegrini LF (2009) Analysis and thermo-economic and environmental optimization applied to the combined production of sugar, alcohol and electricity. Ph.D. Thesis, Polytechnic School of the University of São Paulo, São Paulo, Brazil (in Portuguese)
41. Velásquez-Arredondo HI (2009) Exergy and exergo-environmental analysis of the biofuels production. Ph.D. Thesis, Polytechnic School of the University of São Paulo, São Paulo, Brazil (in Portuguese)
42. Malça J, Freire F (2006) Renewability and life-cycle energy efficiency of bioethanol and bio-ethyl tertiary butyl ether (bioETBE): assessing the implications of allocation. Energy 31:3362–3380
43. Shapouri H, Duffield JA, Wang M (2002) United States Department of Agriculture. USDA, The Energy Balance of Corn Ethanol: An Update: In: http://www.transportation.anl.gov/pdfs/AF/265.pdf. Accessed 15 jan 2008
44. Kaltschmitt M, Reinhardt GA, Stelzer T (1997) Life cycle analysis of biofuels under different environmental aspects. Biomass Bioenergy 12:121–134
45. Velásquez-Arredondo HI, Pellegrini LF, Oliveira S Jr (2008) Ethanol and sugar production process from sugar cane: renewability evaluation: In: Proceeding of the 12th brazilian congress of thermal science and engineering, Belo Horizonte
46. Velásquez-Arredondo HI, Ruiz Colorado AA, Oliveira S Jr (2009) Ethanol production from banana fruit and its lignocellulosic residues: exergy and renewability analysis. Int J Thermodyn 12:155–162

Chapter 8
Exergy Method for Conception and Assessment of Aircraft Systems

Symbols

AMS	Air management system
B	Exergy flowrate/rate (kW)
$B_{\text{t-Fuel}}$	Total fuel exergy consumed, kJ
$B_{\text{t-inlet air}}$	Total inlet air exergy, kJ
$B_{\text{t-dest,mission}}$	Total destroyed exergy during the mission, kJ
C	Specific cost (US$/kJ)
C	Cost rate (US$/s)
C_{equip}	Equipment/System cost rate (US$/s)
ECU	Environmental control unit
F	Generic function (fuel consumption rate, exergy efficiency, SEC, SFC)
f_u	Annual utilization factor
MEA	More electric airplane
SEC	Specific exergy consumption
SFC	Specific fuel consumption, lbm/(h lbf)
W	Power (kW)
Δt_{phase}	Flight phase duration (min)
η_{b}	Exergy efficiency

Subscripts

Air	Air at the inlet of the engine
Anti-Ice	Anti-ice system
Anti-ice, inlet	Anti-ice air at the inlet of the system
B	Exergy
Bleed	Extracted bleed air from the engine, bleed system
Cabin	Cabin
Cabin,Air,out	Outlet cabin air
Cc	Combustor
Col	Collector

S. de Oliveira Jr., *Exergy*, Green Energy and Technology,
DOI: 10.1007/978-1-4471-4165-5_8, © Springer-Verlag London 2013

Comp	Compressor
Comp, bleed	Bleed air from the compressor
Dest, destroyed	Destroyed
Destr, Engine	Destroyed in the engine
Dest, Bleed	Destroyed in the bleed system
Dest, AMS (conventional)	Destroyed in the AMS of conventional airplane
Dest, AMS (MEA)	Destroyed in the AMS of more electric airplane
Dest, Anti-Ice	Destroyed in the anti-ice system
Dest, ECU	Destroyed in the environmental control unit
Dest, Cabin	Destroyed in the cabin
Dest, ElectricSystem	Destroyed in the electric system
Dest, mission	Total destroyed exergy of the mission
EAI_Air (inlet)	Engine anti-ice inlet
ECU	ECU
ECU_Air (inlet)	ECU anti-ice inlet
ECU, Inlet	High pressure air at the inlet of the ECU
ECU, Outlet	Outlet ECU air
Electric, el	Electric system
Engine	Engine
Equip	Equipment
Ex	Extracted
Fan	Fan
Fan, Air	Extracted fan air from the engine
Fan, Air, out	Extracted fan air from the engine
Fan, bleed	Bleed air from the fan
Fuel	Fuel
Norm	Normalization of function F
Gases	Gases leaving the engine
Generator	Electric generator
Global	Global
HPT	High pressure turbine
HX, air, in	Inlet ram air of the ECU heat exchanger
HX, air, out	Outlet ram air of the ECU heat exchanger
Hydraulic	Hydraulic system
I	Useful output flows
J	Input flows
Lost	Exergy loss rate
LPT	Low pressure turbine
MEA	More electric airplane
Mec_Hydraulic	Mechanical power extracted from engine to hydraulic system
Mission	Mission
Mix	Mixer
Noz	Nozzle

Q, Leading_Edge	Leading edge heat transfer of anti-ice system
Q, Heat_Transfer	Cabin heat transfer
RAM_Air	RAM air
SAI	Stabilizer anti-ice
SAI_Air (inlet)	Stabilizer anti-ice inlet
T, Thrust	Thrust
T-fuel	Total exergy of fuel
T-inlet air	Total exergy of inlet air
WAI	Wing anti-ice
WAI_Air (inlet)	Wing anti-ice inlet

8.1 Introduction

Modern aircraft are designed to attend the client needs, minimizing fuel consumption, and inefficiencies—this approach is named "traditional optimization". Most of these studies rely on rules-of-thumb, individual experience, and non-integrated, non-interdisciplinary approach of basic calculations, i.e., simple trade-off analysis [1]. Such analysis relies on cost–benefit studies among different options, but not in the same basis. Thus, it may lead to suboptimal solutions.

Many authors have criticized such analysis [2–6]. These authors agree that there must be a common basis to compare different alternatives in aircraft design, especially when developing new systems.

Moorhouse [7] states that it is possible to define aircraft design parameters (for instance, weight, fuel consumption, drag) as energy functions. Thus, the initial optimization problem may be characterized as one of minimizing additional weight and losses constrained by the energy requirements of the client (weight to be transported). In this way, the project becomes committed to maximum efficiency and minimum waste of useful energy, considering the adequate constraints—parameters of an energy optimization problem.

Energy systems become increasingly more complex, leading to higher levels of interaction between each subsystem. Consequently, there is a need to develop tools that would allow the design/synthesis of the system in complete integrated way, allowing all demands of each subsystem to be reached in the best possible way. The aeronautical industry also has evolved to design extremely complex aircraft, with highly integrated systems, requiring more information in order to evaluate the whole system [1, 7].

Exergy analysis and minimization of exergy destruction can be used mainly in areas where the total cost of the installation is dominated by the cost due to Thermodynamic irreversibilities [8]. Paulus and Gaggioli [2] state that the design of vehicles do not have the same goal of energy systems designed to provide mass flow, heat, or work transfer. Instead, they are designed to attend performance goals, and these are related to Thermodynamic inefficiencies.

Some authors [1, 2, 5, 8] suggest the use of decomposition methods, based on the work of El-Sayed and Evans [9], in order to overcome some drawbacks related to the design/synthesis of aircraft based on exergy considerations. Other authors [10–12] follow the "Entropy Generation Minimization" approach [13], a more phenomenological one. However, the demonstration of a completely optimized design of an aircraft system using exergy methods have not been documented so far [3].

8.2 Exergy Analysis in Aerospace Industry

The exergy approach has been used for some years in aerospace industry involving isolated systems and integrated aircraft-systems analysis, since several systems and processes on an aircraft contribute to the destruction of all the exergy supplied by the fuel. Analyzing the direction of fuel flow to illustrate the method, the first exergy loss (about 30 %) is due to the combustion process [10]. Associating this with the loss due to the irreversible operation of the engine, the exergy destruction in the engine can reach values up to 50 % of the total exergy destroyed [12]. In addition to propulsion system losses, aerodynamic drag and electric, pneumatic, and hydraulic power user systems generate irreversibilities. Paulus and Gaggioli [2] used exergy for the decomposition and detailed design of the subsystems and devices of a light experimental aircraft (alternator and engine). The method proposed used exergy analysis and decomposition as well as thermoeconomics to the selection of an optimum design.

Figliola and co-workers [3] developed a model for the environmental control system (ECS) of an advanced military aircraft, which includes seven integrated subsystems: (1) cold poly-alpha-olefin (PAO) loop; (2) vapor cycle system; (3) air cycle system; (4) hot PAO loop; (5) oil loop; (6) hydraulic system; (7) fuel loop. The closed-circuit liquid loop subsystem uses PAO as the liquid coolant. The model includes a methodology to analyze each subsystem, component by component, for each separate mission aspect, or as integrated over the mission. The main objectives are to reduce the entropy generation for the system, while satisfying the mission performance requirements of the ECS and takeoff weight (fuel penalty), this one as a result of the energy analysis. In order to do that, the main variables such as evaporator effectiveness, cold PAO mass flow rate, etc., were expressed in terms of ground takeoff weight. A multi-objective, optimization decision-based approach was applied in a tradeoff study between these two objectives (minimization of takeoff weight and entropy), and provided similar results.

At the same direction, Roth [4] introduces the exergy methodology and shows the exergy usage during an F-5E subsonic area intercept mission. In this case, 90 % of the total exergy is destroyed in the propulsion system and the vast majority of these propulsive losses consist of exhaust gases, irreversible combustion, and residual kinetic energy of the jet efflux left in the wake of the vehicle. The remaining 10 % of the exergy is converted into thrust work and used to overcome vehicle drag. Moreover, the author establishes the bridge between aero-thermo performance and vehicle weight as it is possible to quantify losses incurred during the mission (such as drag work, engine inefficiencies, etc.) in terms of the fuel weight required to offset those losses, over an entire mission. In order to illustrate that, it is shown a comparison of conventional gross weight breakdown for the F-5E versus the chargeable gross weight breakdown (measured based on losses in thrust work potential, instead of exergy). Also, it is possible to allocate fuel cost chargeability to the aerothermodynamics loss mechanisms. At the end, the author introduces the modeling structure at aircraft level to identify and quantify the losses in a vehicle design, assigning the losses during the entire mission.

Moorhouse [7] improves the idea of allocating losses during the mission when introduces the methodology of quantifying the mission requirements in terms of energy. The method must consider the work necessary to takeoff, climb, cruise, descent, weight variations, speed variations, etc. An exergy breakdown can be done to quantify the exergy destruction in each mission phase.

Muñoz [1] developed different decomposition strategies and applied them to the integrated optimization of two subsystems, which are part of an advanced military aircraft. The idea was to carry out the conceptual design of a low-bypass turbofan engine with afterburning and the full synthesis/design optimization of an air-cycle ECS.

Rancruel [14] applied the methodology developed by Muñoz and Spakovsky [15] to the synthesis/design and operational optimization of the advanced tactical fighter aircraft. The total system was decomposed into five subsystems as follows: propulsion system, ECS, fuel loop system, vapor compressor and PAO loops system, and airframe system of which the latter is a non-energy based subsystem.

Markell [16] compared an exergy-based methodology to a more traditional based measure by applying both to the synthesis/design and operational optimization of a hypersonic vehicle configuration comprised of an airframe subsystem and a propulsion subsystem consisting of inlet, combustor, and nozzle components. Results of these optimizations show that the exergy method performs well when compared to the standard performance measure and, in a number of cases, leads to more optimal synthesis/designs in terms of the fuel mass flow rate required for a given task.

Butt [17] applied energy and exergy-based methods to the integrated synthesis/design of an air-to-air fighter (AAF) aircraft with and without wing-morphing capability. The morphing-wing and fixed-wing aircraft models were optimized using four different objective functions: the minimization of fuel consumed, the minimization of total exergy destroyed and lost, the minimization of the exergy destroyed and lost by the propulsion subsystem (PS), and the maximization of the thrust efficiency. Out of these four objective functions, the minimization of fuel consumption and total amount of exergy destroyed and lost produced the best performance results from the optimization runs. This is due to the fact that both objective functions in essence are seeking to minimize the use of the highest quality energy present on board the aircraft, i.e., the fuel exergy.

Brewer [18] applied exergy to synthesis/design of generic hypersonic vehicles, in which tradeoffs between dissimilar technologies are observed, is proposed, and measured against traditional methods of assessing highly integrated systems.

A quasi-one dimensional hypersonic vehicle simulation model was designed and optimized over a formal mission with three objective functions: maximized thrust efficiency, minimized fuel consumption, and minimized exergy destruction plus fuel exergy loss. For the three objective functions, the minimum fuel mass, and minimum exergy destruction, and exergy fuel loss proved to be able to design and operate a vehicle which meets the mission constraints using nearly identical amounts of fuel. The optimized thrust efficiency objective vehicle consumed significantly larger amounts of fuel and destroyed more exergy while promoting a broad, thrust maximizing body.

Periannan [6] studied different objective functions based on energy and exergy considerations on the optimization of three subsystems: PS, ECS, and airframe subsystem-aerodynamics (AFS). The author concluded that an exergy-based approach is not only able to pinpoint where the greatest inefficiencies in the system occur but produces a superior optimum vehicle as well as by accounting for irreversibilities losses in subsystems only indirectly tied to fuel usage.

8.3 Exergy and Thermoeconomic Analysis of a Turbofan During a Typical Commercial Flight

8.3.1 Introduction

The power plant is the aircraft power source. It provides thrust to the airframe as well as electric and hydraulic power to drive all the aircraft equipment and subsystems, and bleed air to the air management system (AMS), which is composed of the pneumatic system, air conditioning system, ice protection system, engine starting system, and pressurization system. Depending on the configuration, the power plant consists of a piston or turbine engines, propellers or fan as well as all engine subsystems and utilities. From this point on, engine and power plant will be used interchangeably.

Most recently, the so-called 'more-electric' architectures has brought a new benchmark to energy-optimized aircraft. This architecture increases the electric energy use in order to replace the mix of hydraulic, pneumatic, and electric energy to fulfill the requirements of systems, augmenting the energy efficiency of non-propulsive systems [19].

Different studies involving commercial aircraft systems have been developed to apply exergy analysis to evaluate the replacement of the conventional architecture by 'more-electric' ones [20, 21]. Since the system energy flows are generated in the power plant, and the useful impact is expressed in terms of fuel consumption reduction, the analysis of the aircraft engine is of fundamental interest.

The focus of this study is the development of a model to evaluate the performance of the energy conversion processes inside the engine using exergy and thermoeconomic analysis as assessment tools. Previous studies have dealt with

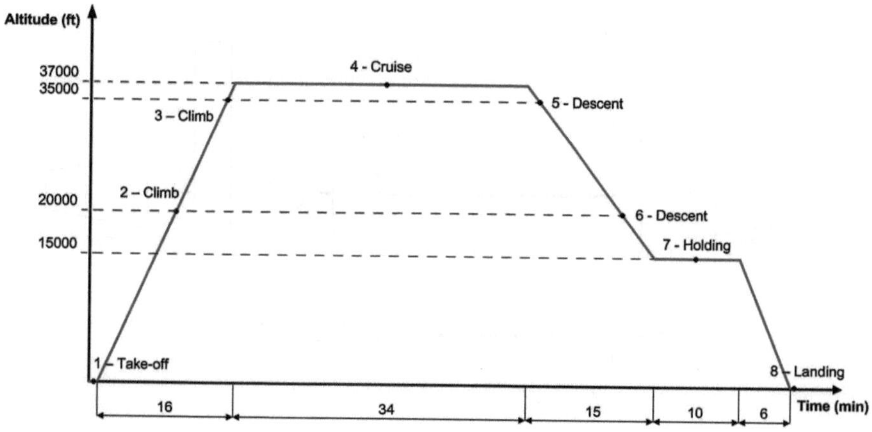

Fig. 8.1 Flight mission profile [43]

exergy analysis applied to scramjet propulsion systems [22] and turbofan with an after burner [23]; however, these studies considered one or two operational points (at sea level or in cruise conditions). Roth et al. [24] propose the use of the work potential concept to evaluate the performance of propulsion systems.

Therefore, the main goals of this approach are:

- To characterize the power plant operation evaluating the input and output exergy rates and the useful flow cost over a complete mission;
- To define the performance of engine components in each flight phase.

8.3.2 System Description and Modeling

The mission profile used in this study represents a simplified typical commercial flight as shown in Fig. 8.1. The operational points considered are takeoff, climbing, cruise, descent, holding, and landing.

To develop the engine numerical model, the GSP (Gas Turbine Simulation Program) software [25] was used in order to calculate the fuel consumption, thrust and inlet air, bleed, fan air and outlet gases thermodynamic characteristics. The power plant is a classical turbofan with separated flows, a high bypass ratio, and two shafts. Figure 8.2 shows a schematic representation of turbofan analyzed and implemented in GSP.

Bleed Air and Extracted Power Controls were implemented to simulate the different energy demands of the board systems during the whole mission. Bleed controls were located on the fan and compressor, and hydraulic, and electric power extractions were located on the turbine stages.

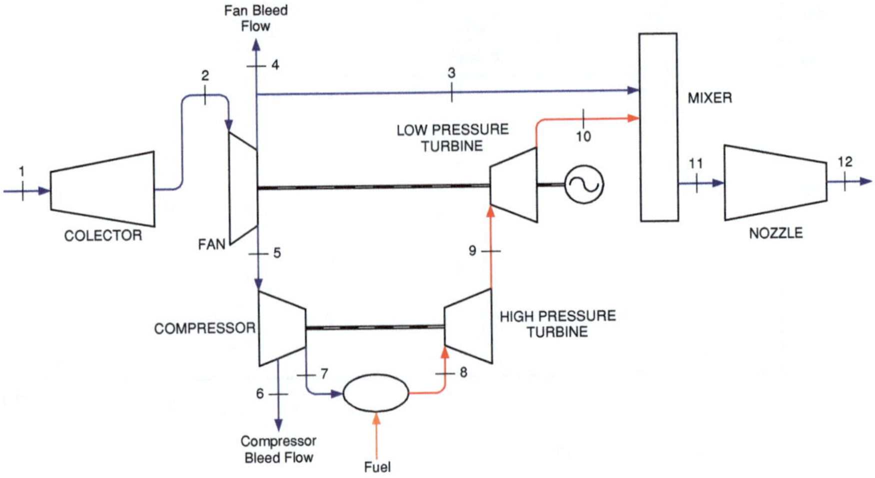

Fig. 8.2 Turbofan model [43]

Table 8.1 Flight phases characteristics with anti-ice OFF [43]

Flight phase	Height (m)	Mach number	Time duration (min)	Thrust (N)
Take-off	0	0.20	1	28,268
Climb	6,096	0.57	12	13,652
Climb	10,668	0.66	4	8,634
Cruise	11,277	0.77	34	7,713
Descent	10,668	0.77	4	2,478
Descent	6,096	0.62	11	2,509
Holding	4,572	0.40	10	13,193
Landing	0	0.20	0.5	2,095

The compressor has two bleed air conditions in order to meet the demands of the anti-ice system:

- anti-ice ON: air is extracted from the 14th stage of the compressor;
- anti-ice OFF: air is extracted from the 9th stage of the compressor.

The design point for the engine was set on cruise phase: altitude 37,000 (ft), Mach 0.77. All other phases were calculated considering an off-design simulation of the engine.

The numerical results were validated in the design point using real reference data and a mathematical model developed for this engine with Matlab® [26] software. The average error based on real reference data is 3.3 %, and the maximum error is 8.2 %. Table 8.1 shows some characteristics for each one of the flight phases.

8.3.3 Exergy Analysis

Two cases and models were considered in this study:

1. Global model: the power plant is considered a unique open system; this model allows the evaluation of the exergy destruction and loss rates for the different flight conditions.
2. Local model: the power plant is studied analyzing its equipment one by one, using local balances for each component in order to evaluate its performance and thermoeconomic cost.

To direct the study, the following assumptions were made:

1. All gases were modeled as ideal gases;
2. The fuel is kerosene (JET-A1), and its chemical formulae is $C_{12}H_{23}$;
3. The fuel physical exergy is not considered;
4. All engine equipment are considered adiabatic;
5. Heat exchangers (air–fuel, fuel–oil) are not considered.

EES® software [27] was used to implement the analysis.

8.3.3.1 Global Model

The first step to evaluate the exergy content of an energy flow is the definition of a reference state. Generally this choice is easy, because the system under evaluation is fixed upon the Earth or its altitude is constant. Analyzing a flight vehicle or a jet engine, operating in high altitude, and different environmental conditions during the flight, makes it more difficult to choose a unique reference [28]. Szargut's environment gives negative values for some exergy flows in these conditions [29]. For this reason, preliminary analysis were performed using two different reference systems, one fixed on the ground and one attached to the power plant. Thus, when the reference is fixed on the ground, Szargut's environment was considered, and, when attached to engine, the reference state was set by the stagnation outside air temperature and pressure for each flight phase. In addition, as a first approach to the exergy calculations, standard chemical exergies given by Szargut et al. [29] were used for both references. Regarding this aspect, Etele and Rosen [30] showed that the variation of the fuel chemical exergy for a ground fixed reference environment and an environment reference at 15,000 m is only 0.6 %.

Figure 8.3 presents the control volume considered in the global model for a turbofan engine.

The exergy balance for this control volume is shown in Eq. 8.1:

$$B_1 + B_{\text{fuel}} = B_{12,\text{lost}} + W_{\text{ex}} + B_{\text{comp,bleed}} + B_{\text{fan,bleed}} + B_T + B_{\text{destroyed}} \tag{8.1}$$

Table 8.2 presents exergy rates for the different flows during cruise phase for both reference systems. It can be seen that taking the reference state on the Earth

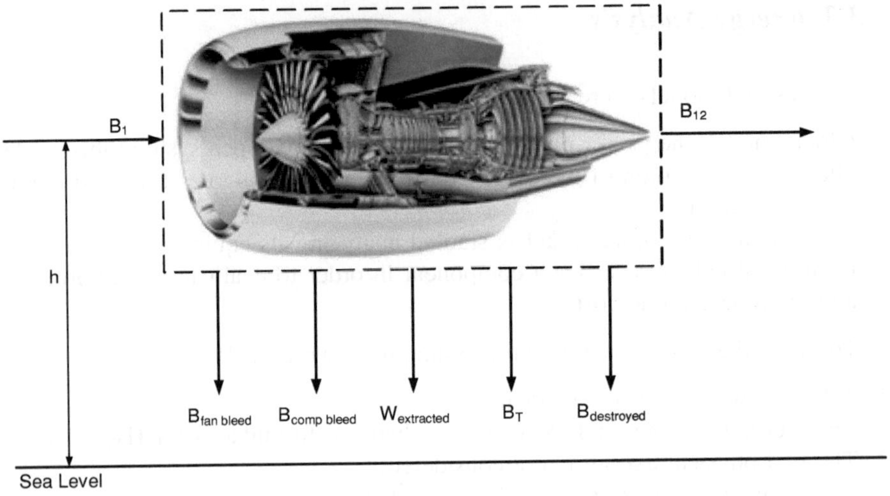

Fig. 8.3 Global model control volume [43]

Table 8.2 Exergy rates for the different flows during cruise [43]

#	Description	Exergy flow rate (kW)	
		Ground reference	Engine reference
1	Collector inlet	−202.8	588.6
2	Fan inlet	792.0	423.6
3	Fan outlet to mixer	2114.6	1862.0
4	Fan bleed	42.3	37.3
5	Fan outlet to compressor	392.8	340.4
6	Compressor bleed	43.7	42.5
7	Compressor outlet	3456.3	3490.0
8	High pressure turbine inlet	8303.6	8734.0
9	Low pressure turbine inlet	4611.1	5055.0
10	Low pressure turbine outlet	2421.0	2875.0
11	Mixer outlet	2984.5	3477.0
12	Nozzle outlet	349.4	3385.0
–	Fuel	7186.0	7168.8
–	Thrust	1745.5	1745.5
–	Fan power	2071.6	2071.6
–	Compressor power	3546.6	3546.6
–	Extracted power to hydraulic and electric systems	16.4	16.4

gives a negative value for the inlet air exergy, due to the smaller pressure of this flow compared to the reference pressure.

Furthermore, it is important to mention that, when the reference is attached to the engine, the exergy of the thrust is related to the kinetic exergy of the exhaust gases, i.e., the kinetic exergy of the exhaust gases is equal to the exergy of the

Table 8.3 Exergy rate distribution in anti-ice OFF flight conditions [43]

Phase	Destroyed exergy (%)	Lost exergy (%)	Thrust exergy (%)	Bleed and extracted exergy (%)
1	64	26	9	1
2	68	11	20	1
3	71	5	23	1
4	69	5	25	1
7	71	13	15	1
8	83	11	4	2

thrust plus an exergy destruction rate associated to the energy conversion process and an exergy loss rate associated to the exhaust gases. The term $B_{12,\text{lost}}$ in Eq. 8.1 represents this exergy loss rate. When the reference is fixed on the ground, the term $B_{12,\text{lost}}$ accounts for the exergy of stream 12 that is no further used.

8.3.3.2 Local Model

As for the local model, exergy balances are developed for each component, defining its efficiency and performance related to the power plant global values. Such analysis allows the identification of the components that contribute most for the exergy destruction inside the engine. Hence, these results might define optimization priorities among the system and direct investment to where solutions will be more cost-effective or exergy efficient.

In this analysis, the reference state for the exergy calculations was set by the stagnation air temperature and pressure for each flight phase [28]. Again, as a first approach, standard chemical exergies given by Szargut et al. [29] were used.

8.3.4 Exergy Analysis Results

8.3.4.1 Global Balance

Table 8.3 shows the distribution of output exergy flows for all the flight phases. The exergy efficiency is defined as the ratio of sum of the thrust and bleed exergies (formed by air bleed and extracted power) to the fuel exergy (Eq. 8.2):

$$\eta_b = \frac{\sum_i B_i}{\sum_j B_j} \tag{8.2}$$

The results for the complete mission are presented in Table 8.4, including anti-ice OFF and ON conditions and the reference state fixed on the ground.

Table 8.4 Exergy efficiency in each flight phase [43]

Phase	Anti-ice system	Exergy efficiency (%)
1	OFF	10.06
2	OFF	20.71
3	OFF	24.07
4	OFF	26.46
5	OFF	20.37
6	OFF	13.94
7	OFF	16.34
8	OFF	5.87
1	ON	10.55
2	ON	21.01
6	ON	14.86
7	ON	17.44
8	ON	8.14

Table 8.4 and also Fig. 8.1 show that:

- Cruise is the design and longest phase. It presents the maximum exergy efficiency (26 %), demonstrating that the engine design point is the most efficient one;
- The exergy efficiency, lower than 10 % during takeoff, increases during the mission reaching its maximum value in cruise. After the cruise, the efficiency decreases reaching its minimum value during landing (6 %);
- Exergy losses are a function of the aircraft speed and thrust. This leads to an interesting result: a better efficiency is obtained mainly by reducing exergy losses and not by decreasing the engine destroyed exergy;
- During landing phase with anti-ice system ON the exergy bleed flow value is near to thrust exergy value.

Considering the reference state attached to the engine, and that the thermochemical and part of the kinetic exergy of the outlet gases as an exergy loss, the results are very similar to the ones in Tables 8.3 and 8.4. Hence, the use of different environment references does not give any misleading results regarding the engine efficiency. However, the change in the environment reference leads to a different distribution of the irreversibilities between destroyed and lost exergy. Whenever the reference is set to the engine, the thermochemical exergy of the exhaust gases is high, increasing the exergy loss associated to this flow (since it includes the conversion of kinetic energy into thrust). On the other hand, using the reference fixed on the ground gives smaller values for the thermochemical exergy of the exhaust gases, leading to higher values of exergy destruction. All in all, the relation of total irreversibilities to the exergy input in both cases is very similar, independently of the reference chosen.

To investigate the complete flight, an integral analysis was considered. This study allows relating the destroyed exergy (proportional to thrust and fuel flow) for each phase to its time duration. Thus, it is possible to identify the critical phases

Table 8.5 Total destroyed exergy in the various flight phases with anti-ice OFF [43]

Phase	Total destroyed exergy (%)
1	4
2	25
3	5
4	36
5	3
6	9
7	19
8	<1

during the mission. These results are presented in Table 8.5, which shows that cruise is the flight condition of maximum exergy destruction since it is the longest, followed by takeoff (maximum thrust required) and holding phases.

8.3.4.2 Local Balance

For each flight phase, it was calculated the distribution of irreversibilities among components. In the nozzle, the exergy balance considered the variation between the inlet exergy (stream 11 in Fig. 8.2) and the total outlet exergy (stream 12 in Fig. 8.3).

$$B_{11} = B_{12} + B_{\text{destroyed}} \tag{8.3}$$

Another exergy balance was set by the difference of stream 12 exergy and the thrust exergy to calculate the destroyed exergy associated to the thrust generation from the kinetic exergy of stream 12 and the lost exergy associated to thermo-chemical and the residual kinetic exergy (relative velocity of the exhaust gases) of stream 12.

$$B_{12} = B_T + B_{\text{destroyed}} + B_{12,\text{lost}} \tag{8.4}$$

In this way, it is possible to evaluate the exergy destruction of the energy conversion process associated to the thrust generation.

Table 8.6 shows the total destroyed and lost exergy in each component for the complete flight.

Components that destroy more exergy are those related to the more irreversible processes, such as combustor and mixer. Their contribution to the total exergy destruction is between 50 and 66 % depending on the flight phase. Also, the destroyed and lost exergy associated to the generation of thrust is very significant. The sum of the combustor, mixer, and thrust generation losses represents up to 83 % of the total irreversibilities, being 80 % on average. Besides, Table 8.6 shows that the main distribution of irreversibilities does not change considerably during the mission.

Table 8.6 Relative destroyed and lost exergy in anti-ice OFF conditions [43]

Destroyed/lost exergy (%)	Flight phase							
	1	2	3	4	5	6	7	8
Inlet	2	3	3	3	3	3	2	< 1
Fan	5	5	4	4	6	6	5	3
Compressor	6	6	6	6	7	8	6	10
Fan duct	1	1	1	1	3	3	1	1
Combustor	32	33	31	33	34	37	33	44
HP turbine	<1	<1	1	1	1	1	1	1
LP turbine	<1	<1	<1	<1	<1	<1	<1	<1
Turbine duct	<1	<1	<1	<1	<1	<1	<1	<1
Mixer	19	20	19	20	20	21	19	22
Nozzle	<1	2	1	2	2	2	1	<1
Shaft friction	2	2	2	2	2	2	2	1
Thrust generation	32	27	31	28	21	16	28	15

Moreover, the following remarks can be drawn:

- Engine destroyed exergy is inversely proportional to thrust and represents between 70 and 80 % of the total irreversibilities. During phases of high and medium thrust, this value is approximately 70 %, while during low thrust phases it can reach 85 %;
- For the whole flight, the components that destroy more exergy are the combustion chamber and the mixer;
- The appraisal of optimization studies involving engine components must consider technological and economics aspects. Thus, not necessarily the most irreversible components are the better focus of an optimization process. For instance, considering the 'more-electric' architectures it seems important to validate the variation on the exergy performance of equipment involved on bleed air and extracted power: fan, compressor, and turbines;
- Fan and compressor destroy more exergy than turbines;
- The lost exergy is directly proportional to thrust (or to relative speed of exhaust gases). The higher the lost exergy, the higher the thrust is (or the relative speed of exhaust gases).

Figure 8.4 presents the Grassmann diagram in cruise phase, using the reference attached to the engine to avoid negative values for exergy. This diagram shows that 22.5 % of the inlet exergy is used for thrust and 1.3 % as input to systems based on compressor air bleed.

8.3.5 Thermoeconomic Analysis

In order to determine the utilities exergy based production costs, a complete evaluation of the cost of different exergy rates and flow rates of the engine was

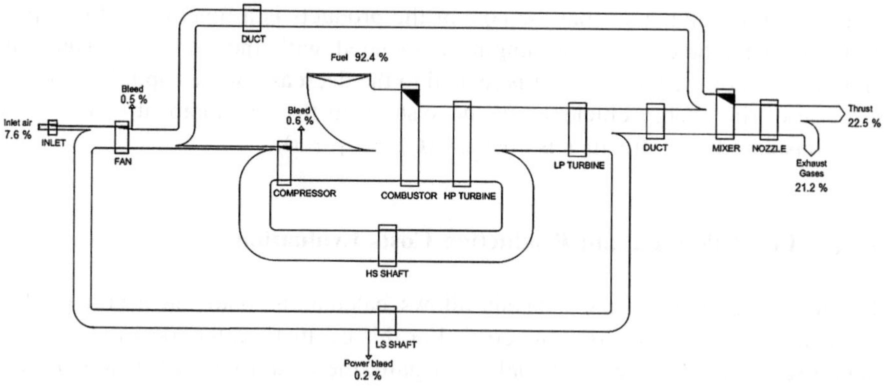

Fig. 8.4 Grassmann diagram during cruise phase [43]

developed based on the results of the exergy analysis. Two different analyses were performed:

- Global power plant production cost evaluation, considering only the inlet and outlet flows (air, fuel, exhaust gases, thrust and bleed exergy rate and extracted power), as shown in Fig. 8.3;
- Local power plant production cost evaluation, taking into account each power plant component and its relative input and output exergy rates and flowrates, as shown in Fig. 8.2.

8.3.5.1 Global Power Plant Production Costs Evaluation

Following the same approach used in the exergy evaluation of the power plant, two reference systems were adopted: one fixed upon the Earth and the other fixed to the engine.

For both reference systems, the cost balance is:

$$C_{\text{fan,bleed}} + C_{\text{comp,bleed}} + C_{\text{ex}} + C_{\text{T}} = C_1 + C_{\text{fuel}} + C_{\text{equip}} \tag{8.5}$$

No cost is assigned to stream 12, since part of it is converted into thrust and part is either destroyed or lost in the exhaust, as discussed previously.

Making use of the thermoeconomic specific cost ($c = C/B$), the equality criterion was used in this analysis, considering the same thermoeconomic specific costs for all the useful products [31]:

$$c_T = c_{\text{comp,bleed}} = c_{\text{fan,bleed}} = c_{\text{ex}} = c \tag{8.6}$$

$$\begin{aligned} c &= \frac{C_{\text{equip}} + c_{\text{fuel}} B_{\text{fuel}}}{B_T + B_{\text{comp,bleed}} + B_{\text{fan,bleed}} + W_{\text{ex}}} \\ &= \frac{C_{\text{equip}}}{B_T + B_{\text{comp,bleed}} + B_{\text{fan,bleed}} + W_{\text{ex}}} + \frac{c_{\text{fuel}}}{\eta_{\text{Global}}} \end{aligned} \tag{8.7}$$

Equation 8.7 indicates that the cost of the products is a function of the capital cost and the efficiency of the engine associated with the fuel cost. Thus, the minimization of the product cost is related to the decrease of the capital cost or the increase of the exergy efficiency of the system. In the development of Eq. 8.7 it was considered that stream 1 is costless ($C_1 = c_1 = 0$).

8.3.5.2 Local Power Plant Production Costs Evaluation

The local thermoeconomic analysis allows calculating both the useful product costs and the internal exergy rate costs. For this evaluation, the local model is the same used in the local exergy analysis. Again, the equality criterion was used to attribute cost for more than one stream leaving each control volume.

Cost balances applied to components shown in Fig. 8.2 are as follows:

Collector:

$$c_1 B_1 + C_{\text{col}} = c_2 B_2 \tag{8.8}$$

Fan:

$$c_2 B_2 + C_{\text{fan}} + W_{\text{fan}} c_{\text{fan}} = c_3 B_3 + c_4 B_4 + c_5 B_5 \tag{8.9}$$

$$c_3 = c_4 = c_5 = c_{\text{fan,bleed}} \tag{8.10}$$

Compressor:

$$c_5 B_5 + C_{\text{comp}} + W_{\text{comp}} c_{\text{comp}} = c_6 B_6 + c_7 B_7 \tag{8.11}$$

$$c_6 = c_7 = c_{\text{comp,bleed}} \tag{8.12}$$

Combustor:

$$c_7 B_7 + c_{\text{fuel}} B_{\text{fuel}} + C_{\text{cc}} = c_8 B_8 \tag{8.13}$$

High Pressure Turbine:

$$c_8 B_8 + C_{\text{HPT}} = c_{\text{comp}} W_{\text{comp}} + c_9 B_9 \tag{8.14}$$

$$c_{\text{comp}} = c_9 \tag{8.15}$$

Low Pressure Turbine:

$$c_9 B_9 + C_{\text{LPT}} = c_{\text{fan}} W_{\text{fan}} + c_{\text{ex}} W_{\text{ex}} + c_{10} B_{10} \tag{8.16}$$

$$c_{\text{fan}} = c_{10} = c_{\text{ex}} \tag{8.17}$$

Mixer:

$$c_{10} B_{10} + c_3 B_3 + C_{\text{mix}} = c_{11} B_{11} \tag{8.18}$$

Table 8.7 Data used for the exergy based costs analysis [43]

Fuel specific initial cost (US$/GJ)	9.022
Power plant economic life (years)	25
Annual utilization factor	0.285
Total capital cost (US$)	2,000,000

Table 8.8 Components initial costs [43]

Component	Capital cost (US$)
Power plant	2,000,000
Collector	100,000
Fan	340,000
Compressor	440,000
Combustor	280,000
High pressure turbine	360,000
Low pressure turbine	240,000
Mixer	80,000
Nozzle	160,000

Nozzle:

$$c_{11}B_{11} + C_{\mathrm{noz}} = c_{\mathrm{T}}\mathrm{B}_{T} \tag{8.19}$$

The solution of this set of equations gives the values of each one of the production costs of the utilities of the power plant.

8.3.5.3 Fuel and Power Plant Components Costs

The fuel considered is aeronautical kerosene JET-A1. Data used for the economic model are presented in Table 8.7. The fuel specific initial cost includes logistic costs and was evaluated as 50 US$/t [21] Capital cost of the power plant and its components were evaluated using DAPCA method [32]. The considered components costs are evaluated as a total engine cost percentage and are presented in Table 8.8.

The engine annual operating cost was calculated as the sum of the equipment cost rate and the fuel cost rate multiplied by the annual utilization factor, f_{u}:

$$C_{\mathrm{total}} = \left(\sum_{\mathrm{i}} C_{\mathrm{equip}} + c_{\mathrm{fuel}}\mathrm{B}_{\mathrm{fuel}}\right)f_{\mathrm{u}} = 1,694,000(\mathrm{US\$/year}) \tag{8.20}$$

8.3.5.4 Results

Table 8.9 presents the average annual production costs of the power plant products for both global evaluation approaches. It is interesting to verify that the obtained results for electrical, hydraulic, and thrust are quite close, since the change in the

Table 8.9 Global exergy based costs analysis [43]

Product	Ground reference	Engine reference
Fan bleed (US$/t)	5.74	5.07
Compressor bleed (US$/t)	27.21	26.52
Electrical power (US$/MWh)	367.92	368.64
Hydraulic power (US$/MWh)	367.92	368.64
Thrust (US$/MWh)	367.92	368.64

Table 8.10 Local exergy based costs [43]

Flow	Cost
Fan inlet (US$/t)	0.06
Compressor inlet (US$/t)	1.69
Combustor inlet (US$/t)	18.87
High pressure turbine inlet (US$/t)	36.39
Low pressure turbine inlet (US$/t)	22.12
Mixer inlet (primary flow) (US$/t)	13.29
Mixer inlet (secondary flow) (US$/t)	1.94
Exhaust nozzle inlet (US$/t)	3.96
Fan bleed air (US$/t)	1.94
Compressor bleed air (US$/t)	10.86
Total extracted power (US$/MWh)	131.70
Thrust (US$/MWh)	380.88

reference system does not alter the exergy efficiency of the engine, which is directly related to the cost. On the other hand, exergy flow rates are more sensitive to changes in the reference systems (Table 8.2).

Table 8.10 presents the results obtained with the local evaluation approach. In this case, as the cost formation process is more accurately followed by cost balance equations applied to components of the power plant, the results are more representative than those shown in Table 8.9 for the same products. Besides that, the set of cost balance equations allows one to better understand the energy conversion processes.

Comparing the results of Table 8.9 to those of Table 8.10, it is possible to verify that the global balance overestimates (more than doubles) the cost of bleed air from the fan and the compressor, although the cost of thrust is only 3.2 % lower than that of local model. This is a direct consequence of the contribution of the thrust exergy to the total useful exergy from the engine (95 %).

8.3.6 Closing Comments

The presented power plant exergy and thermoeconomic analysis allows evaluating the performance of this important aircraft component over a complete flight mission.

For the global analysis, two reference systems were proposed and compared: one fixed on the ground and one attached to the engine. In the first case, negative values for exergy were obtained, as reported also in Szargut et al. [29] while for the second case it does not happen as the reference varies according to the stagnation temperature and pressure at each flight phase. Although, as a first approach, the chemical exergies used were those reported in Szargut et al. [29], the use of a variable reference seems to be thermodynamically better.

In the global analysis, the overall engine performance was evaluated over a complete flight mission, characterizing the destroyed, lost and useful exergy trends. The maximum exergy efficiency was found during cruise phase, 26.5 %, but this value decreased to 6 % during landing, considering the reference fixed on the ground. With the reference attached to the engine, the maximum efficiency is 23.7 %, since the exergy of the incoming air is no longer negative (an 11 % decrease). Hence, as the specific thermoeconomic cost of the products is related to the capital cost and the engine exergy efficiency associated with the fuel cost, when the reference is fixed on the ground this cost should be smaller. However, the decrease in the cost is not as high as the efficiency due to the capital cost, meaning that the capital cost has a higher impact on the cost than fuel cost on the final products, 71 and 29 %, respectively for the cruise phase. This justifies the use of local balances to determine the product cost as they associate the correct participation of the capital cost on the cost formation of the products.

Regarding the irreversibilities in each flight phase, cruise has the highest total exergy destruction (36 %), even though it has the higher exergy efficiency. This is a consequence of the duration of the phase (the most inefficient phase is landing, which lasts 0.5 min).

The local exergy analysis showed the distribution of irreversibilities among the engine equipment, defining that the most inefficient components are the combustion chamber and the mixer, responsible for 50–66 % of the total destroyed exergy. Furthermore, it was shown that the thrust generation accounts for between 15 and 35 % of the total irreversibilities, being higher for phases that demand more thrust.

8.4 Aircraft Air Management Systems Trade-off Study Using Exergy Analysis as a Design Comparison Tool

8.4.1 Introduction

In this section, exergy analysis is used in order to compare different airplane architectures, conventional, and 'more electric' one, of the air management system (AMS) of a commercial aircraft. The analysis evaluates the exergy efficiency, exergy destroyed rate, and exergy destroyed cost for each architecture. Also, it shows the economic impact (related to fuel consumption savings) of the change from the conventional architecture to the 'more eletric' one.

8.4.2 Systems Description

8.4.2.1 Engine Bleed Air Architecture

The air management systems (AMS) of commercial aircraft are mainly composed of: (a) pneumatic system; (b) air conditioning system; (c) ice protection system; (d) engine starting system; (e) pressurization system. In conventional architectures, the pneumatic system distributes and controls bleed air to the pneumatic users (air conditioning, engine start, ice protection, pressurization), as shown in Fig. 8.5.

Bleed pressures from low stages of the engine compressor may be insufficient at lower engine power settings. Bleed pressures from high stages of the engine compressor may be too elevated at higher engine power settings. One function of the engine bleed system is to alternately select between low pressure (LP) and high pressure (HP) bleed supply sources. This is necessary to maintain adequate and safe bleed supply pressures at all engine operating conditions.

The aircraft air conditioning system may be composed by one or more machines, which usually are vapor or air cycles (reverse Brayton cycle). The machines provide cooling and heating for flight deck and cabin, filtered cabin air recirculation, conditioned air supply for gaspers, cooling air for avionics and emergency ram air ventilation for flight deck and cabin. In case of air cycle, the air supplied is compressed to keep cabin pressurized at safe and comfort levels. On the other hand, the vapor cycle requires an external compressor to pressurize the cabin. Pellegrini et al. [20] provide an exergy analysis of an air cycle environmental control unit (ECU) pack.

The ice accretion on aircraft wings and stabilizers causes aerodynamic performance degradation, weight increase, control and maneuverability difficulties, which may decrease the operational safety margin. When an aircraft flies through a cloud formed by super cooled water droplets, the ice will accrete on the non-protected surfaces subjected to impingement. As the water droplets are in a metastable equilibrium, any perturbation of thermal, mechanic or chemical nature can trigger the freezing process.

The specification of the protected regions of the wing is essential in ice protection system design. For example, it will affect fuel consumption and engine thrust, if engine bleed air is used to supply a hot air anti-ice system. An appropriate design should be a compromise between the ice protection system performance maximization and the impact minimization in aircraft overall operational performance. In order to protect the airfoils and allow safe flight in icing conditions, commercial and some military aircraft have ice protection systems, which can be classified in de-ice and anti-ice types. The de-ice system operates cyclically to remove the ice layer accreted after some exposition time. When the system is not actuated, the ice is allowed to grow on the airfoil; when it is actuated, the system removes the ice from the airfoil. On the other hand, the anti-ice system prevents the ice accretion on airfoils and continuously works while the aircraft flies in icing condition. Most commercial aircraft has a hot air anti-ice system for airfoil leading

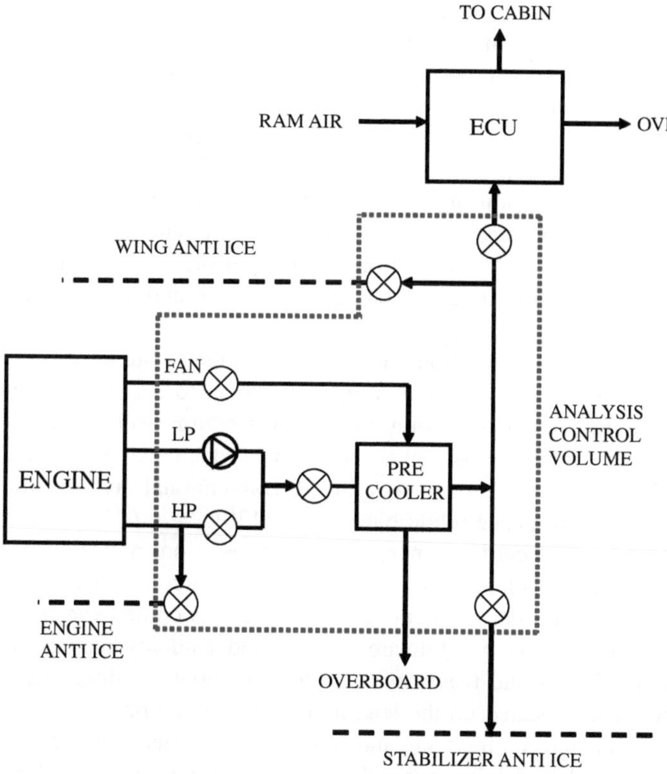

Fig. 8.5 Integrated pneumatic system-conventional architecture [21]

edges (wings and stabilizers) and engine nacelle lips protection. It transfers heat to protected surfaces using engine hot and pressurized bleed air as a thermal source.

To estimate the electrical power or the engine bleed air exergy rate demand, it is necessary to use a mathematical model to simulate the operation airfoil of a thermal anti-ice system as the one developed by Silva et al. [33, 34]. An adequate thermal anti-ice numerical code shall be used for conception, integrated optimization of aircraft systems, architecture definition, ice protection system sizing, and system development. In addition, during the certification phase, the code shall support critical cases matrix definition and test campaign planning.

8.4.2.2 More Electric Architecture

Currently, one of the most researched solutions for systems energy optimization is the 'more or all electric' aircraft, whose architecture does not involve bleed air as a source of anti-ice power and cabin/cockpit air source (Fig. 8.6). In this type of architecture, the engine drives an electric generator that supplies electric power to all or most systems, including air management systems. Due to minimization of losses and optimum control of equipment (allowed by electronics), the electric power system can deliver, approximately, what is demanded by the users in that particular operational condition. However, the main gain in the electric architecture, with no engine air extraction, is the extension of engine life and economy, as tapping off the air may increase the turbine inlet temperature and disturb the flow field in the engine core. In addition, a more electric engine can be designed for an optimum condition with no bleed air, i.e., compressor pressure and mass flow can be what is required for the best combustion, turbine performance, shaft power and maximum thrust, not requiring the usual excess of air and pressure for ECS operation and pneumatic system losses.

One of the more electric solutions to air management systems is an electric engine driven cabin compressor, which pumps outside air to the ECU using a high efficient air intake (ram air). In this solution, the cabin compressor uses less energy because the compression rate is lower and more adequated to the ECU air requirements. There is a waste of energy in the conventional architecture and a decrease in engine performance while bleeding air [20].

Regarding the 'more electric' concept for ice protection, a system based on electrical heating was developed to remove ice from aircraft surfaces, using a graphite heating element. Such element can be heated and cooled very rapidly. In this system, small areas of the graphite are strongly and suddenly heated, so the ice over that section debonds and leaves with airflow, without melting. The heating panels are strategically located on the leading edges and can be heated on a time basis, alternating locations on the wing and stabilizers, in order to minimize power consumption. A complete deicing cycle does not take long, and very thin accretions of ice can be shed without damaging the aircraft.

Yet, there are other more recent technologies for a more electric ice protection system, such as electro-mechanical expulsion deicing system, sonic pulse electro-expulsive deicer, shape memory alloy, piezoeletric among others [35–37].

The engine anti-ice solution for the 'more electric' alternative presents no changes comparing to the conventional architecture. Hot bleed air can be chosen to protect the engine lip because of the low power requirement (less protected area), the convenient location and for installation simplicity.

8.4.3 Modeling and Simulation

Two architectures were chosen for the present analysis: (1) conventional design (engine bleed air driven and anti-ice); (2) more electric design (electric compressor, electric heating anti-ice and engine bleed for engine anti-ice). The ECU is identical in the two cases (same ducting, valves, heat exchangers, turbine, compressor and fan).

It was considered different systems weights (conventional and more electric architectures) and flight mechanics coupled with an engine numerical simulation

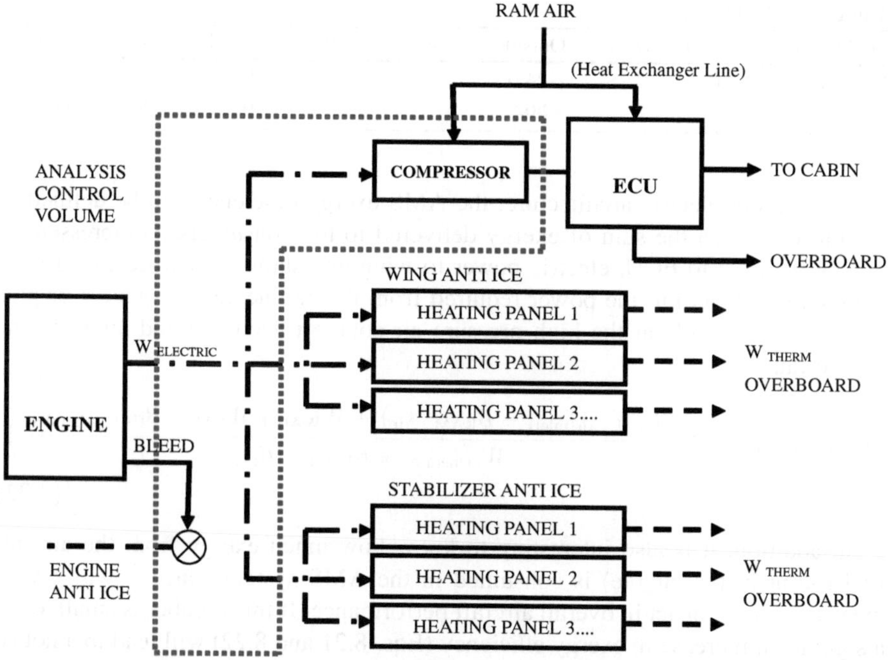

Fig. 8.6 Air management system—more electric architecture [21]

tool (evaluation of the thrust, aircraft speed and associated fuel consumption). The engine anti-ice system was not considered in the analysis, since there is no change between engine anti-ice architectures of both aircraft. This analysis also considers an increase electric system weight, i.e., electric generation and distribution, due to the higher electric power requirements of the more electric airplane. On the other hand, the more electric architecture presents a larger air intake (ram air scoop), whose drag penalty increase was not evaluated.

The case study considers one operational point during cruise and one point during climb at maximum thrust, which are relevant flight phases in terms of fuel consumption. The operational parameters used in present analysis are shown in Table 8.11.

In the conventional architecture, the AMS exergy efficiency can be defined as the ratio between the sum of the exergy delivered to the consumers (ECU, wing anti-ice, stabilizer and engine anti-ice) to the exergy available at the high pressure air source (bleed air) plus the exergy of the mass flow rate extracted from engine fan:

$$\eta_{\mathrm{b,AMS(conventional)}} = \left(\frac{B_{\mathrm{ECU\_Air(inlet)}} + B_{\mathrm{WAI\_Air(inlet)}} + B_{\mathrm{SAI\_Air(inlet)}} + B_{\mathrm{EAI\_Air(inlet)}}}{B_{\mathrm{Bleed}} + B_{\mathrm{Fan,Air}}} \right) \tag{8.21}$$

Table 8.11 Input data [21]

Flight phase	Altitude (kft)	Outside air temperature (°C)	Mach	ECU	Anti-ice
Climbing	20	−24.6	0.57	ON	ON
Cruise	37	−56.5	0.77	ON	OFF

In the more electric architecture, the AMS exergy efficiency can be defined as the ratio between the sum of exergy delivered to the consumers (compressed air exergy increase to ECU, electric power to wing and stabilizer anti-ice, exergy to the engine anti-ice) to the power required from the engine electric generator plus the exergy available at the high pressure air source (bleed air) and air from the engine fan:

$$\eta_{b,AMS(electric)} = \left(\frac{\left(B_{ECU\_Air(inlet)} - B_{RAM\_Air}\right) + W_{WAI} + W_{SAI} + B_{EAI\_Air(inlet)}}{W_{Generator} + B_{Bleed} + B_{Fan,Air}}\right) \tag{8.22}$$

In addition, it is also interesting to know how much exergy from the aircraft fuel (burnt in the engine) is consumed in the AMS system, since it will give a measure of the impact in overall aircraft performance. If this number is small, even a significant increase in exergy efficiency (Eqs. 8.21 and 8.22) will lead to a not so important decrease in aircraft fuel consumption. This index, named specific exergy consumption, similar to factor f_i defined in Chap. 3, is given by:

$$SEC_{AMS(conventional)} = \left(B_{Bleed} + B_{Fan,Air}\right)/B_{Fuel} \tag{8.23}$$

$$SEC_{AMS(electric)} = \left(W_{Generator} + B_{Bleed} + B_{Fan,Air}\right)/B_{Fuel} \tag{8.24}$$

Equations 8.23 and 8.24 require the knowledge of fuel mass flow rate consumption that is obtained from the engine performance numerical tool.

A common performance index used in the aeronautical industry is the specific fuel consumption (SFC). This index is a measure of fuel consumption to provide a given power for a given period (fuel burnt per thrust), usually used for gas turbines with a mechanical shaft output (i.e., turboprops and turbo shafts). The term is used to describe turbine efficiency; therefore, a lower number indicates better efficiency.

The normalization of fuel consumption rate, exergy efficiency, specific exergy consumption, and specific fuel consumption are performed as:

$$\Delta F_{norm} = (F_{electric} - F_{conventional})/F_{conventional} \tag{8.25}$$

where F is the generic function to be analyzed, i.e., it can be fuel consumption rate, $\eta_{Ex,AMS}$, SEC or SFC.

In order to compare both architectures in a thermoeconomic basis, it is important to evaluate the exergy destroyed, defined as:

$$B_{\text{Dest,AMS(conventional)}} = B_{\text{Bleed}} + B_{\text{Fan,Air}} - \left(B_{\text{WAI\_Air(inlet)}} + B_{\text{SAI\_Air(inlet)}} + B_{\text{EAI\_Air(inlet)}} + B_{\text{ECU\_Air(inlet)}}\right) \quad (8.26)$$

$$B_{\text{Dest,AMS(electric)}} = B_{\text{Bleed}} + B_{\text{Fan,Air}} + W_{\text{Generator}} + B_{\text{RAM\_Air}} - \left(W_{\text{WAI}} + W_{\text{SAI}} + B_{\text{ECU\_Air(inlet)}} + B_{\text{EAI\_Air(inlet)}}\right) \quad (8.27)$$

The exergy destruction is related to the inefficiencies that are inherent to each architecture. The cost associated with the exergy destruction in a process is a 'hidden' cost, but a very important one, that can be revealed only through a thermoeconomic analysis. The cost flow rate associated with the exergy destruction in a given component can be represented either by the cost of the additional fuel that needs to be supplied to this component to cover the exergy destruction and generate the same exergy flow rate of the product, or by the monetary flow rate lost at the product side because of the exergy destruction. In practical applications, the use of the fuel cost is more common [38].

In this way, the cost evaluation of thermodynamic losses (exergy destroyed cost) is given by the exergy-based cost of aircraft fuel times the exergy destroyed in each flight phase:

$$C_{b(\text{Dest})} = c_{b(\text{fuel})} \cdot B_{\text{Dest}} \quad (8.28)$$

Besides thermodynamic losses (which will have an influence on fuel consumption), a complete thermoeconomic analysis should consider other operational costs like maintenance, equipment, flight crew, etc. The analysis presented here aims at the evaluation of the exergy destruction cost only.

8.4.4 Results

Tables 8.12 and 8.13 present the input and output exergy flow rates for both architectures, based on the control volumes shown in Figs. 8.5 and 8.6. It was considered that during the climbing phase the anti-ice system is always ON. Such hypothesis is conservative and a simplification of the real operation of the system.

For the conventional architecture, the bleed exergy requirements were calculated at the engine bleed ports (high pressure port when anti-ice is ON and low pressure port when anti-ice is OFF)—Fig. 8.5. Furthermore, the inlet air requirements for ECU, wing anti-ice and stabilizer anti-ice were also evaluated. In this case, it is important to evaluate the engine fan air exergy, which cools the bleed air in order to meet the requirements of the consumers.

For the 'more electric' architecture, it was considered the bleed air requirements for the engine anti-ice system calculated at the engine bleed ports, the electricity generated to attend the ECU compressor, and electric anti-ice systems (wing and stabilizer). Also, in this case it is necessary to evaluate the RAM air exergy, which is the exergy of external air flow that is compressed in the electric compressor and sent to the ECU. The electrical demand for the anti-ice systems

Table 8.12 Engine bleed port exergy and inlet exergy requirements for conventional architecture (results shown in kW for one pack, half wing, half stabilizer, one engine) [21]

	AMS input		AMS output			
Flight phase	Bleed	Fan air	ECU (inlet)	Engine AI (inlet)	Wing AI (inlet)	Stab AI (inlet)
Climbing	459.70	272.30	97.97	34.06	128.40	71.56
Cruise	76.47	155.20	62.96	0	0.00	0.00

were calculated based on available data from a supplier of anti-ice systems, the protected area was calculated according to the best practice recommended by the aeronautical industry.

It is interesting to notice a considerable decrease in exergy consumption of the electric anti-ice systems. The reason is that the conventional architecture uses excess of hot bleed air, by forcing the air to the internal leading edge area. In such systems, the air mass flow rate is usually controlled either by altitude or by airfoil skin temperature. This may cause the system to provide more enthalpy than required to prevent significative ice accretion in critical areas. On the other hand, the 'more electric' anti-ice system optimizes the power requirement because it does not heat all panels at the same time and applies the electrical power to remove cyclically the ice. There is also an exergy decrease in the requirements of the more electric architecture ECU, since in this case the cabin compressor can operate more efficiently and attend the ECU requirements more effectively.

Table 8.14 presents the values of the exergy efficiencies and specific exergy consumption, calculated according to Eqs. 8.21–8.24.

It is worth to note that for both architectures, the climbing phase is more efficient than cruise, although the impact on the consumption of fuel is higher during climbing.

During cruise, when only the ECU is ON, the MEA architecture is more efficient and has a lower impact on the exergy consumption, in accordance to Pellegrini et al. [20].

More energy efficient systems reflect in lower fuel consumption, as shown in Table 8.15. Climbing is the flight phase that improvements are more evident, also because it was considered anti-ice ON in this phase. This can be significant during winter in regions that icing conditions are frequently forecasted, for example North America and Europe which represent about 59 % of the new airplane deliveries in the next 20 years [39].

The ΔSEC parameter indicates a better use of exergy, as it shows great improvements for the MEA (Table 8.15). To complement this statement, the exergy destroyed (shown in Table 8.16) also evidences the MEA more efficient systems. In this case, the exergy destroyed is significantly lower basically because the bleed waste is much smaller in this configuration.

In addition, it is interesting to note that exergy destroyed rate during climbing phase, for the conventional architecture, is almost 140 % higher than during cruise phase. Such difference is much lower for the MEA architecture. The values

Table 8.13 Exergy requirements for MEA architecture (results show in kW for one pack, half wing, half stabilizer, one engine) [21]

AMS INPUT					AMS OUTPUT			
Flight phase	Bleed	Fan air	$W_{Generator}$	RAM air	ECU (inlet)	Engine AI (inlet)	W_{WAI}	W_{SAI}
Climb	48.75	0	65.80	52.61	83.82	34.06	21.80	7.10
Cruise	0	0	33.30	38.19	60.88	0	0.00	0.00

Table 8.14 Bleed exergy efficiency and specific exergy consumption [21]

	Conventional (%)		MEA (%)	
Flight Phase	η_{AMS}	SEC	η_{AMS}	SEC
Climbing	45.36	5.99	88.45	0.95
Cruise	27.18	3.61	68.14	0.52

Table 8.15 Electric architecture improvement results [21]

	Consumption relative difference (%)			
Flight phase	ΔFuel	ΔSFC	$\Delta\eta_{AMS}$	ΔSEC
Climbing	−8.16	−7.73	95.00	−84.16
Cruise	−0.82	−4.22	150.70	−85.51

presented in Table 8.16 show the great impact on exergy destroyed rate of the conventional anti-ice system. Regarding the MEA architecture, the compressor is the main source of exergy destruction in the AMS system.

The determination of the exergy destroyed costs considers that Jet-A fuel price is US$5.00/gallon (US$1.32/L or, with average density, US$1.65/kg) and the exergy-based cost of Jet-A fuel is US$3.62e-5/kJ (the calculated exergy of the kerosene type jet fuel is 45,673 kJ/kg according to Szargut et al. [29]). Table 8.17 indicates the costs of exergy destruction by operation hour

It is considered a mission with 15 min climbing and 40 min cruise, without descent and holding, and considering anti-ice on during climbing. The cost of exergy destroyed, calculated based on time values shown before, is US$27.69 for the conventional airplane, and US$1.32 for the MEA. This indicates that the MEA saves US$26.37 per mission due to fewer thermodynamic losses. This number should be higher if one considers the complete mission profile. This may represent US$21,096.00 in savings in a year per aircraft, while operating eight flight legs a day, 25 days a month and 4 months a year with anti-ice ON (winter time).

The total economic savings per aircraft, related to fuel consumption decrease, represent US$29,576.00, calculated with the same considerations made before.

Table 8.16 Comparison between conventional and MEA exergy destroyed rates [21]

Flight phase	Exergy destroyed rate (kW) Conventional	MEA
Climbing	400.01	12.30
Cruise	168.71	10.61

Table 8.17 Costs associated to exergy losses [21]

Flight phase	Exergy destroyed costs (US$/h) Conventional	MEA
Climbing	52.13	1.60
Cruise	21.99	1.38

8.4.5 Closing Comments

Exergy analysis may bring different aspects of aircraft design into a common basis. However, some of these must be further studied in order to understand how they could be addressed in the analysis. Still, the indexes based on exergy do show the location of the main sources of irreversibilities, allowing a comparison of the increase/decrease of them because of a change in the architecture of the system. The combined use of aeronautical traditional indexes (such as fuel burnt, SFC and takeoff weight) and exergy-based ones proved to be more interesting since it provides more information regarding the whole system. Such integrated analysis allows the comparison of different architectures, helps the engineer to find the equipment to be optimized in the plant, and provides understanding on how much the system is important for the aircraft and compared to other systems.

The comparative study presented allowed a validation of the conclusions above. The results showed a better use of the exergy supplied to the ECU, wing and stabilizer anti-ice systems in the 'more-electric' architecture. Even with the simplification assumed in this present case study (same drag for both architectures), the fuel burnt rate and SFC index presented reduction. In addition, there is a significant reduction in exergy consumption of the wing and stabilizer anti-ice systems, which led to almost 8 % fuel reduction impact in engine performance.

In spite of adding a source of irreversibilities (electric motor and air compressor), the electric architecture reduces the exergy destroyed since there is no pre-cooler and bleed valves, which causes exergy destruction and are located upstream the environmental control unit. The electric anti-ice systems present significant lower exergy destruction; however, the anti-ice systems remain OFF the majority of the flights. These results may indicate that the new engine generation, which is adopted in more or all-electric aircraft, must have additional net gain on fuel consumption by providing only shaft power and not bleeding air. Moreover, due to high exergy destruction rate inherent to current jet propulsion systems, the engine performance gains may not be sufficient to reach significant fuel consumption reduction values. Therefore, an integrated exergy analysis for all or more

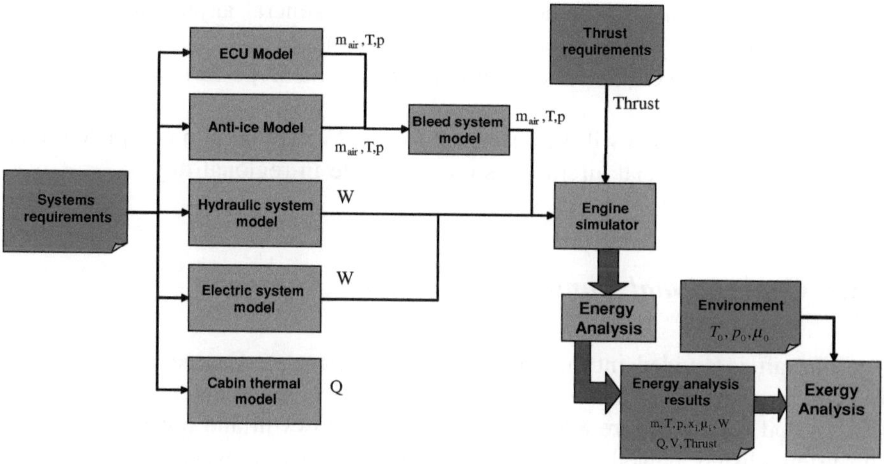

Fig. 8.7 Airplane systems modeling and simulation [41]

electric aircraft must consider the effects of the new generation of engines and its new user systems plus weight decrease due to new materials, structure optimization, and breakthrough electric system architecture as well as less drag due to improved aerodynamics.

Regarding the exergy destroyed cost, results indicate that the 'more electric' architecture has a better performance compared to the conventional one. Furthermore, considering the impact of the change of the AMS architecture, it represents 71 % of the total economic savings from the reduction of the fuel consumption.

A complete thermoeconomic analysis must be performed, in order to validate the results presented. Also, such analysis will indicate if the optimization procedure should be made in order to reduce the exergy destroyed rate (cost related to fuel consumption) or on the use of different equipment (related to capital cost).

8.5 Exergy Method for Conception and Assessment of Aircraft Systems

8.5.1 Introduction

This section describes a methodology, based on the exergy analysis, to assess the performance of airplane conceptions, configurations and systems. This approach allows the comparison of different airplane architectures by means of the mission exergy efficiency, as well as the exergy destroyed rate and its distribution among the several components in which the energy conversion process take place inside the aircraft.

Figure 8.7 presents a scheme of the steps of a general aircraft modeling and simulation behavior procedure, indicating that the exergy analysis is the last activity of the procedure, whose product is a global performance evaluation standard.

The exergy approach will be described for the evaluation of a complete typical mission of a commercial aircraft design to operate in regional flights.

8.5.2 Conventional Commercial Aircraft Approach

The aircraft is divided into eight subsystems: Airframe, Engine, Bleed System, Electric System, Hydraulic and FC Systems, Anti-ice, environmental control unity (ECU), and Cabin. Figure 8.8 shows the exergy flows in and out of each of the subsystems listed as well as the exergy destroyed rates in every subsystems. Exergy flows in blue represent useful exergy flows, while those in red stand for exergy destruction or losses.

The methodology allows one to evaluate how the fuel exergy is consumed by each of the subsystems, and also to quantify their thermodynamic losses. The definition of a mission exergy efficiency that accounts for the ratio of useful exergy to the input of exergy (inlet air and fuel) may provide an adequate comparison between different architectures or designs for the same flight mission. The distribution of irreversibilities among systems and/or flight phases makes it possible to assess the major sources of thermodynamic losses. Such information is valuable when considering a future optimization methodology application to aircraft design process.

8.5.3 Airplane Systems

As already stated, the powerplant is the aicraft power source. Usually in aeronautics, airframe is the structure of an aircraft exclusive of its powerplant. Airframe is represented by an aircraft flight mechanics, i.e., a mathematical model that applies the force balance to a steady state and leveled flight. The force balance considers altitude, outside air temperature, phase of flight, flaps configuration, weight, aerodynamic forces (lift and drag), and engine thrust in order to find, for instance, the equilibrium true air speed and the angle of attack. Depending on ambient temperature and altitude, the engine can provide a range of thrust within a range of speed. Therefore, the pilot is allowed to equilibrate the flight by adjusting the thrust lever angle.

Figure 8.8 presents an example of exergy analysis that considers the exergy demand and penalties imposed by systems to aircraft. Basically, it is necessary to evaluate the flight mechanics coupled with an engine numerical simulation tool to find the thrust, aircraft speed and associated fuel consumption.

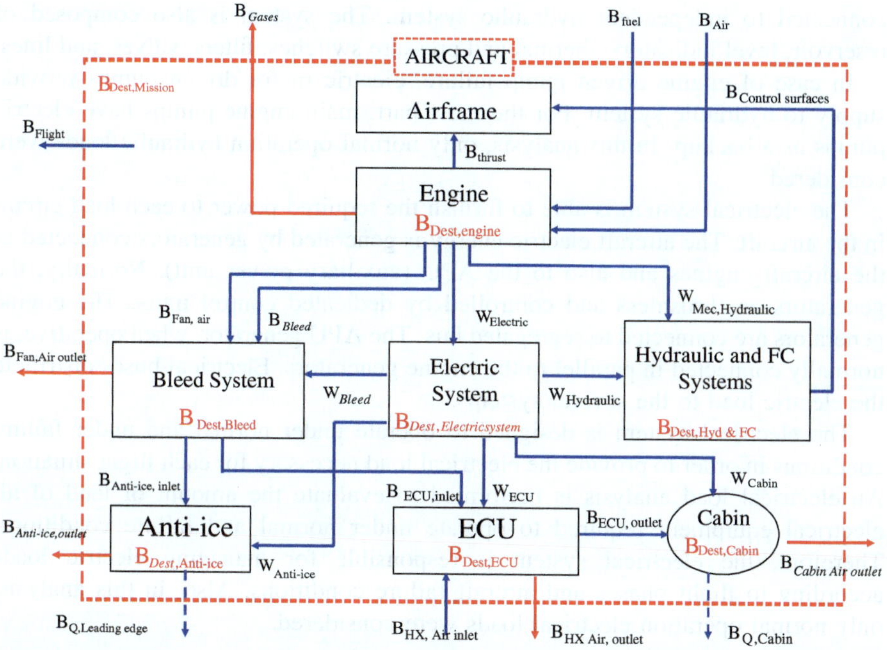

Fig. 8.8 Demand and penalties imposed by aircraft systems to engine and airframe [41]

Hence, an engine simulator such as the GSP—Gas turbine simulation program [25] can be used to evaluate fuel consumption, thrust and inlet air, bleed, fan air, and outlet gases thermodynamic characteristics (as described in Sect. 8.3). Also, the power that is required to run the hydraulic pumps and electric generators must be evaluated and used as an input of the engine simulation tool.

The air management system (see description in Sect. 8.4) is decomposed into 4 subsystems (Fig. 8.8):

1. Bleed system—Responsible for extracting hot air and fan air from the engine, and adjusting pressure and temperature before delivering to the consumers (anti-ice and ECU).
2. ECU—Provides cooling and heating for flight deck and cabin
3. Anti-ice—Prevents the ice accretion on airfoils
4. Cabin—Provides filtered air recirculation, conditioned air supply for gaspers, cooling air for avionics and emergency ram air ventilation for flight and cabin.

The hydraulic power of an aircraft is supplied by two independent systems which provide constant pressure and variable flow according to services demand. Hydraulic power system provides rudder, aileron, elevator, spoilers, landing gear, brakes, nose wheel steering, thrust reverser actuators with power extracted from the engine. Also, auxiliary systems are powered by this system. Usually, the hydraulic power is generated by engine driven hydraulic pumps, which are connected to independent hydraulic system. The system is also composed of reservoir, level indicators, thermal and pressure switches, filters, valves, and lines.

In case of engine driven pump failure, electric motor driven pumps provide supply to hydraulic system. For the most part, main engine pumps have electric pumps as a backup. In this analysis, only normal operation hydraulic loads were considered.

The electrical system is able to furnish the required power to each load circuit in the aircraft. The aircraft electric energy is generated by generators connected to the aircraft engines and also to the APU (auxiliary power unit). Normally, the generators are brushless and controlled by dedicated control units. The engine generators are connected to segregated bus. The APU generator, when operative, is normally connected in parallel to the engine generators. Electrical buses distribute the electric load to the several systems.

The electrical system is designed to operate under normal and under failure conditions in order to provide the electrical load necessary for each flight situation. An electrical load analysis is performed to evaluate the amount of load of all electrical equipment required to operate under normal and failure conditions. Therefore, the electrical system is responsible for managing electric loads according to flight phases and aircraft failure conditions. Also, in this analysis, only normal operation electrical loads were considered.

8.5.4 Exergy Analysis of an Airplane Mission

The exergy destruction is related to the inefficiencies that are inherent to each system. The exergy losses are associated to exergy flows that leave the system without any further use (for instance, exhaust gases from a power plant). In aircraft systems, the exergy losses might be related to the thermal exergy of the gases leaving the engine, ram air leaving the heat exchanger of the ECU and Fan air after the pre-cooler. Since these flows are no longer used they are considered as a source of destruction of exergy. Thus, with the products of each system known, as well as its exergy inputs, the destroyed exergy rate may be calculated by Eqs. 8.29–8.34.

$$B_{\mathrm{Dest,Engine}} = B_{\mathrm{Fuel}} + B_{\mathrm{Air}} - \left(B_{\mathrm{Thrust}} + B_{\mathrm{Bleed}} + B_{\mathrm{Fan,Air}} + W_{\mathrm{Electric}} + W_{\mathrm{Mec\_Hydraulic}}\right) \quad (8.29)$$

$$B_{\mathrm{Dest,Bleed}} = B_{\mathrm{Bleed}} + B_{\mathrm{Fan,Air}} + W_{\mathrm{Bleed}} - \left(B_{\mathrm{Anti-ice,Inlet}} + B_{\mathrm{ECU,Inlet}}\right) \quad (8.30)$$

$$B_{\mathrm{Dest,ECU}} = B_{\mathrm{ECU,inlet}} + B_{\mathrm{HX,Air,in}} + W_{\mathrm{ECU}} - \left(B_{\mathrm{ECU,outlet}}\right) \quad (8.31)$$

$$B_{\mathrm{Dest,Anti-ice}} = B_{\mathrm{Anti-ice,Inlet}} + W_{\mathrm{Anti-ice}} - \left(B_{Q,\mathrm{Leading\_Edge}}\right) \quad (8.32)$$

$$B_{\mathrm{Dest,Cabin}} = B_{\mathrm{ECU,outlet}} + W_{\mathrm{Cabin}} - \left(B_{Q,\mathrm{Heat\_Transfer}}\right) \quad (8.33)$$

$$B_{\text{Dest,Electric System}} = W_{\text{Electric}} - \left(W_{\text{Bleed}} + W_{\text{ECU}} + W_{\text{Cabin}} + W_{\text{Hydraulic}} + W_{\text{Anti-ice}}\right) \tag{8.34}$$

The exergy efficiency of a system can be defined as the ratio between the exergy of the net effect in the system to the exergy consumed in it. The exergy efficiencies of the systems considered in the analysis are:

$$\eta_{\text{b,Engine}} = \left(\frac{B_{\text{Bleed}} + B_{\text{Fan,Air}} + W_{\text{Electric}} + W_{\text{Mec\_Hydraulic}} + B_{\text{Thrust}}}{B_{\text{Fuel}} + B_{\text{Air}}}\right) \tag{8.35}$$

$$\eta_{\text{b,Bleed}} = \left(\frac{B_{\text{ECU,Inlet}} + B_{\text{Anti-ice,Inlet}}}{B_{\text{Bleed}} + B_{\text{Fan,Air}} + W_{\text{Bleed}}}\right) \tag{8.36}$$

$$\eta_{\text{b,ECU}} = \left(\frac{B_{\text{ECU,Outlet}}}{B_{\text{ECU,Inlet}} + B_{\text{HX,Air,in}} + W_{\text{ECU}}}\right) \tag{8.37}$$

$$\eta_{\text{b,Anti-ice}} = \left(\frac{B_{Q,\text{Leading\_Edge}}}{B_{\text{Anti-ice,inlet}} + W_{\text{Anti-ice}}}\right) \tag{8.38}$$

$$\eta_{\text{b,Cabin}} = \left(\frac{B_{Q,\text{Heat\_Transfer}}}{B_{\text{ECU,Outlet}} + W_{\text{Cabin}}}\right) \tag{8.39}$$

The exergy efficiency of electric and hydraulic systems were not considered in the analysis, since these are responsible mainly for supplying power to other systems, with no energy conversion process involved. However, electric and hydraulic powers extracted from the engine were evaluated. It was also considered in the analysis the exergy flows related to electric energy that each system requires for functioning and control (not shown in Fig. 8.8).

The specific exergy consumption (SEC) of the analyzed systems are given by:

$$\text{SEC}_{\text{Bleed}} = \left(B_{\text{Bleed}} + B_{\text{Fan,Air}} + W_{\text{Bleed}}\right)/B_{\text{Fuel}} \tag{8.40}$$

$$\text{SEC}_{\text{ECU}} = \left(B_{\text{ECU,Inlet}} + W_{\text{ECU}}\right)/B_{\text{Fuel}} \tag{8.41}$$

$$\text{SEC}_{\text{Anti-ice}} = \left(B_{\text{Anti-ice,Inlet}} + W_{\text{Anti-ice}}\right)/B_{\text{Fuel}} \tag{8.42}$$

$$\text{SEC}_{\text{Electric}} = \left(W_{\text{Electric}}\right)/B_{\text{Fuel}} \tag{8.43}$$

$$\text{SEC}_{\text{Hydraulic}} = \left(W_{\text{Mec\_Hydraulic}} + W_{\text{Hydraulic}}\right)/B_{\text{Fuel}} \tag{8.44}$$

$$\text{SEC}_{\text{Cabin}} = \left(B_{\text{ECU,outlet}}\right)/B_{\text{Fuel}} \tag{8.45}$$

With the destroyed exergy rates determined for each system in each flight phase, it is possible to estimate the total destroyed exergy of the mission, using the time duration of the phases specified in Table 8.18.

Table 8.18 Input data (ice protection simulated only on ice accretion probable phases—1, 2, 7, 8, and 9) [42]

Flight phase	Altitude (kft)	Outside air temperature (°C)	Mach	Time duration between phases (min.)
1. Run and Takeoff	0	15.0	0.20	2.0
2. Climb	20	−24.6	0.57	10.0
3. Climb	35	−54.3	0.66	7.5
4. Cruise	37	−56.5	0.77	1.0
5. Cruise	41	−56.5	0.77	10.0
6. Descent	35	−54.3	0.77	2.4
7. Descent	20	−24.6	0.62	6.0
8. Holding	15	−14.7	0.40	20.0
9. Approach and Landing	0	15.0	0.20	6.0

$$B_{\text{dest,mission}} = \sum_{\substack{\text{flight}\\\text{phase}}} \left(\sum_{\text{system}} B_{\text{dest,system}} \right) \cdot \Delta t_{\text{phase}} \tag{8.46}$$

Since the exergy efficiency might be calculated using the destroyed exergy, the exergy efficiency of the complete mission is defined as:

$$\eta_{\text{mission}} = 1 - \frac{B_{\text{dest, mission}}}{B_{t\text{-fuel}} + B_{t\text{-inlet air}}} \tag{8.47}$$

The total exergy of the fuel and inlet air are calculated in the same way as the total destroyed exergy.

8.5.5 *Modeling and Simulating the Flying Mission*

Figure 8.9 presents a flight mission profile for a commercial aircraft. The case study that will be described considers operational points during climbing, cruise, descent, holding, and landing. The operational parameters used in the analysis are shown in Table 8.18. As a complement, the study also considers anti-ice ON, however, only on specific phases that are susceptible of ice formation.

The studied airplane is a typical commercial one employed in regional companies to transport passengers, with the following mains characteristics:

- range: 2,200 km with full capacity and fuel reserve of 10 % of the mission fuel;
- payload: 52 passengers;
- cruise speed: 0.78 Mach;
- maximum operational altitude: 12,500 m
- taking off distance at sea level: 1,500 m (standard environment conditions);
- landing distance at sea level and maximum landing weight: 1,350 m;

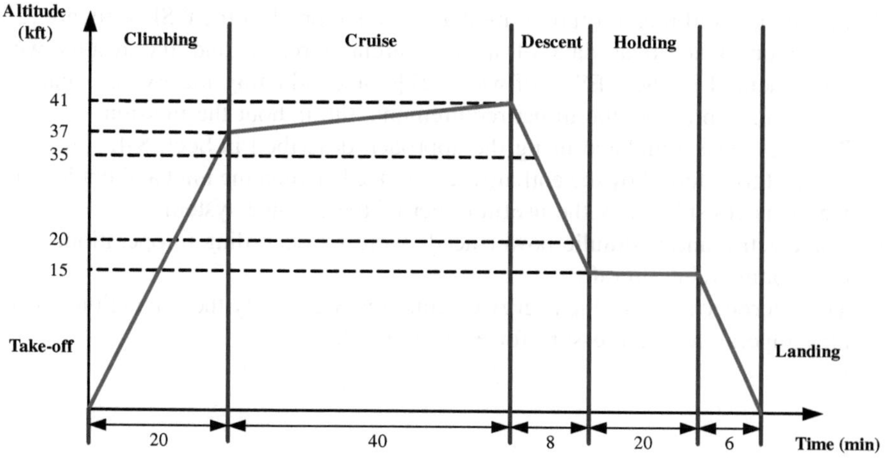

Fig. 8.9 Flight mission profile [42]

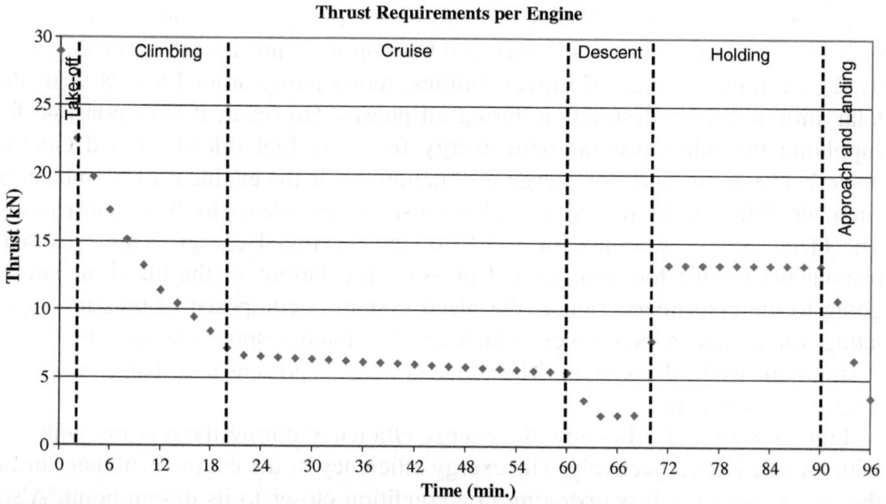

Fig. 8.10 Thrust requirements per engine throughout the mission [41]

- crew: two pilots and one flight attendant
- length: about 25 m
- span: about 20 m

The exergy analysis of each flight phase was performed with the following hypotheses:

1. For each flight phase, the systems were simulated in a steady-state condition;
2. All gas flows were considered as an ideal mixture of ideal gases;

3. For the gas turbine, a turbofan model was developed in the GSP software [25] based on operational data of a commercial aircraft, and the results were implemented in the EES® software [27] for conducting the exergy analysis. Figure 8.10 presents the thrust requirements throughout the mission;
4. The ECU was simulated using the approach described in Sect. 8.4;
5. All heat transferred by the enthalpy difference between the inlet and out let anti-ice air is considered as the useful effect of the anti-ice system;
6. The electric and hydraulic power needs were set according to operational data of a commercial aircraft;
7. The reference state for the exergy calculations was set by the stagnation outside air temperature and pressure for each flight phase.

8.5.6 Results

The results of the analysis can be seen in Figs. 8.11, 8.12, 8.13, 8.14, 8.15 and 8.16. Figures 8.11 and 8.12 show the distribution of the irreversibilities among the systems for two flight phases (cruise and holding with anti-ice ON). The engine is by far the higher source of irreversibilities, representing around 96–98 % of the total aircraft exergy destruction during all phases. However, it is responsible for supplying the other systems with exergy from the fuel (bleed air and electric power). The main source of exergy destruction inside the engine is the combustion chamber, followed by the mixer and the lost exergy related to the exhaust gases. The bleed system accounts for 1–3 % of the destroyed exergy. This system is responsible for the temperature and pressure regulations of the bleed air before going to the consumers. Hence, the bleed system is composed of pressure regulating valves and a pre-cooler, which are the main responsible for the exergy destruction inside the system. The other systems represent less than 1 % of the total irreversibilities.

Figures 8.13 and 8.14 show the exergy efficiency during the mission with and without anti-ice, respectively. The exergy efficiency of the engine is higher during the cruise, meaning it is operating in a condition closer to its design point. Also, the ECU efficiency has its greatest values during cruise, approximately 30 %. This higher efficiency is a consequence of the lower temperature of the RAM air used in the heat exchanger in high altitudes. This analysis is in accordance with the results obtained by Conceição et al. [40].

On the other hand, the exergy efficiency of the cabin has an opposite behavior to the ECU. The exergy destruction associated with the cabin increases during flight, which decreases the exergy efficiency. This is explained by the increase of the exergy associated to the heat transfer. At higher altitudes, the heat transfer through the fuselage increases because of the higher temperature gradient between cabin and outside environments, despite the fact that at higher altitudes, skin heating is higher. Also, the mechanical and thermal exergy associated to the outlet

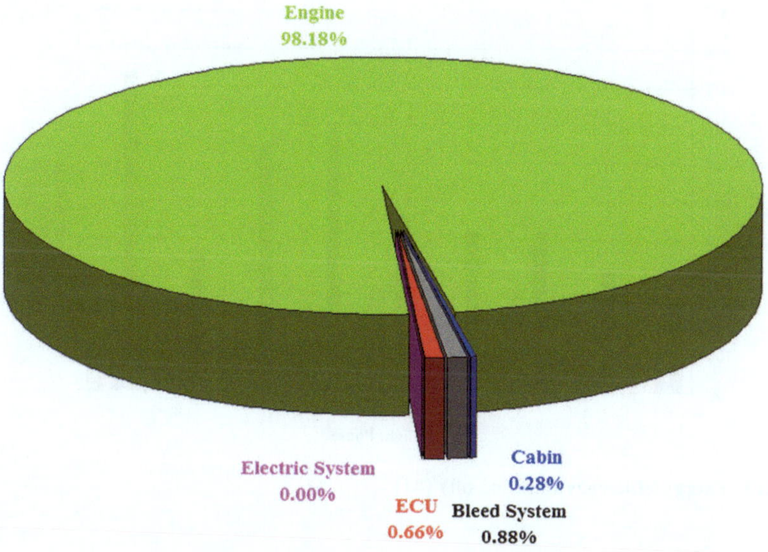

Fig. 8.11 Distribution of irreversibilities during cruise@41 kft [41]

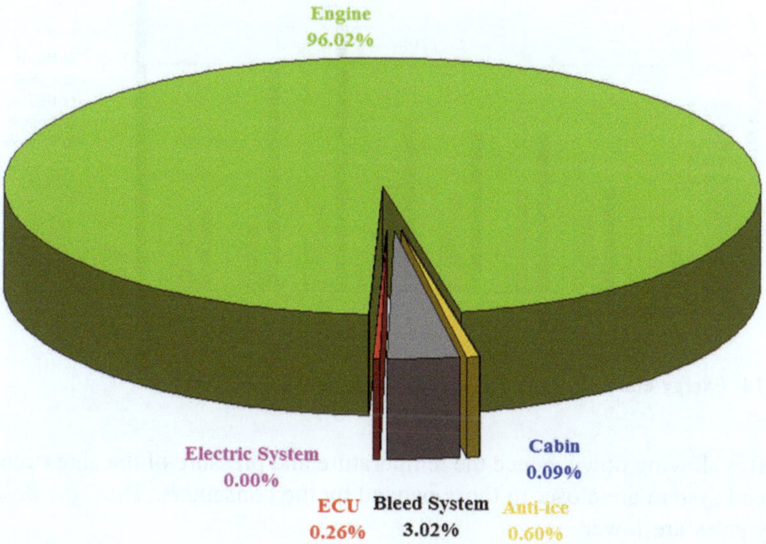

Fig. 8.12 Distribution of irreversibilities during holding@15 kft with anti-ice on [41]

cabin air are no longer used, hence are considered as destroyed exergy. These values are higher for higher altitudes.

Figures 8.13 and 8.14 also indicate that the exergy efficiencies of the bleed system are higher for descent and approach phases, since in these phases the

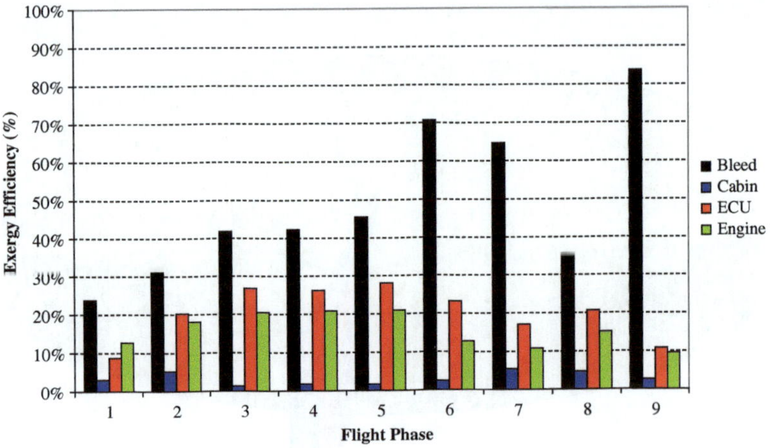

Fig. 8.13 Exergy efficiency (anti-ice off) [41]

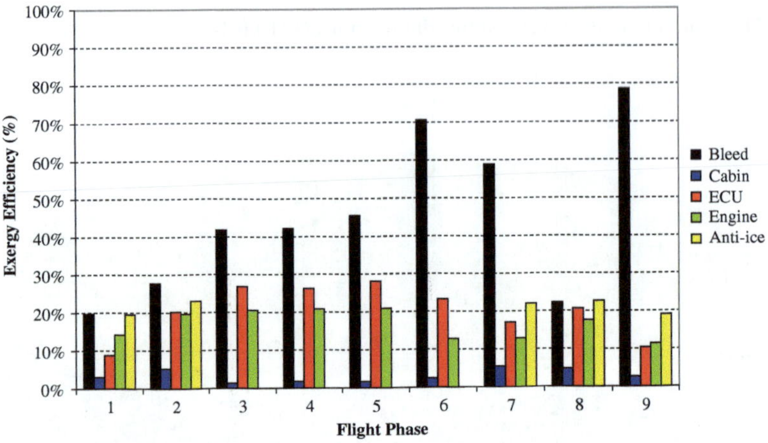

Fig. 8.14 Exergy efficiency (anti-ice on) [41]

aircraft is slowing down, hence the temperature and pressure of the air extracted by the bleed system are closer to those needed by the consumers. Thus, the destroyed exergy rates are lower.

The exergy efficiency of the anti-ice system is almost the same for different flight phases in which this system is turned on. This is a consequence that such system is not optimized for different icing conditions and flight altitudes. It is dimensioned for the most severe icing condition.

The SEC index is a value that indicates how much of the fuel exergy is consumed by the different systems. According to Fig. 8.15, the bleed system is the higher consumer of exergy, followed by the anti-ice system and ECU. Hence,

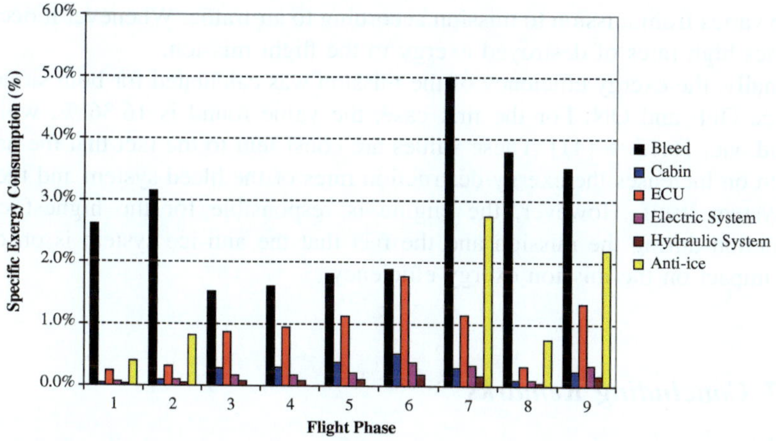

Fig. 8.15 SEC (anti-ice on) [41]

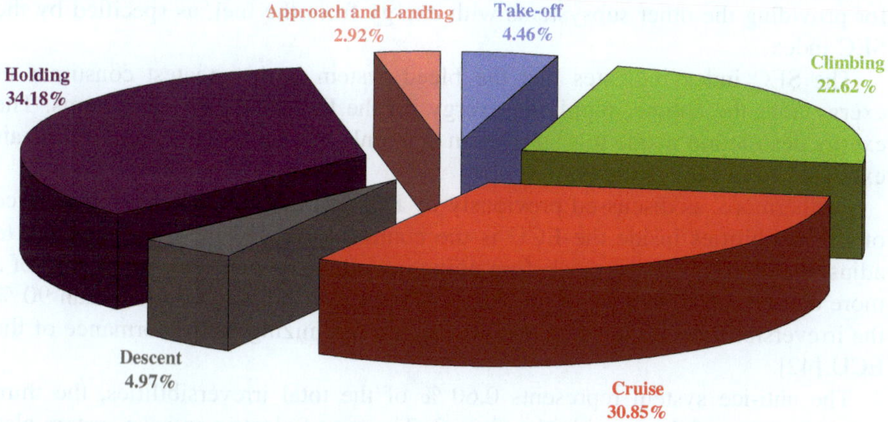

Fig. 8.16 Distribution of irreversibilities among flight phases (anti-ice on) [41]

measures to optimize the operation of the bleed system will have a greater impact on fuel consumption than an optimization of the other systems.

Exergy destruction can be analyzed in each flight phase with an integral approach that considers total systems rate of irreversibilities multiplied by each flight phase time (Eq. 8.46). Figure 8.16 shows the distribution of irreversibilities among each flight phase, considering anti-ice turned on. Holding is responsible for the highest total irreversibilities (34 %), while climbing and cruise are responsible for 23 and 31 %, respectively. In a destroyed exergy rate basis, takeoff is the phase that presents the highest exergy destruction rate, followed by holding, climb, cruise, descent and approach and landing. It is important to notice that the holding phase varies from mission to mission according to air traffic. Whenever it occurs, it imposes high rates of destroyed exergy to the flight mission.

Finally, the exergy efficiency of the mission was calculated for both situations: anti-ice OFF and ON. For the first case, the value found is 16.36 %, while for second one, 16.15 % [41]. These values are consistent to the fact that the anti-ice system on increases the exergy destruction rates of the bleed system and the anti-ice system itself. However, the engine is responsible for the highest exergy destruction during the mission and the fact that the anti-ice system is on causes little impact on the mission exergy efficiency.

8.5.7 Concluding Remarks

An exergy analysis of a commercial aircraft during a complete mission has been proposed. The results show that the engine is responsible for almost all exergy destruction and losses inside the aircraft. However, the engine is also responsible for providing the other subsystems with exergy from the fuel, as specified by the SEC index.

The SEC index indicates that the bleed system is the greatest consumer of exergy after the engine, supplying exergy for the ECU and anti-ice systems. The exergy destruction inside this subsystem is mainly related to the cooling of the air extracted from the engine compressor.

Furthermore, as discussed previously by Pellegrini et al. [20], the main source of irreversibilities inside the ECU is the control valve and pre-cooler needed to adjust the bleed air properties before going into the ECU machine. In the case of a more electric ECU, the use of an electric compressor reduces by more than 90 % the irreversibilities in the valve and pre-cooler, optimizing the performance of the ECU [42].

The anti-ice system represents 0.60 % of the total irreversibilities, the third greatest source of them inside the aircraft. The use of electric anti-ice system also reduces the exergy destruction and losses, since the operation of the anti-ice system is optimized through the use of electric panels. The heating panels are strategically located on the leading edges and can be heated on a time basis, alternating locations on the wing and stabilizers, in order to minimize power consumption. In contrast, when bleed air is used, it is sent continuously to the anti-ice system.

Considering the complete mission, the cruise, climbing, and holding phases represent 88 % of the total losses for the mission. Since the holding phase varies according to air traffic, it should be minimized as much as possible as it imposes high destroyed exergy rates.

An integral analysis was performed, and the exergy efficiency of the mission was defined. For a complete mission without the use of the anti-ice system, the exergy efficiency was 16.36 %. With the anti-ice ON, decreases to 16.15 %. The exergy efficiency of a complete mission might provide a fair comparison among competitive designs for aircraft, and also the exergy destruction and losses distribution allows one to identify correctly the impact of each subsystem on the performance of the aircraft during the mission.

It is also important to point out that different airplane missions and architecture, for instance, military airplanes and intercontinental ones, can present quite different exergy destruction rates among flight phases, as well as a different hierarchy of SEC values. Nevertheless the use of exergy based evaluation parameters allows the overall performance comparison in a rational and same basis.

References

1. Muñoz JD (2000) Optimization strategies for the synthesis/design of highly coupled, highly dynamic energy systems. Ph.D. Thesis, Faculty of Virginia Polytechnic Institute and State University, Blacksburg
2. Paulus D, Gaggioli R (2000) Rational objective functions for vehicles. In: Proceedings of 8th AIAA/USAF/NASA/ISSMO symposium on multidisciplinary analysis and optimization, Long Beach
3. Figliola RS, Tipton R, Li H (2003) Exergy approach to decision-based design of integrated aircraft thermal systems. J Aircraft 40:49–55
4. Roth B (2003) The role of thermodynamic work potential in aerospace vehicle design. In: Proceedings of the 16th international symposium on air breathing engines (ISABE), Cleveland
5. Rancruel DF, von Spakovsky MR (2003) Decomposition with thermoeconomic isolation applied to the optimal synthesis/design of an advanced fighter aircraft system. I J Thermodyn 6:93–105
6. Periannan V (2005) Investigation of the effects of various energy and exergy-based objectives/figures of merit on the optimal design of high performance aircraft system. Master Dissertation, Faculty of Virginia Polytechnic Institute and State University, Blacksburg
7. Moorhouse DJ (2003) Proposed system-level multidisciplinary analysis technique based on exergy methods. J Aircraft 40:11–15
8. Bejan A, Siems DL (2001) The need for exergy analysis and thermodynamic optimization in aircraft development. Exergy 1:14–24
9. El-Sayed YM, Evans RB (1970) Thermoeconomics and the design of heat systems. J Eng Power Trans ASME 92:17–26
10. Vargas JVC, Bejan (2001) A integrative thermodynamic optimization of the environmental control system of an aircraft. Int J Heat Mass Tran 44:3907–3917
11. Shiba T, Bejan A (2001) Thermodynamic optimization of geometric structure in the counterflow heat exchanger for an environmental control system. Energy 26:493–511
12. Ordonez JC, Bejan A (2003) Minimum power requirement for environmental control of aircraft. Energy 28:1183–1202
13. Bejan A (1996) Entropy generation minimization. CRC Press, New York
14. Rancruel DF (2002) A decomposition strategy based on thermoeconomic isolation applied to the optimal synthesis/design and operation of an advanced fighter aircraft system. Master Dissertation, Faculty of Virginia Polytechnic Institute and State University, Blacksburg
15. Muñoz JR, von Spakovsky MR (2003) Decomposition in energy system synthesis/design optimization for stationary and aerospace applications. J Aircraft 40:35–42
16. Markell KC (2005) Exergy methods for the generic analysis and optimization of hypersonic vehicle concepts. Master dissertation, Faculty of Virginia Polytechnic Institute and State University, Blacksburg

17. Butt JR (2005) A study of morphing wing effectiveness in fighter aircraft using exergy analysis and global optimization techniques. Master dissertation. Faculty of Virginia Polytechnic Institute and State University, Blacksburg
18. Brewer KM (2006) Exergy methods for the mission level analysis and optimization of generic hypersonic vehicles. Master Dissertation, Faculty of Virginia Polytechnic Institute and State University, Blacksburg
19. Ensign TR (2007) Performance and weight impact of electric environmental control system and more electric engine on citation CJ2. In: Proceeding of the 45th AIAA aerospace science meeting and exhibit, Reno
20. Pellegrini LF, Gandolfi R, Silva GAL et al. (2007) Exergy analysis as a tool for decision making in aircraft systems design In: Proceedings of 45th AIAA, Reno
21. Gandolfi R, Pellegrini LF, Silva GAL et al. (2007) Aircraft air management systems trade-off study using exergy analysis as design comparison tool. In: Proceedings of 19th congress of mechanical engineering, Brasilia
22. Amati V, Bruno C, Simone D et al. (2006) Development of a novel modular simulation tool for the exergy analysis of a scramjet engine at cruise condition. Int J Thermodyn 9:1–11
23. Turgut ET, Karakoc TH, Hepbasli A (2007) Exergetic analysis of an aircraft turbofan engine. Int J Energ Res 31:1383–1397
24. Roth BA, McDonald R, Mavris D (2002) A method for thermodynamic work potential analysis of aircraft engines. In: proceedings of the 38th AIAA/ASME/SAE/ASEE joint propulsion conference and exhibit, Indianapolis
25. GSP Development Team (2004) GSP 10 User manual. National Aerospace Laboratory NRL, Amsterdam, The Netherlands
26. MathWorks Inc., MATLAB, The language of technical computing. 1994–2007
27. Klein SA (2011) Engineering equation solver—EES, F-Chart software. www.fChart.com
28. Göğüs YA, Çamdali U, Kavsaoğlu MS (2002) Exergy balance of a general system with variation of environmental conditions and some applications. Energy 27:625–646
29. Szargut J, David RM, Steward F (1988) Exergy analysis of thermal, chemical, and metallurgical processes. Hemisphere Publishing, New York
30. Etele J, Rosen AR (2001) Sensitivity of exergy efficiencies of aerospace engines to reference environment selection. Int J Exergy 1:2001
31. Gaggioli RA, Wepfer WJ (1980) Exergy economics: I. Cost accounting applications II. Benefit-cost conservation. Energy 5:823–837
32. Raymer DP (1992) Aircraft design: a conceptual approach, American institute of aeronautics and astronautics
33. Silva GAL, Silvares OM, Zerbini EJGJ (2007) Numerical simulation of airfoil thermal anti-ice operation. Part 1: mathematical modeling. J Aircraft 44:627–633
34. Silva GAL, Silvares OM, Zerbini EJGJ (2007) Numerical simulation of airfoil thermal anti-ice operation. Part 2: implementation and results. J Aircraft 44:634–641
35. Goraj Z (2004) An overview of the deicing and anti-icing technologies with prospects for the future. In: Proceedings of 24th international congress of the aeronautical sciences (ICAS), Yokohama
36. Lawson CP (2006) Electrically powered ice protection systems for male uavs—requirements and integration challenges. In: Proceedings of 25th international congress of the aeronautical sciences, Hamburg
37. Venna SV, Lin Y, Botura G (2007) Piezoelectric transducer actuated leading edge de-icing with simultaneous shear and impulse forces. J Aircraft 44:509–515
38. Tsatsaronis G (1993) Thermoeconomic analysis and optimization of energy systems. Prog Energ Combust 19:227–257
39. Current market outlook (2006) In: Boeing. Available via DIALOG http://www.boeing.com/commercial/cmo/. Cited in 2006
40. Conceição ST, Zaparoli EL, Turcio WHL (2007) Thermodynamic study of aircraft air cycle machine: 3-wheel x 4-wheel. In: Proceedings of 16th SAE Brazil international mobility technology congress and exposition, São Paulo

41. Gandolfi R (2010) Exergy method for conception and performance evaluation of aeronautical systems. Ph.D. Thesis, Polytechnic School of the University of São Paulo, São Paulo, Brazil (In Portuguese)
42. Gandolfi R, Pellegrini LF, Silva G, Oliveira Jr S (2008) Exergy analysis applied to a complete flight mission of commercial aircraft. In: 46th AIAA Aerospace science meeting and exhibit, Reno, Nevada, 7–10 Jan 2008
43. Tona C, Raviolo PA, Pellegrini LF (2010) Exergy and thermoeconomic analysis of a turbofan engine during a typical commercial flight. Energy 35:952–959

Chapter 9
Exergy Analysis and Environmental Impact

Symbols

a_i	activity of species i
b	specific chemical exergy (kJ/kmol)
B	exergy rate/flow rate
n_i	number of moles of species i
$\eta_{\mathrm{b,env}}$	environmental exergy efficiency
η_{d}	exergy index for contaminant destruction
η_{p}	exergy index for waste converted product
μ_i	chemical potential of species i
$\mu_{o,i}$	chemical potential of species i at the reference state

Subscripts

contaminant	related to contaminants of a given process
deact	additional natural resources during waste deactivation
destroyed	destroyed
disp	related to waste disposal of the process
materials/utilities	related to materials and utilities
nat.res	natural resources consumed by the processes
prep	required for extraction and preparation of the natural resources
process	related to a given energy conversion process
product	useful effect of a process
reject	related to a reject
waste	related to wastes

Abbreviations

AEnC	accumulated energy consumption
AExC	accumulated exergy consumption
COD	chemical oxygen demand

S. de Oliveira Jr., *Exergy*, Green Energy and Technology,
DOI: 10.1007/978-1-4471-4165-5_9, © Springer-Verlag London 2013

BOD biological oxygen demand
TOC total organic carbon
UASB upflow anaerobic sludge blanket
WTP wastewater treatment plant

9.1 Introduction

The quantification of environmental impact has been done in many different ways, involving technical, economic, and social aspects. Even a technical approach to environmental problems ends up in the need for a complex analysis, involving various scientific disciplines and methodologies [1, 2]

Also, treatment processes and measures with a view to maintain emissions within legally established limits have been evaluated, almost exclusively, with respect to their emissions abatement efficiency and economic aspects.

Since environmental issues inevitably require a multidisciplinary analysis, definition of acceptable legal limits regarding the release of waste materials into the environment has been effected through an approach that is not highly systematic, considering physical, chemical, biological, ecological, and toxicological parameters, among others. Furthermore, it should be emphasized that, in general, legal limits are normally dictated by the most recent developments in treatment technology.

Although suited to the pluralistic nature of environmental problems, the kind of analytical tool normally used ends up relegating the comparison of environmental solutions that attain the same emissions abatement targets, to an economic assessment. Very often, aspects that provide evidence of a lesser global environmental impact of the adopted solution end up being neglected, to the detriment of a local analysis.

In this context, the concept of exergy arises as a powerful tool for analysis not only of environmental impact, but also the measures and processes necessary for mitigating this impact. Since it is a measure of the potential for carrying out work contained in the material (fuel, food, or any kind of material), exergy becomes the natural choice for assessing the quantity and quality of resources, instead of other parameters [3, 4, 5].

Some authors have suggested that the quantification of the environmental impact can be better driven by the use of the exergy concept [3, 6, 7]. Others calculated that impact based on the exergy [8–13]. Makarytchev [11] presented an evaluation of the environmental impact of a fuel gas cogeneration and the electric power from coal, using data generated by the exergy analysis to quantify that impact in terms of environmental efficiency and risks indexes.

The exergy concept has been utilized in the ecologic and environmental field by Jorgensen [14, 15] and Fuliu [16], and as an ecological indicator and objective function in the modeling of aquatic systems Bendoriccio and Jorgensen [17, 18].

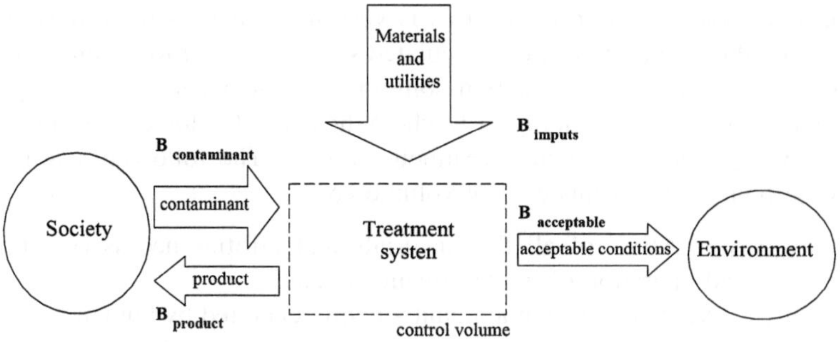

Fig. 9.1 Boundaries of the considered problem [46]

Exergy can be defined as a sustainable development registration that emphasizes the connection between generated services/products and resources consumption. This fact makes exergy a better measure of the damage and a good ecological index since a high exergy efficiency means less exergy wastes to the environment or less environmental damage Gong [3, 7].

Also, by using life cycle analysis methodology on an exergy basis, it is possible to evaluate an environmental impact mitigation process, with the same energy dimension, in a given facility, with respect to its products and input materials/utilities, irrespective of whether these are fuels or not Gong [3, 4].

9.2 Exergy Analysis of Environmental Impact Mitigation Processes

9.2.1 Exergy Indexes

All the individual stages of a given production process, from the raw material, taken from the environment, to the end product, result in exergy destruction. Very often, the final production stage is characterized by a relatively high degree of exergy efficiency, although the production of intermediary products may occur with very low rates of exergy efficiency, with resulting wastes that are discharged into the environment. The mitigation of the impacts generated by these wastes requires the adoption of treatment process. Therefore, it is both interesting and useful to use the accumulated exergy consumption (AExC) proposed by Szargut [19] and Szargut and Morris [20], which expresses the sum consumption of the exergy of natural resources throughout the production chain.

A similar method is that of accumulated energy consumption (AEnC), which has already been developed to a significant extent during the 1970s. Nevertheless, the calculation based on exergy is more informative, as it considers the exergy of non-energy raw materials taken from nature Szargut [19] .

The definition of system boundaries is very important, as these have a direct influence on the sum exergy of the calculated stages. In order to conduct an exergy analysis of environmental mitigation solutions, consider here a generic problem with the boundaries shown in Fig. 9.1, where the control volume considered is that which envelops the contaminant treatment process. The following exergy flows across the boundaries of this control volume are:

- $B_{materials/utilities}$: AExC of all the materials and utilities necessary for implementation and operation of the treatment system;
- $B_{contaminant}$: Exergy of the contaminant stream generated by human activity and the object of treatment;
- B_{waste}: Exergy of the waste stream produced by the treatment process, which are discarded into the environment; and
- $B_{product}$: Exergy of products useful to society, obtained through the treatment process.

In this sense, this study is limited to the analysis of the adopted treatment process and/or environmental solution. In this case, one has the following exergy balance for the treatment system:

$$B_{\text{contaminant}} + B_{\text{materials/utilities}} = B_{\text{products}} + B_{\text{waste}} + B_{\text{destroyed}} \tag{9.1}$$

From this, it can be seen that one has moved away from traditional focus on exergy balance of the production process, to assessment of the system for abatement of the emission, waste, and/or contaminant that the process generates.

With given or estimated chemical composition of the wastes and contaminants, the specific exergy streams are calculated according to the equations presented in Chap. 2.

With regard to the exergy of input materials/utilities, the methodology employed was that of exergy life cycle analysis, proposed by Szargut [19], according to which (AExC) expresses the exergy sum of natural resources consumed throughout the implementation and operation of this system. The difficulty associated with the use of this methodology lies in calculating the exergy of non-fuel materials, resources, and utilities taken from the environment.

Szargut et al. [19] and Dewulf et al. [5] tabulated several values for the specific exergy content of a variety of materials and utilities, ranging from extraction of raw materials from the environment, handling, manufacture, and transport, to their construction and operation.

Therefore, the methodology comprised the following steps:

- Identification and characterization regarding the chemical composition of the environmental contaminant to be assessed;
- Calculation of the exergy of this stream;
- Calculation of the AExC of several treatment, decontamination, or recycling processes, determining the materials and necessary utilities for their implementation;
- Calculation of the exergy of the output stream (s) after treatment; and

- Evaluation of yield, in exergy terms, of the treatment process given by the methodology detailed below.

Depending on the objective of the treatment process, one can have two different situations that require particular evaluation criteria. The treatment processes can be:

- processes involving the destruction of contaminant exergy;
- processes to recover the exergy of the contaminant.

The aim of first type of processes is to reduce the exergy contained in the contaminant stream to a minimum. These are processes used in the treatment of air emissions and liquid effluents, or decontamination of soil and groundwater, in which there is minimal or even negligible product recovery. With respect to these processes, according to the nomenclature shown in Fig. 9.1, an exergy index (η_{d}) is defined to evaluate the destruction of the contaminant:

$$\eta_d = \frac{B_{\text{contaminant}} - B_{\text{waste}} - B_{\text{product}}}{B_{\text{materials/utilities}}} \tag{9.2}$$

The aim of the second type of processes is to maximize the exergy that can be obtained from a given contaminant, through the use of treatment processes that mitigate the impact caused on the environment and produce some kind of product that can be useful to society. Such processes include the final disposal of solid or liquid waste materials, which have an elevated specific exergy that can be partially recovered by society. Considering these processes, according to the nomenclature shown in Fig. 9.1, an exergy index (η_p) can be defined with regard to making the best use of the converted product:

$$\eta_p = \frac{B_{\text{waste}} + B_{\text{product}}}{B_{\text{contaminant}} + B_{\text{materials/utilities}}} \tag{9.3}$$

In the forthcoming sections, three case studies are presented in which the methodology described was applied. The considered case studies deal with the problems of air emissions treatment, soil and groundwater contamination, and the final disposal of solid waste materials.

9.2.2 Air Emissions Treatment

This study was developed for the treatment of the effluent gases of an enclosed painting compartment, in which metallic parts are painted using a solvent-based paint. A radial fan removes the air contaminated by volatile organic compounds (VOCs). The chemical composition of the exhaust gases is shown in Table 9.1. For a flow rate of 141 $\mathrm{Nm^3/h}$, and an exhaust gas temperature of 315 K, the total exergy rate of the contaminant was calculated as $B_{\text{contaminant}} = 20.06$ kW.

Table 9.1 Composition of exhaust gases [46]

Contaminant	Concentration (μg/Nm^3)
Benzene	24
Toluene	60,162
Xylene	39,015
n-butyl acetate	41,958
Ethyl alcohol	27,972
Acetone	23,310

Table 9.2 Characteristics of air emissions treatment processes [46]

Process	Characteristics
After-burner	Auxiliary fuel: CH_4
	Excess air: 1150 %
	Temp. of gases at outlet: 1295 K
	Fuel consumption: 154.9 Nm^3/h
	Electrical power consumption: one 5.6 kW axial fan
	100 % thermal oxidation of contaminants
Activated carbon	Number of columns: 3
	Mass absorption capacity: 0.25 kg of contaminant/kg of activated carbon
	Electrical power consumption: one 11.2 kW axial fan
	100% absorption of contaminants
Biofilter	Dimensions: 20 × 30 × 2.5 m
	Volume of substrate: 1250 m^3
	Electrical power consumption: one 11.2 kW axial fan
	90 % of contaminants metabolized

Three treatment processes were assessed, namely: contaminant incineration, using an afterburner; adsorption onto columns of activated carbon, and biodegradation using a biofilter. The technical characteristics of these processes are summarized in Table 9.2.

AExC calculations were made for each alternative, considering an operational period of 20 years. Also the exergy content of waste materials is determined, as well as the exergy yield regarding the destruction of the contaminant, η_d. The results obtained are shown in Table 9.3.

Since the afterburner alternative increases the exergy content of the emission, due to temperature elevation ($B_{\text{waste}} > B_{\text{contaminant}}$), η_d has a negative value in this case.

Table 9.3 Exergy index (η_d) calculation [46]

Alternatives				
Parameter	Units	After-burner	Activ. carbon	Biofilter
$B_{contaminant}$	MJ	2.98E + 06	2.98E + 06	2.98E + 06
B_{waste}	MJ	1.76E + 07	0	3.05E + 05
$B_{product}$	MJ	0	0	0
$B_{materials/util}$	MJ	7.97E + 09	2.06E + 08	8.40E + 06
η_d	%	−0.18	1.44	31.90

Table 9.4 Distribution of 8,000 L of fuel in an aquifer [46]

Medium	Phase	Volume of contaminant (L)	% of total	Contaminated volume	% of total
Soil	Free	5,120	64	673	1
Soil	Residual	2,800	35	13,464	20
Water	Dissolved	80	1	53,183	79
Total		8,000	100	67,320	100

9.2.3 Soil and Groundwater Remediation

This case study was also based on a real situation, involving soil and groundwater contamination by petroleum derivative hydrocarbons. In this occasion, a pipeline assembly fault resulted in a leak of approximately 8,000 L of diesel oil into the soil from an underground storage tank at a gas station located in the interior of the state of São Paulo—Brazil. Table 9.4 shows an estimate of the distribution of fuel throughout the saturated and non-saturated zones of the soil, according to the fraction distribution model developed by the United States Environmental Protection Agency [21].

Based on information regarding the average composition of diesel oil, and the specific exergy values for each hydrocarbon drawn up in Dewulf and Van Langenhoven [22], using Eq. 9.2, the exergy of the contaminant was calculated as $\mathbf{B}_{contaminant} = 328{,}628$ MJ, for the 8,000 L of leaked fuel, this being the object of elimination (destruction) by the treatment processes.

Three treatment processes were evaluated, namely: pump and treat (*P* and *T*), multi-phase extraction (MPE), and removal and incineration of contaminated soil. The technical characteristics of these processes are summarized in Table 9.5.

The AExC value was calculated for each alternative, considering an operating period of 10 months in the case of pumping, 18 months in the case of MPE, and 1 month for soil removal and incineration. Since the processes act on different portions of the contamination, a remediation efficiency value was also calculated, in order to assess the rate of exergy consumption per unit volume decontaminated. This parameter, as well as the exergy yield of destruction of the contamination, η_d, are shown in Table 9.6.

Table 9.5 Characteristics of soil and groundwater decontamination processes [46]

Process	Characteristics
Pump and treat	Action: soil, free and residual phases Components: three intrinsically safe pneumatic pumps, oil/water separation tank, and an adsorption system for the phase dissolved in water Installed power: 3.73 kW Efficiency: 64 % removal of contamination
MPE	Action: soil, free and residual phases and water (dissolved phase). Promotes in situ bioremediation due to soil oxygenation. Components: vacuum pump, vacuum tank, emulsion breaking system, oil/water separation tank, and system for adsorption of phase dissolved in water and air emissions Installed power: 20 kW Efficiency: 100 % removal of contamination (considering the levels of decontamination that should be attained according to the risk assessment conducted previously)
Removal and incineration	Action: soil, free phase and partially in residual phase Removal: Excavation of 26,896 m^3 of soil, transport of 3,026 tons of contaminated soil by truck to an incineration facility Incineration: incinerator with capacity of 100 t/day of hazardous waste material, consuming 6.25 kW_{el}, 60 Nm^3/h of natural gas, and 10 m^3/h of process water Efficiency: 35 % removal of contamination

Table 9.6 Exergy index (η_d) calculation [46]

Alternatives				
Parameter	Units	*P* and *T*	MPE	Incineration
Volume decontaminated	m^3	673	67.320	2,522
% of total	%	1	100	4
AExC	MJ	1.02E + 06	9.06E + 05	3.03E + 06
Remediation efficiency	MJ/m^3	1,516	13	1,201
$B_{contaminant}$	MJ	3.29E + 05	3.29E + 05	3.29E + 05
B_{waste}	MJ	1.18E + 05	0.00E + 00	2.14E + 05
$B_{product}$	MJ	0	0	0
$B_{materials/util}$	MJ	1.02E + 06	9.06E + 05	3.03E + 06
η_d	%	21	36	4

9.2.4 Final Disposal of Urban Solid Waste Materials

For the purposes of this study, consideration was given to a specific assessment according to the mass of the following materials present in domestic refuse: paper, cardboard, polyethylene (PE), polypropylene (PP), polyvinyl chloride (PVC). Chemical exergy, as well as the (AExC) of the analyzed materials are shown in Table 9.7 Szargut [19, 22].

Table 9.7 Characteristics of the analyzed materials [46]

Product	Exergy of raw material (MJ/kg)	AExC (MJ/kg)
Cardboard	19.50	70.84
Paper	16.50	69.16
PE	46.50	86.00
PP	46.40	85.20
PVC	19.70	67.00

Table 9.8 Characteristics of urban solid waste treatment processes [46]

Process	Characteristics
Landfill	Capacity of landfill: 100 t/day
	Average density of waste: 8 t/m^3
	Installed power of systems for treating percolate and making use of biogas: 63.04 kW
	Work demand: 420,000 t km
	Quantity of soil moved: 495,000 m^3
	Cogeneration: 1.1 MW$_{el}$ + 109 kWth [23]
Recycling triage	Capacity: 2 t/day manual segregation system
	Installed power: 23.5 kW
Plastics recycling	Installed power: 25.36 kW
	Gas consumption: 30 m^3/t
Paper/cardboard recycling	Installed power: 34.69 kW
	Gas consumption: 30 m^3/t
Incineration	Capacity: 100 t/day
	Air consumption: 653,240 kg/day (100 % excess)
	Energy input: 8 × 10^5 MJ/day
	Enthalpy of exhaust gases: 6.3 × 10^5 MJ/day
	Ash production: 25.26 t/day
	Installed power: 91 kW

With regard to solid waste treatment, three alternatives were considered, namely: sanitary landfill, recycling, and incineration. Table 9.8 shows the main characteristics of the processes that were analyzed.

The AExC value was calculated for each alternative, considering an operating period of 20 years. Also calculated were the exergy content of the waste materials and the exergy yield for obtaining the products, η_P. The results are shown in Table 9.9.

Table 9.9 Exergy index (η_p) calculation [46]

Alternatives/product	$B_{materials}$ (MJ/kg)	Product	$B_{product}$ (MJ/kg)	η_p (%)
Landfill				
Cardboard	1.15	Heat and Electricity	1.04	5
Paper	0.47	Heat and Electricity	1.04	6
PE	0.21	Heat and Electricity	0.00	0
PP	0.23	Heat and Electricity	0.00	0
PVC	0.15	Heat and Electricity	0.00	0
Recycling				
Cardboard	18.09	Cardboard	19.50	52
Paper	18.09	Paper	16.50	48
PE	12.62	PE	46.50	79
PP	12.62	PP	46.40	79
PVC	12.62	PVC	19.70	61
Incineration				
Cardboard	0.54	Heat and Electricity	7.12	36
Paper	0.54	Heat and Electricity	6.88	40
PE	0.54	Heat and Electricity	18.91	40
PP	0.54	Heat and Electricity	18.64	40
PVC	0.54	Heat and Electricity	9.87	49

9.2.5 Comments on the Exergy Indexes for Mitigating Environmental Impacts

The use of the exergy indexes for destruction of the contaminant and conversion into product, showed results consistent with those technological options that are the most notable in terms of ensuring the sustainability of environmental solutions.

With respect to all the alternatives evaluated, the input materials/utilities consumed during the operation had a preponderant part to play in calculating accumulated exergy in each process, with special emphasis on the utilities gas and electricity. The exergy consumed in installing the processes had less significant orders of magnitude.

Case studies involving the treatment of air emissions and the remediation of contaminated soil and groundwater, respectively, were assessed according to η_d, as the main focus of these processes was that of eliminating the exergy contained in the contaminants. None of the processes in question showed any kind of by-product that is of use to society. Although in some of them good use was made of thermal wastes, as would be possible in the afterburner process, or pyrolysis for activation of the saturated activated charcoal, involved in both studies, the resultant products would be of less importance when compared with the main objective of these processes. Nevertheless, the methodology used in calculating η_d would show the distortion of providing lesser yields for those processes in which some kind of useful product is returned to society.

In situations where the objective was that of destroying the exergy of the contaminant, processes involving biological activity were those that showed the highest yields, for example, use of a biofilter (32.7 %) and MPE (36.0 %). On the other hand, those involving thermal destruction through the consumption of input materials/utilities (mainly natural gas) showed the lowest values of the exergy index, such as the removal and incineration of soil (4 %, second case study), use of the afterburner (0.18 %), and adsorption onto activated carbon (1.48 % first case study).

The second case study showed a unique situation, in which the available processes that were evaluated act on a different medium and portion of the contamination. Therefore Table 9.6 shows the specific remediation efficiency per volume decontaminated. The index η_d calculated in the same table confirms the actual raking (at the time the study was developed) that is observed, which has MPE ($\eta_d = 36$ %), as the most sophisticated and efficient technology, concerning decontamination levels, followed by *P* and *T* ($\eta_d = 21$ %). Soil removal ($\eta_d = 4$ %) shows to be the less usual alternative, due to its partial solution of the problem, and its high costs.

Urban solid waste treatment processes were the only ones assessed based on product index, η_p, in view of the fact that these processes manage to return products that are of use to society, such as process heat, electrical power, and recycled materials.

The differences obtained using η_p, which identifies recycling as the best alternative and sanitary landfill as the worst, was highly consistent with the notion of sustainability that environmentalists and government institutions have been trying to disseminate. The highest exergy content and consequent added value of recycled products explain the high figure obtained, even if this process involves a higher consumption of exergy. At the opposite extreme, landfill seems to be an alternative of low specific exergy consumption; however, on the other hand, it provides little or no useful product to society.

Currently, choices of alternatives regarding processes for the mitigation of environmental impact are principally focused on cost, very often ignoring the sustainability of the technologies employed. With the use of exergy, the concept of sustainability gains a very valuable quantitative nature for conducting more technical, rational, and universal analyses of environmental solutions.

9.3 Exergoenvironmental Evaluation of Wastewater Treatment Processes

9.3.1 Introduction

Improvement in global health, sanitation, and consequent reduction in the spread of disease depends largely on good hygiene practices, availability of health facilities, and reliable collection and treatment of wastewater. The World Health

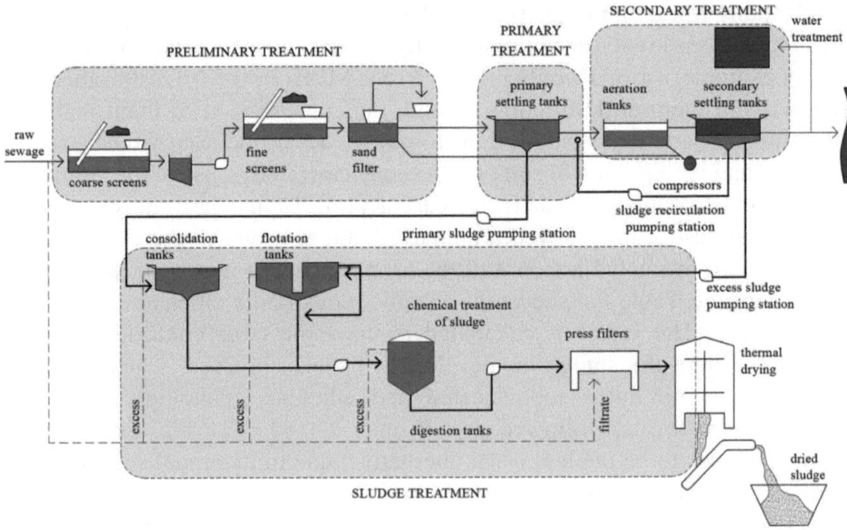

Fig. 9.2 Barueri wastewater treatment plant [45]

Organization estimates that 2.4 billion people lack access to any type of sanitation equipment [24].

There is an increasing demand for more sustainable wastewater treatment systems. However, the criteria needed to characterize the environmental performance of such a system are not fully developed and it is a challenge designing sustainable wastewater treatment systems that address the positive effects to the environment, society, and economy. The more advanced conceptions of wastewater treatment systems value the level of ability of the operation staff of the plant, the jobs in the community, aesthetics of the physical structure of the plant, the minimization of atmospherics emissions, operations costs, and the energy use, thus, the treatment system performance maximization. Several authors have proposed different wastewater treatment systems sustainability indexes, which include the exergy concept [24–27]. Hellström [28] showed how an exergy analysis could be used to estimate the consumption of physical resources in a wastewater treatment plant.

The exergy concept has been used for water quality evaluation, elucidating the relation between exergy, and the water quality parameters as COD (Chemical Oxygen Demand), BOD (Biochemical Oxygen Demand), TOC (Total Organic Carbon) ([13, 29–35]). Other authors presented renewability exergy indexes for different processes ([36–38], Torio et al. [39]).

The following sections describe a comparative exergy and renewability analysis of three wastewater treatment plants: a conventional plant with secondary activated sludge treatment, a facultative lagoon—upflow anaerobic sludge blanket reactor (UASB) system, and a chemically enhanced primary treatment plant.

9.3.2 Configurations of Wastewater Treatment Plants

Figure 9.2 illustrates a conventional plant with secondary activated sludge treatment with organic material removal of 90 % BOD and about 7 m^3s^{-1} treatment capacity (Barueri Wastewater Treatment Plant). The treatment process that takes place in the plant consists of the following stages:

1. The Preliminary Treatment consists of two phases: screening and sand removal. Screening removes large solids, which are retained by the screens. The main reasons for the screening are to protect the pumps and tubes, later treatment units and the tanks. The sand is removed by sedimentation.
   The aims of sand removal are to protect the equipment from wear and turbulence, eliminate or reduce the risk of blockages in pipes, tanks, siphons, and passages, and simplify the liquid transportation, especially transfer of sludge (see Fig. 9.2).
2. The Primary Treatment consists of primary settling tanks which are rectangular or round. Sewage flows slowly through the tanks, allowing suspended solids to gradually settle to the bottom of the tanks. This solid mass, called primary sludge, can be consolidated at the bottom of the tank and sent directly for digestion, or can be sent to the consolidation tanks. A large part of these solids is made up of organic matter. Depending on the nature and size of the suspended solids, rotating sieves may be used instead of the screening system or the primary settling tanks. The aim is to separate the larger suspended solids, by means of flowing them through the moving sieves, from the center to the outside. The retained solids are continuously removed in buckets.
3. The Secondary Treatment is made of three phases. In the aeration tank (phase one), organic matter is removed by biochemical reaction, using microorganisms (bacteria, protozoan, fungi). This process relies on contact between the microorganisms and the organic material in the sewage, which forms their food. They convert the organic material into carbon dioxide, water, and their own cell structure. The secondary settling tanks perform an important function in the activated sludge process (phase two), being responsible for the separation of the suspended solids present in the aeration tank, and allowing a clarified liquid to flow out, leaving sediments solids at the base of the tank, which can be returned in a higher concentration. The effluent from the aeration tanks is settled, so that the activated sludge is separated and returns to the aeration tanks. The return of this sludge is necessary to supply the aeration tanks with a sufficient quantity of microorganisms to keep the feeding process going in sufficient strength to decompose the organic material efficiently. The liquid effluent from the secondary settling tanks is either released directly or conveyed for treatment so that it can be reused internally or sold for uses such as washing streets and watering gardens. In the pumping station the excess sludge is sent to the third stage of the secondary treatment: the sludge formed from the suspended solids by means of the alimentation of microorganisms must be removed to maintain

equilibrium in the system (solids in = solids out). The sludge is extracted and sent for treatment (see Fig. 9.2).

4. The Sludge Treatment consists of five phases: (a) Consolidation: this stage takes place in consolidation and flotation tanks. As the sludge still contains large quantities of water, its volume must be reduced. The consolidation process increases the solid content in the sludge, reducing its volume. This process can increase the proportion of solids from 1 to 5 %. In this way, subsequent units, such as digester tanks and drying units have less work to do. The most common methods include gravity consolidation and flotation. Gravity consolidation is based on the principle of zone sedimentation, as in the conventional settling tanks. The consolidated sludge is removed from the base of the tank. Flotation involves the introduction of air in a compression chamber. When the solution is depressurized, the dissolved air forms micro bubbles that carry the clumps of sludge to the surface, where they are removed. (b) Anaerobic Digestion: digestion has the following aims: to destroy dangerous microorganisms, to stabilize unstable substances and organic material present in the crude sludge, reduce the volume of the sludge through liquefaction, gasification, and consolidation, to enable the sludge to reduce its liquid level, and to allow the use of the sludge—after stabilization—as a fertilizer or soil conditioner. Without oxygen, only anaerobic bacteria survive, which are able to use combined oxygen. Acidogenic bacteria breakdown carbohydrates, proteins, and lipids, turning them into volatile acids. Methanogenic bacteria convert a large part of these acids into gases, mainly methane. The stabilization of these substances can also be performed by addition of chemicals, a process known as chemical stabilization. (c) Chemical Conditioning: chemical conditioning results in the coagulation of solids and the freeing of absorbed water. Conditioning is used before the mechanical drying systems, such as filtration, centrifuging, etc. The chemicals used include iron chloride, lime, aluminum sulfate, and organic polymers. (d) Press Filters: drying in the press filters occurs under high pressure. The advantages of this system include: high concentration of solids in the sludge cake, low turbidity in the filtrate and high solid retention. The resulting proportion of solids is between 30 and 40 % for a 2–5 h filtration cycle—the time needed to fill the press, maintains it under pressure, open it, remove the cake, and close the press. (e) Thermal Drying: thermal drying of the sludge is the process of reduction through evaporation of water into the atmosphere by means of heat, resulting in a proportion of solids between 90 and 95 %. This reduces the final volume of the sludge significantly [45].

Figure 9.3 illustrates an upflow anaerobic sludge blanket (UASB) reactor system, with 8 L/s treatment capacity and 66 % BOD efficiency of organic matter removal, which is an anaerobic treatment system wherein the organic matter is digested, absorbed, and metabolized into bacterial cell mass and biogas. Anaerobic digestion is the degradation of organic material without the aid of oxygen.

The UASB process is a combination of physical and biological processes. The main feature of physical process is the separation of solids and gases from the

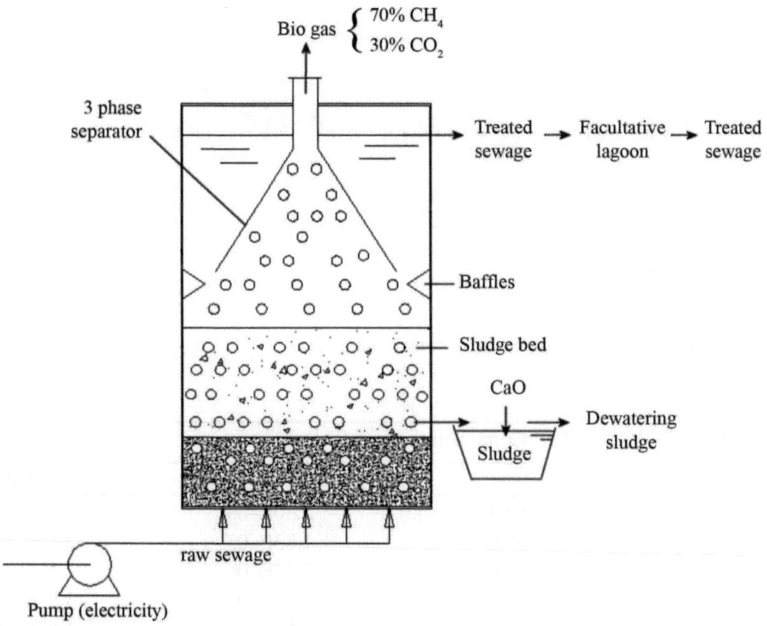

Fig. 9.3 Facultative lagoon–UASB reactor system Mora [42]

liquid, and that of biological process is the degradation of decomposable organic material under anaerobic conditions.

In the UASB treatment concept, the treatment tank consists of an upflow reactor with feed distribution internal system at the bottom of the reactor and a three-phase separator (gas, liquid, solid) at the top. The wastewater is evenly distributed over the reactor bottom through feed inlet pipes and flows upwards through a bed of anaerobic sludge in the lower part of the reactor called the digestion compartment. During the passage through the sludge bed, particulate matter is entrapped and the degradable matter is completely or partially digested. Dissolved organic matter is removed from the solution by the anaerobic bacteria and converted into biogas and a small fraction into new bacterial biomass. The biogas provides a gentle mixing in the sludge bed. In the upper part of the reactor, a three-phase separator is installed.

The biogas produced is collected in a gas collector (gas holder) from where it is withdrawn. The remaining water sludge mixture enters a settling compartment where the sludge can settle and flow back into the digestion compartment. After settling, the water is collected in the effluent gutters and discharged out of the reactor to the final polishing unit (FPU) to meet discharge standards.

The treated sewage in UASB reactor is disposed after polishing in a facultative lagoon [40].

Figure 9.4 shows the processes of the WTP Cañaveralejo, that is a chemically enhanced primary treatment (CEPT), with organic material removal of 47 % BOD and 3.849 m^3/s treatment capacity. The description of the primary treatment is similar to that presented for the WTP Barueri. In physical–chemical treatment, the

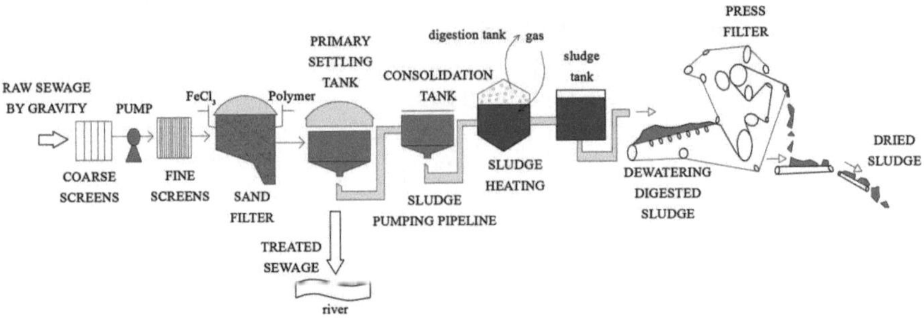

Fig. 9.4 Cañaveralejo wastewater treatment plant [44]

Table 9.10 Composition of the raw and treated sewage for Barueri WTP [42]

Composition (mol L^{-1})		
	Raw sewage	Treated sewage
COD	2.30E–03	3.01E–04
NH_3	1.83E–03	5.80E–04
NO_3	2.42E–06	1.11E–04
NO_2	2.01E–07	3.50E–06
S_2	1.60E–05	1.60E–05
SO_4	4.53E–04	3.30E–04
Cd	6.23E–08	4.00E–08
Ni	1.21E–06	6.81E–07
Ag	1.11E–07	3.71E–08
Zn	6.73E–06	1.62E–06
Mg	1.70E–06	1.30E–06
Mo	2.08E–07	2.08E–07
Pb	1.26E–07	7.40E–08
Cu	1.42E–06	2.72E–07
Cr	2.40E–06	6.15E–07
Fe	5.91E–05	8.43E–06
Alcohol	2.34E–06	3.61E–07
P	1.65E–04	7.75E–05
Detergent	5.40E–04	3.25E–05
Sn	1.20E–06	1.10E–06

solid matter is removed by means of coagulation, flocculation, and sedimentation processes. In coagulation, low concentrations iron salts are employed, alone, or in combination with cationic polymers. Flocculation is achieved after adding anionic polymers and the action of electrostatic forces that promote the formation of flakes of coagulated larger particles. During sedimentation there is a raise in the speed of sedimentation of particles due to the increase of their size. The settlement unit is similar to the conventional decanting unit, adding only the system of dosage and application of coagulants and polymers [41].

Table 9.11 Sludge composition of Barueri WTP [42]

Components	Composition (mol kg^{-1})
COD (mol L^{-1})	1.14E–01
Cd	1.30E–04
Pb	9.70E–04
Cu	9.50E–03
Cr	1.40E–02
Mg	5.50E–03
Fe	6.00E–01
Ni	5.30E–03
Zn	3.54E–02
Ag	5.60E–04
Mo	2.10E–04

Table 9.12 Composition of the raw and treated sewage of the Facultative Lagoon–UASB reactor system [42]

Composition (mol L^{-1})		
	Raw sewage	Treated sewage
COD	3.34E–03	1.09E–03
$CaCO_3$	5.88E–03	5.50E–03
NO_2	1.47E–07	7.92E–07
Cl	1.77E–03	1.62E–03
SO_4	1.94E–03	–

Table 9.13 Sludge composition of UASB reactor [42]

Components	Composition (mol kg^{-1})
COD (mol L^{-1})	4.16E–01
Ca	1.10E–04
Mg	3.81E–05
K	3.99E–04
Na	1.02E–03
P	5.62E–03
B	1.32E–04
Cu	4.68E–03
Zn	5.37E–03
Mn	3.20E–03
Fe	7.89E–03

Tables 9.10, 9.11, 9.12, 9.13, 9.14 and 9.15 present the raw and treated sewage composition as well as the sludge composition of three analyzed wastewater treatment plants: the Wastewater Treatment Plant Barueri located in the Metropolitan Area of Sao Paulo (Brazil); a facultative lagoon—UASB reactor system, located in the rural area of Ginebra (Colombia), and the Wastewater Treatment Plant Cañaveralejo, located in the Urban area of Cali (Colombia).

Table 9.14 Composition of the raw and treated sewage of Cañaveralejo WTP [42]

Composition (mol L^{-1})		
	Raw sewage	Treated sewage
COD	2.20E–03	1.39E–03
$CaCO_3$	1.93E–03	1.86E–03
Cl	1.51E–03	1.52E–03
Cd	5.67E–08	5.34E–08
Ni	7.86E–07	6.01E–07
Ag	1.11E–07	1.11E–07
Zn	2.36E–06	1.24E–06
Pb	5.31E–07	5.31E–07
Cu	4.64E–07	2.36E–07
Cr	1.16E–06	9.14E–07
Fe	4.95E–05	4.85E–05
P	1.76E–04	1.29E–04
Detergent	1.62E–04	1.25E–04
Hg	8.47E–10	8.47E–10

Table 9.15 Sludge composition of cañaveralejo WTP [42]

Components	Composition (mol kg^{-1})
COD (mol L^{-1})	1.73E–01
Cd	9.79E–05
Pb	1.69E–03
Cu	4.07E–03
Cr	2.16E–03
Fe	8.20E–01
Ni	1.69E–03
Zn	1.38E–02
Ag	2.23E–04
Hg	2.38E–07

9.3.3 Exergy Evaluation of the Environmental Performance and Renewability of the WTP

The environmental performance and renewability of the wastewater treatment process is done by means of evaluating the environmental exergy efficiency ($\eta_{b,env}$) and the renewability exergy index (λ).

The analysis of the environmental and renewability performance was carried out for the three described WTP considering operation in steady state conditions and using annual average data of each process. The chemical exergy of organic matter in the wastewater was calculated according to Eq. 9.4, proposed by Tai et al. [29] that established a relation between chemical exergy of organic substance and the COD:

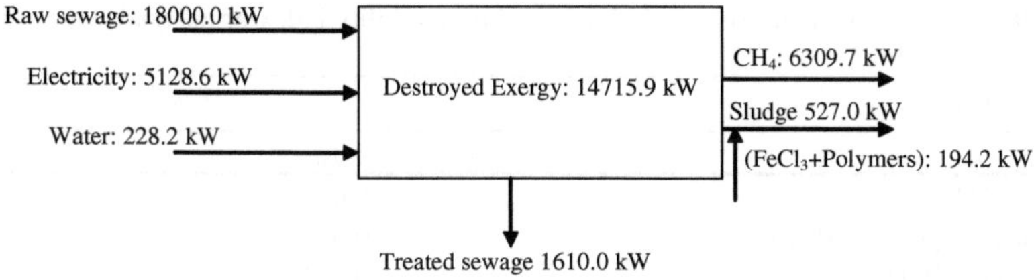

Fig. 9.5 Exergy balance of Barueri WTP [42]

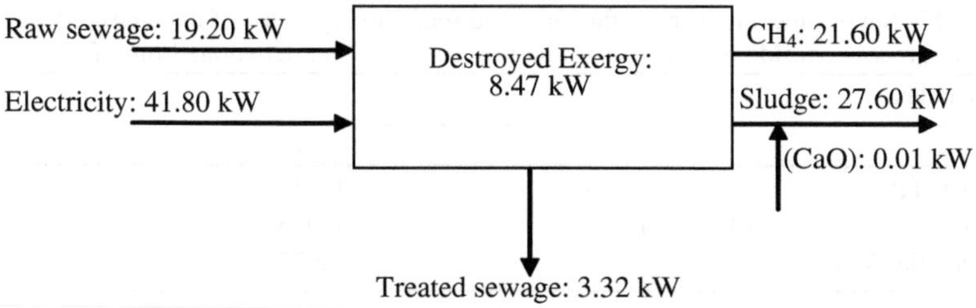

Fig. 9.6 Exergy balance of the UASB reactor [42]

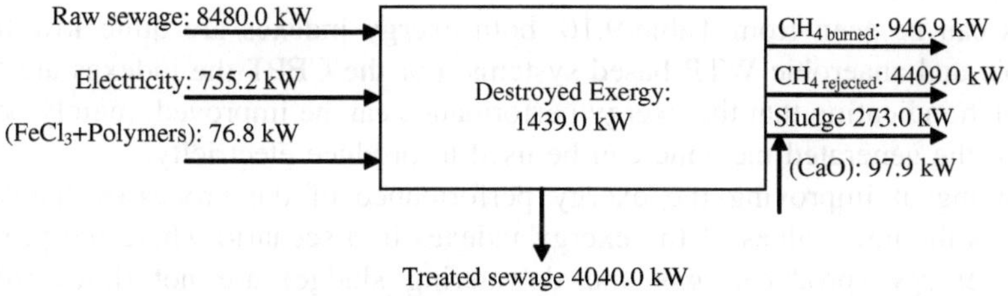

Fig. 9.7 Exergy balance of Cañaveralejo WTP [42]

$$B_{\text{org,mat.}} = 13.6 \text{ COD} \tag{9.4}$$

The molecular mass of sewage was assumed to be that of the substance $C_{10}H_{18}O_3N$; the exergy of inorganic substances for the raw and treated sewage was calculated considering real mixture (activity $\neq$ molar fraction) and 298.15 K as reference temperature. The exergy of sludge was calculated considering ideal mixture (activity = molar fraction), and the exergy flows due to biogas and chemicals were calculated according to standard chemical exergy data presented by Szargut et al. [19]. With the information generated by this exergy analysis, the environmental exergy efficiency, and renewability exergy index were determined and compared. A detailed description of exergy calculations is shown in Mora [42].

Figures 9.5, 9.6 and 9.7 present the exergy balances for the three analyzed wastewater treatment processes.

Table 9.16 Environmental exergy efficiency and renewability index of the analyzed wastewater treatment plants [42]

Exergy index		
Process	$\eta_{b,\text{env}}$	λ
Barueri WTP	0.070	0.060
Facultative–lagoon UASB reactor	0.054	0.057
Cañaveralejo WTP	0.394	0.770

Table 9.17 Environmental exergy efficiency and renewability index of the analyzed wastewater treatment process considering the use of the produced gas and dehydrated mud [42]

Exergy index		
Plant	$\eta_{b,\text{env}}$	λ
Barueri WTP	0.348	0.410
Facultative lagoon—UASB reactor	0.983	7.060
Cañaveralejo WTP	0.673	4.200

Table 9.16 shows the calculated exergy indexes based on the results of the exergy balances.

As can be seen from Table 9.16, both exergy indexes are quite low for the aerobic and anaerobic WTP-based systems. For the CEPT the indexes are better, although indicating that the exergy performance can be improved, mainly because part of the generated methane can be used to produce electricity.

Aiming at improving the exergy performance of the processes, Table 9.17 presents the new values of the exergy indexes in a scenario where the processes waste exergy (produced gas and dewatering sludge) are not destroyed. For instance, if all the methane is used as fuel in an internal combustion engine with 30 % thermal efficiency and sludge exergy was used for methanol production, as proposed by Ptasinski et al. [43]. These procedures increase significantly the values of the exergy indexes, as shown in Table 9.17.

The values obtained for the renewability exergy index greater than 1 (see Table 9.17), mean that the exergy of the products of sewage treatment system (UASB reactor-Facultative Lagoon and WTP Cañaveralejo), could be used to restore the previous environmental conditions and still have a net exergy flow rate for another purpose.

The restoration of the environment in this case can be related to the production of electricity from the produced biogas, which replaces the effect caused on the environment by the process power network electricity consumption, and the net exergy flow rate can be represented by an excess of produced electricity and by the use of dewatering mud for agricultural purposes or still as an input in the methanol production.

9.3.4 Concluding Remarks

The exergoecology analysis, supplemented with the exergy indexes, is a scientific methodology with well-defined criteria to assess and quantify the environmental performance of sewage treatment processes on a single basis: the exergy concept. With the application of this methodology it is possible to compare and characterize the environmental exergy performance and renewability of WTP technologies.

The environmental exergy efficiency identifies the technical inefficiencies in the conversion of organic matter in sewage flows, and highlights clearly that the technology used to utilize the organic matter in sewage is far from being optimized. This is because the technical solutions have not considered the recovery of exergy from organic matter as an important aspect. According to Hellström [35, if a urine separation system was included in the wastewater treatment plant, the nutrients exergy recovery could be improved with a consequent increase of the environmental exergy efficiency.

The global comparison of the three analyzed sewage treatment processes indicates that the process with the higher environmental performance and renewability values, considering the methane and sludge of process as useful effect, was the Facultative Lagoon—UASB Reactor system, with values, respectively, of 0.983 and 7.060 (see Table 9.17). That is, environmental performance is better as greater is the potential for recovery the by-products of the process.

As the values of the exergy indexes are influenced by the definition of the boundaries of the considered control volume, it is important to observe the size and compatibility of the control volumes in order to avoid distortions in a comparative analysis.

References

1. Ayres RU, Ayres LW, Martinàs K (1995) Waste potential entropy: the ultimate ecotoxic. Économique appliquè. INSEAD, Fontainebleau
2. Ayres RU, Ayres LW, Martinàs K (1996) Eco-thermodynamics: exergy and life cycle analysis. INSEAD, Fontainebleau
3. Gong M (1999) On exergy as an environmental indicator. Chalmers University of Technology, Gothenburg
4. Wall G (1993) Exergy, ecology and democracy—concepts of a vital society or a proposal for an exergy tax. In: international conference on energy systems and ecology, Krakow
5. Dewulf J, Van Langenhoven H, Dirckx J (2001) Exergy analysis in the assessment of the sustainability of waste gas treatment systems. Sci Total Environ 273:41–52
6. Rosen MA, Dincer I (1997) On exergy and environmental impact. Int J Energy Res 21(7):643–654
7. Wall G, Gong M (2001) On exergy and sustainable development–part 1: conditions and concepts. Exergy 1:128–145
8. Bastianoni SA (1998) Definition of 'pollution' based on thermodynamic goal functions. Ecol Model 113:163–166

9. Botero EA (2000) Rating exergy of natural resources, minerals, water and fossil fuels. Ph.D. Thesis. Department of Mechanical Engineering, University de Zaragoza, Zaragoza, (Spanish)
10. Creyts JC, Carey VP (1997) Use of extended exergy analysis as a tool for assessment of the environmental impact of industrial processes. Adv Energy Syst Div 37:129–137
11. Makarytchev SV (1997) Environmental impact analysis of ACFB-based gas and power cogeneration. Energy 23(9):711–717
12. Szargut J (2002) Application of exergy for the determination of the pro-ecological tax replacing the actual personal taxes. Energy 27:379–389
13. Valero A, Uche J, Valero A et al (2006) Physical hydronomics: application of the exergy analysis to the assessment of environmental costs of water bodies. The case of the inland basins of catalonia. Department of Mechanical Engineering, University de Zaragoza, Zaragoza. Avaliable via DIALOG. http://teide.cps.unizar.es:8080/pub/publicir.nsf/codigos/o436/$FILE/cp0436.pdf. Cited in 7 Mar 2008
14. Jorgensen SE (1988) Use of models as experimental tool to show that structural changes are accompanied by increased exergy. Ecol Model 41:117–126
15. Jorgensen SE (1992) Exergy and ecology. Ecol Model 63:185–214
16. Fuliu X (1997) Exergy and structural exergy as ecological indicators for the development state of the lake chaohu ecosystem. Ecol Model 99:41–49
17. Bendoricchio G, Jorgensen SE (1997) Exergy as goal function of ecosystems dynamic. Ecol Model 102:5–15
18. Jorgensen SE, Nielsen SN (2007) Application of exergy as thermodynamic indicator in ecology. Energy 32:673–685
19. Szargut J, Morris DR, Steward FR (1988) Exergy analysis of thermal, chemical, and metallurgical processes. Hemisphere Publishing Corporation, New York
20. Szargut J, Morris DR (1987) Cumulative exergy consuption and cumulative degree of perfection of chemical process. Int J Energ Res 11:245–261
21. EPA US (1993) An overview of underground storage tank remediation options, solid waste and emergency response, 5403 W EPA 510-F-93-029. U S Government Printing Office, Washington
22. Dewulf J, Van Langenhoven H (2001) Assessment of the sustainability of technology by means of a thermodynamically based life cycle analysis. Environ Sci Technol 9(267):273
23. Rettenberger G (1987) Gas haushalt von deponien: ablagerung von abfällen. Verlag für Energie und Umwelt, Berlin
24. Muga HE, Mihelcic JR (2008) Sustainability of wastewater treatment technologies. J Environ Manage 88:437–447
25. Lundin M, Molander S, Morrinson GM (1997) Indicators for the development of sustainable water and wastewater systems. In: Proceedings of the 3rd annual International sustainable development research conference, Manchester
26. Balkema AJ, Preisig HA, Otterpohl R et al (2002) Indicators for the sustainability assessment of wastewater treatment systems. Urban Water 4:153–161
27. Palme U, Lundin M, Tillman AM et al (2005) Sustainable development indicators for wastewater systems—researchers and indicator user in a co-operative case study. Resour Conserv Recy 43:293–311
28. Hellström D (1997) An exergy analysis for a wastewater treatment plant: an estimation of the consumption of physical resources. Water Environ Res 69:44–51
29. Tai S, Matsushige K, Goda T (1986) Chemical exergy of organic matter in wastewater. Int J Environ Stud 27:301–315
30. Chen GQ, JI X (2007) Chemical exergy based evaluation of water quality. Ecol Model 200:259–268
31. Huang LQ, Chen GQ, Zhang Y et al (2007) Exergy as a unified measure of water quality. Commun Nonlinear Sci Numer Simul 12:663–672
32. Chen B, Chen GQ, Hao FH et al (2007) The water resources assessment based on resource exergy for the mainstream yellow river. Commun Nonlinear Sci Numer Simul 14:331–344

33. Zaleta-Aguilar A, Ranz L, Valero A (1998) Towards a unified measure of renewable resources availability: the exergy method applied to the water of a river. Energ Convers Manag 39:1911–1917
34. Gallegos-Muñoz A, Zaleta-Aguilar A, Gonzalez-Rolón B et al (2003) On an exergy efficiency definition of a wastewater treatment plant. Int J Thermodyn 6:169–176
35. Hellström D (2003) Exergy analysis of nutrient recovery processes. Water Environ Res 48:27–36
36. Dewulf J, Van Langenhove H (2005) Integrating industrial ecology principles into a set of environmental sustainability indicators for technology assessment. Resour Conserv Recy 43:419–432
37. Manish S, Indu RP, Rangan B (2006) Sustainability analysis of renewables for climate change mitigation. Energy for Sustain Dev 10:25–36
38. Chen GQ, Jiang MM, Yang ZF et al (2009) Exergy assessment for ecological economic system: chinese agriculture. Ecol Model 220:397–410
39. Torío H, Angelotti A, Schmidt D (2009) Exergy analysis of renewable energy-based climatisation systems for buildings: a critical view. Energ Buildings 41:248–271
40. National buildings construction corporation limited (2005) Avaliable via DIALOG. http://www.nbccindia.com/nbccindia/public/jsp_pub/technology.jsp. Cited in 5 Mar 2008
41. Cete Poli/UFRJ (2001) Experimental Center Sewage Treatment UFRJ. Avaliable via DIALOG. http://www.saneamento.poli.ufrj.br/cete/main_cete_missao.htm. Cited in 27 Jan 2009
42. Mora-Bejarano CH (2009) Exergoecological evaluation process wastewater treatment. Ph.D. Thesis—Polytechnic School of the University of São Paulo, São Paulo, Brazil (in Portuguese)
43. Ptasinski KJ, Hamelinck C, Kerkhof PJAM (2002) Exergy analysis of methanol from the sewage sludge process. Energ Convers Manage 43:1445–1457
44. Empresas Municipales de Cali (2001) Plan de manejo ambiental PTAR cañaveralejo. EMCALI, Cali
45. Seawage treatment (2004) In: SABESP, Companhia de Saneamento Básico do Estado de São Paulo. Avaliable via DIALOG http://www.sabesp.com.br/english/o_que_fazemos/coleta_e_tratamento/tratamento_de_esgotos/default.htm. Cited in 6 Jul 2004
46. Lattouff R, Oliveira S Jr (2003) Exergy analysis of environmental impact mitigation processes. In: Proceedings of 16th International Conference on Efficiency, Cost, Optimisation, Simulation, and Environmental Impact Of Energy Systems, Copenhagen

Chapter 10
Exergy Analysis and Human Body Behavior

Symbols

A	Body area (m^2)
ATP	Adenosine triphosphate
$\boldsymbol{B}$	Exergy (J)
B	Exergy rate/flow rate (W)
b	Specific exergy (J/kg)
C	Thermal capacity (J/K)
c_p	Specific heat at constant pressure (J/kg K)
f_{cl}	Ratio between the surface area of the clothed and the nude segment
H	Enthalpy flow rate (W)
h	Specific enthalpy (J/kg) and heat transfer coefficient (W/m^2 K)
h_{lv}	Enthalpy of vaporization (J/kg)
M	Metabolism (W)
M	Mass flowrate (kg/s)
P	Pressure, partial pressure (Pa)
Q	Heat transfer rate (W)
R	Gas constant (J/kg K)
R_{cl}	Thermal resistance of clothes (m^2 K/W)
S	Entropy rate/flow rate (W/K)
s	Specific entropy (J/kg K)
s_{lv}	Entropy of vaporization (J/kg K)
U	Internal energy (J)
T	Temperature (°C, K)
t	Time (s)
w	Skin wettedness
W	Power (W)

Greek symbols

η	Second Law efficiency
σ	Steffan–Boltzmann constant ($Wm^{-2}K^{-4}$)

S. de Oliveira Jr., *Exergy*, Green Energy and Technology,
DOI: 10.1007/978-1-4471-4165-5_10, © Springer-Verlag London 2013

φ	Relative humidity
ω	Absolute humidity

Subscripts

0	Reference
1	Lung control volume
2	Tissue control volume
a	Environment air, ambient air
ar	Arterial blood
bl	Blood
c	Convective
cl	Clothing
CO_2	Carbon dioxide
dest	Destroyed
e	Evaporative
ex	Expired, expired air
i	Inlet
liq	Liquid part of the blood
met	Metabolism
out	Outlet
O_2	Oxygen
r	Radiative
res	Respiration
sk	Skin
t, ti	Tissues
ve	Venous blood
w	Water vapor

10.1 Entropy Generation and Life

One of the most fascinating and challenging application of the Second Law of Thermodynamics is the attempt to study and understand the evolution of life and living organisms. Everyone knows that human life has a beginning, a development, and an end. In other words, the evolution of living organisms is a typical example of irreversible process!

One of the first authors who tried to describe life as a function of the physical quantity entropy was Schördinger [1] who suggested that the body extracts negative entropy from the environment to compensate the entropy produced continually. This approach was based on statistical mechanics that relates entropy to the degree of order of the system to affirm that the organism absorbs order continually from the environment.

Two years after, Prigogine and Wiame [2] proposed a principle that a biological organism tends to progress to a state of minimum entropy production (minimum entropy production principle); their theory is based on the foundation that most of entropy production is related to metabolism.

Stoward [3] studied the entropy production in bacterial culture attempting to confirm the minimum entropy principle. Zotin and Zotina [4] examined the validity of Prigogine–Wiame theory in a wide range of organisms and claimed that this theory embraces a very wide of phenomena in biology including the process of growth, aging, death, healing wounds, regeneration, and malignant growth. The conclusions of the study are: (1) during the growth and aging the living organism is continuously approaching the steady-state of entropy production; (2) the deviations of the steady-state occur in the early stages of embryo, in the initial period of regeneration and healing of wounds, and in a malignant growth (cancerous tissues has a high rate of entropy production and they do not pass to the phase when the entropy production decreases).

Balmer [5] studied the fish *Nothobranchius guentheri*, a species with life cycle of 12 months, which lives in intermittent rivers of the East Africa. The results of the study indicate that the entropy generation decreases continuously with increase of time over their lifespan, except in the early stages of life, when the entropy production increases with aging, which differ from Zotin and Zotina [4] who obtained that entropy production always decreases with lifespan.

Aoki [6–9] proposed a method to calculate entropy production in organism by accounting the entropy flow rates and rates exchanged with the environment associated with radiation, convection, respiration, and evaporation. Initially, the entropy generation in a deer [6] was evaluated, then the entropy production in a human under basal conditions [7, 8]. In these two studies, the effect of environmental temperature, forced air currents around the body, clothing, exercises, and chill were quantified and analyzed based on experimental results. It was analyzed [9] the effect of human body growth and aging, demonstrating that minimum entropy production principle is also valid for an adult human. In this work, a three stage hypothesis of entropy production in human life was proposed.

Batato et al. [10] did a pioneer study of the exergy analysis of human body. Based on experimental results, it was found that the metabolism and metabolic exergy have equivalent values. The authors also indicated that although the energy losses to the environment are large, the exergy losses are very small, and they also found that the Second Law efficiency of the human body at rest is almost zero.

Rahman [11] improved the method proposed by Aoki [6–9] by adding to the model some details of the environment conditions such as air current velocities and relative air humidity. Instead of using experimental results, the heat exchange to the environment from the thermal coefficients was estimated (radiation and convection) and clothing was also considered. The results confirmed the minimum entropy principle, and indicated that the physical activities tend to increase the entropy production.

Silva and Annamalai [12, 13] applied the Second Law of Thermodynamics on the biochemical reactions that are typical of human metabolism. They analyzed the entropy generation over the lifespan and their results confirmed the minimum entropy principle. From the entropy generation over life it was predicted that the life expectancy of average U.S. male and female is close to the actual values obtained from statistic data; and it was shown that a reduction in protein diet and in the caloric content of the ingestion yields to a percentage increase in the lifespan. Moreover, it was concluded that physical exercises rises entropy production.

A physiological two-node model to simulate the heat and mass transfer of the human body to the environment was performed by Prek [14, 15]; Prek and Butala [16] aiming to obtain the relation between exergy consumption with a combination of environment conditions and thermal comfort condition. It was found that for a given physiological parameter only a combination of environment conditions ensures the minimum exergy consumption. Ferreira and Yanagihara [17] performed a similar analysis, but with a more detailed model for the thermoregulatory system that led to similar results indicated by Batato et al. [10].

Lems [18] presented an exergy analysis for different energy conversion processes of living systems, including cells metabolism and photosynthesis. This work describes a very detailed methodology to calculate the exergy of the different streams involved in the biochemical processes. The results obtained account for efficiencies up to 60 % considering the conversion of carbon fuels into adenosine triphosphate (ATP) in living cells. However, the way in which this ATP is used was not considered.

Luo [19] shows that the entropy production of tumorous cells is higher than that of normal cells, and therefore, entropy flow is transferred from tumorous toward healthy cells. This process results in information concerning the cancer propagating into the surrounding normal tissue. In his study, it is proposed that the entropy generation mechanism possibly provides a basis for a novel approach to anticancer therapy through the use of ultrasound irradiation because ultrasound absorption results in additional entropy production in normal tissues reducing the difference between the entropy of cancerous and healthy cells.

10.2 Exergy Behavior of the Human Body

10.2.1 Introduction

The application of exergy analysis to human body aims at determining the quality of the energy conversion processes that occur in the several systems which compose the human body. By using this approach, it is possible to quantify the exergy destroyed rate and the exergy efficiency of these process and of the whole body, under different operational and ambient conditions. The information obtained with this analysis allows comparing a healthy body performance with that of a body with a particular disease aiding the realization of the diagnosis of this type of disease and identifying the main sources of the irreversible processes caused by the disease evolution.

10.2.2 Human Body Thermal Model

The human thermal model developed by Ferreira and Yanagihara [20] is composed of the thermoregulatory and passive system. The first system is related to physiological responses to changes in thermal environment or activity level: vasodilatation or constriction, shivering, and sweating. The second system includes heat conduction inside the body, heat transfer by convection because of the blood flow, and heat and mass transfer between the body and the environment. The human body is divided into 15 cylinders with elliptical cross-section representing the head, neck, trunk, arms, forearms, hands, thighs, legs, and feet whose result is a model with realistic dimensions. Each cylinder has a combination of some of these tissues: skin, fat, muscle, bone, brain, viscera, lung, and heart. With the exception of head and trunk that have composition as function of the cross-section, all cylinders have composition independent of the cross-section.

The energy equation with constant density and specific heat and metabolism (source term) is solved for each tissue. The numerical solution is achieved using a coordinate transformation which transforms an elliptical cylinder in Cartesian space in a parallelepiped in the new coordinate system. To calculate the heat transfer between tissues and blood it was necessary to divide vessels in two classes: the small ones (that can be treated as part of continuum) and the big ones. The model is similar to the one proposed by Pennes [21] with the difference that the arterial blood temperature depends on the position inside the tissue and it is not equal to the body core temperature. The big vessels can be modeled as proposed by Wissler [22], using two reservoirs, one of arterial blood and other of venous blood, with the exception of trunk that is modeled with only one reservoir. From this model it is possible to obtain heat and mass transfer to the environment through skin due to radiation, convection, and evaporation, as well as mass flow rates due to respiration (difference of temperature and humidity of inspired and expired air). It is also possible to obtain the transitory response of body due to a variation on environment conditions (temperature dependence with time). A representation of the model is indicated in Fig. 10.1.

The energy balance applied to the human body can be given by Eq. 10.1

$$\frac{\mathrm{d}U}{\mathrm{d}t} = -(Q_\mathrm{c} + Q_\mathrm{r} + H_\mathrm{e} + \Delta H_\mathrm{res}) - W \tag{10.1}$$

where:

H_e = evaporation enthalpy flow rate
Q_c = convective heat transfer rate
Q_r = radiative heat transfer rate
ΔH_res = net respiration enthalpy flow rate
$\mathrm{d}U/\mathrm{d}t$ = total internal energy variation rate
W = power performed

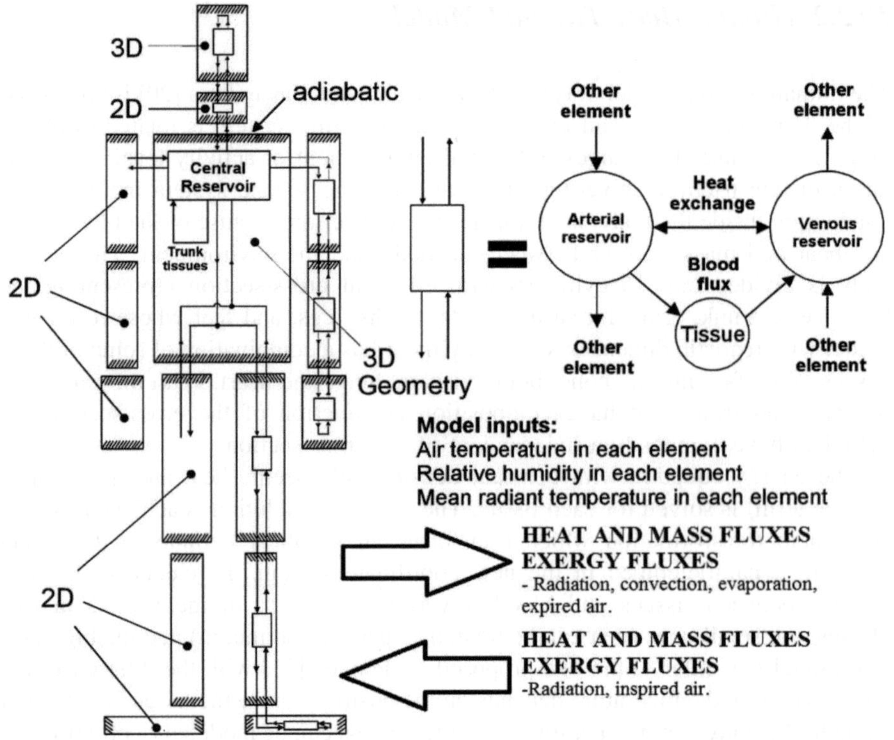

Fig. 10.1 Overview of the passive system model, showing possible boundary conditions, geometry, and circulatory system used in each segment [29]

The total internal energy variation rate of the body was assumed as a sum of the metabolic internal energy rate (M) and the time-dependent internal energy variation rate due to transient environment conditions ($\mathrm{d}U/\mathrm{d}t|_{\Delta T}$), as indicated by Eq. 10.2:

$$\frac{\mathrm{d}U}{\mathrm{d}t} = -M + \left.\frac{\mathrm{d}U}{\mathrm{d}t}\right|_{\Delta T} \tag{10.2}$$

With this distinction, the First Law of Thermodynamics can be rewritten according to Eq. 10.3:

$$\left.\frac{\mathrm{d}U}{\mathrm{d}t}\right|_{\Delta T} = M - (Q_\mathrm{c} + Q_\mathrm{r} + H_\mathrm{e} + \Delta H_\mathrm{res}) - W \tag{10.3}$$

Assuming symmetrical and uniform environment, the combined effect of heat transfer due to radiation and convection is calculated by using Eq. 10.4. A is the body area, T_sk is the skin temperature, T_o is the air temperature, h_o is the combined heat transfer coefficient (convection and radiation) [23], R_cl is the thermal resistance of clothes, and f_cl is the ratio between the surface area of the clothed and nude segment (f_cl becomes one for a nude part of the body).

$$Q_r + Q_c = A\frac{T_{sk} - T_o}{R_{cl} + \frac{1}{f_{cl}h_o}} \tag{10.4}$$

The enthalpy losses by evaporation at the surface of each element can be calculated by Eq. 10.5. $P_{w,sk}$ and $P_{w,a}$ are the water vapor pressure at the skin surface temperature and at environment temperature, w is the skin wettedness (varies from 0.06 when there is only water diffusion to 1.0 when the skin is completely wetted by sweat), and h_e is the equivalent evaporative transfer coefficient of the element.

$$H_e = \mathrm{Aw}\frac{P_{w,sk} - \varphi_a P_{w,a}}{R_{e,cl} + \frac{1}{f_{cl}h_e}} \tag{10.5}$$

The enthalpy flow rates from and to the environment due to respiration can be calculated by Eq. 10.6. Respiration flow rates are divided in 45 and 25 % to neck and head muscles and 30 % to the lung. h_w (T_{ex}) and h_w (T_a) are the vapor-specific enthalpy at expired and inspired air temperatures, respectively.

$$\Delta H_{res} = m_{res}c_{p,a}(T_{ex} - T_{in}) + m_{res}(\omega_{ex}h_w(T_{ex}) - \omega_a h_w(T_a)) \tag{10.6}$$

The pulmonary ventilation rate, m_{res}, is directed related to the metabolic internal energy [24] variation (M—metabolism) as indicated in Eq. 10.7.

$$m_{res} = 1.433 \cdot 10^{-6}\ \mathrm{AM} \tag{10.7}$$

10.2.3 Exergy Analysis

The exergy analysis is applied to the control volumes shown in Fig. 10.1. In order to develop the exergy balances it is necessary to define the environment/reference conditions. In this study, it is considered that environmental parameters are those of the ambient air: $T_0 = T_a$, $P_0 = P_a$, and $\varphi_0 = \varphi_a$. It is also necessary to know a time history of the temperature of the tissues and blood for all nodes of the compartments.

Equation 10.8 presents the general exergy balance and Eq. 10.10 indicates the application of the balance to the model. As done for the internal energy time variation, $\mathbf{dB}/\mathrm{d}t$ is the total exergy variation rate of the body (Eq. 10.8), B_M is the exergy rate associated to the metabolism, and $\mathbf{dB}/\mathrm{dt}|_{\Delta T}$ is the time variation of the exergy of the body due to transient environment conditions (Eq. 10.9).

$$\frac{\mathbf{dB}}{\mathrm{d}t} = \sum B_i - \sum B_o + \sum_k Q_k\left(1 - \frac{T_0}{T_k}\right) - W - B_{dest} \tag{10.8}$$

$$\frac{d\mathbf{B}}{dt} = -B_M + \left.\frac{d\mathbf{B}}{dt}\right|_{\Delta T} \tag{10.9}$$

$$B_{dest} = \left(B_M - \left.\frac{d\mathbf{B}}{dt}\right|_{\Delta T}\right) - (B_e + B_{res} + B_c + B_r) - W \tag{10.10}$$

As Batato et al. [10] achieved that the metabolic internal energy and the metabolic exergy are very close, in this analysis the approximation $B_M \approx M$ is considered.

The thermal exergy rate associated to the heat exchange with the environment by convection and radiation can be determined by Eqs. 10.11 and 10.12.

$$B_c = Q_c\left(1 - \frac{T_0}{T_{sk}}\right) \tag{10.11}$$

$$B_r = Q_r\left(1 - \frac{T_0}{T_{sk}}\right) \tag{10.12}$$

The vaporization exergy through skin is calculated by the Eq. 10.13. The first term of this equation is related to vaporization of the sweat and water on the skin, and the second one is related to the difference in the concentration of saturated vapor near the skin and the concentration of the vapor in the environment.

$$B_e = m_w(h_{lv} - T_0 s_{lv}) + m_w R_w T_0 \ln\left(\frac{P_{w,sk}}{P_{w,0}}\right) \tag{10.13}$$

The exergy flow rates due to respiration are given by Eq. 10.14, and are related to the inspired and expired air rates. Assuming ideal gas behavior, this equation is divided into two, one related to the dry air (Eq. 10.15) and the other one related to the water (Eq. 10.16). It is worth to note that the second term of both equations becomes zero if the adopted reference is equivalent to the environment conditions.

$$\Delta B_{res} = \sum m_o b_o - \sum m_i b_i = \Delta B_a + \Delta B_w \tag{10.14}$$

$$\begin{aligned} \Delta B_a = {} & m_{res}\left[c_{p,a}\left(T_{ex} - T_0 - T_0 \ln\left(\frac{T_{ex}}{T_0}\right)\right) + R_a T_0 \ln\left(\frac{P_{a,ex}}{P_{a,0}}\right)\right] \\ & - m_{res}\left[c_{p,a}\left(T_a - T_0 - T_0 \ln\left(\frac{T_a}{T_0}\right)\right) + R_a T_0 \ln\left(\frac{P_{a,air}}{P_{a,0}}\right)\right] \end{aligned} \tag{10.15}$$

$$\begin{aligned} \Delta B_w = {} & m_{res}\omega_{ex}\left[c_{p,w}\left(T_{ex} - T_0 - T_0 \ln\left(\frac{T_{ex}}{T_0}\right)\right) + R_w T_0 \ln\left(\frac{P_{w,ex}}{P_{w,0}}\right)\right] \\ & - m_{res}\omega_a\left[c_{p,w}\left(T_a - T_0 - T_0 \ln\left(\frac{T_a}{T_0}\right)\right) + R_w T_0 \ln\left(\frac{P_{w,a}}{P_{w,0}}\right)\right] \end{aligned} \tag{10.16}$$

The time variation of the exergy due to transient environment conditions can be calculated by Eq. 10.17, with the hypothesis that the volume of the body does not vary significantly. The first right-side term is the temporal variation of the internal energy and can be calculated from Eq. 10.3. The second one is the temporal variation of the body entropy and can be calculated from each element (hand, arm, forearm, etc.) as indicated by Eq. 10.18.

$$\left.\frac{d\mathbf{B}}{dt}\right|_{\Delta T}=\left.\frac{dU}{dt}\right|_{\Delta T}-T_0\left.\frac{dS}{dt}\right|_{\Delta T} \tag{10.17}$$

$$\left.\frac{dS}{dt}\right|_{\Delta T}=\sum_j\frac{dS_{\text{element,j}}}{dt}=\sum_j\left(\frac{dS_{\text{bl,j}}}{dt}+\frac{dS_{\text{ti,j}}}{dt}\right) \tag{10.18}$$

For each cylinder j, the term $dS_{\text{bl},j}/dt$, the time variation of entropy in the blood compartment is indicated in Eq. 10.19, and $dS_{\text{ti},j}/dt$, the time variation of the entropy in the tissues is indicated in Eq. 10.20 (the blood in the small vessels is already considered in the tissues). In Eq. 10.19, the subscript ar indicates arterial and ve indicates venous blood reservoir. $T_{\text{ar},k}$, $T_{\text{ar},k+1}$, $T_{\text{ve},k}$, and $T_{\text{ve},k+1}$ are the temperatures of the reservoir j in the instant k and $k+1$, respectively, and $C_{\text{bl,j}}$ is the thermal capacity of the blood reservoir (specific heat of blood multiplied by the mass of the reservoir). In Eq. 10.20, i represents the tissue (skin, fat, muscle, etc.), $T_{\text{ti},k}$, $T_{\text{ti},k+1}$ are the temperatures of the tissue i in instants k and $k+1$, respectively, and $C_{\text{ti},i}$ is the thermal capacity of the tissue (specific heat of the tissue multiplied by its mass).

$$\frac{dS_{\text{bl}}}{dt}=\frac{d}{dt}\left[C_{\text{bl,j}}\ln\left(\frac{T_{\text{ar,k+1}}}{T_{\text{ar,k}}}\right)+C_{\text{bl,j}}\ln\left(\frac{T_{\text{ve,k+1}}}{T_{\text{ve,k}}}\right)\right] \tag{10.19}$$

$$\frac{dS_{\text{ti}}}{dt}=\frac{d}{dt}\left[\sum_i C_{\text{ti}i}\ln\left(\frac{T_{\text{tii}},k+1}{}T_{\text{tii}},k\right)\right] \tag{10.20}$$

By using Gibbs relations for each compartment (j = hand, arm, forearm, etc.), it is possible to obtain Eq. 10.21 that gives similar results to Eq. 10.18, but with the only necessity to calculate the mean temperature (T_{m}) of each element (hand, arm, forearm, etc.).

$$\left.\frac{dS}{dt}\right|_{\Delta T}=\sum_j\left(\frac{1}{T_{\text{m}}}\left.\frac{dU}{dt}\right|_{\Delta T}\right)_j \tag{10.21}$$

The exergy efficiency of the human body and its components, η_b, can be calculated according to the definition proposed by Batato et al. [10]:

$$\eta_{\text{b}}=\frac{B_{\text{M}}-B_{\text{dest}}}{B_{\text{M}}} \tag{10.22}$$

Table 10.1 Metabolism as a function of age for a male

Age	M (W)
18	84.3
20	83.6
25	82.0
30	80.3
40	77.0
50	73.7
60	70.4
70	67.1

10.2.4 Results and Discussion

To perform the simulations that correlate exergy destroyed and age, an equation proposed by Harris and Benedict [25] which correlates age, height, and weight with metabolism was used for an adult man. For other simulations the metabolism chosen was 79.1 W (middle aged person) originally used in Ferreira and Yanagihara [20]. Table 10.1 indicates the results of metabolism as a function of age.

Also, as used by Ferreira and Yanagihara [20], the global data of the anatomic model used are: height 1.76 m, weight 67 kg, surface 1.8 m^2, and volume of 0.0627 m^3.

In all studied cases, the air environment parameters were taken as the reference ones. Furthermore, in order to simulate different environment conditions for each age, it is necessary to know the thermal neutrality temperature (T_n). According to ASHRAE [24], the range of operative temperature, from 29 to 31 °C, provides a thermal neutral condition. Figure 10.2 indicates the calculated T_n (for each metabolism) as a function of age, for relative humidity of 50 % and air current of 0.15 m/s. The model was considered to be naked in stand up position. This figure shows that the lowest temperature of thermal neutrality is for the condition of maximum metabolism (18 years) and it tends to increase over the life span.

Figure 10.3 presents the exergy destroyed and exergy efficiency as a function of age, for an air relative humidity of 70 % and operative temperature of 32 °C. The nude model was also considered in stand up position. The principle of minimum entropy generation is valid for an adult (in this case the minimum destroyed exergy, B_{dest}) as indicated by other authors [3–5, 9, 11]. Also, the exergy efficiency tends to decrease over the life span. Therefore, the maximum exergy destroyed and efficiency was obtained for an 18-year-old man.

For an adult middle-aged man (metabolism of 79.1 W), the thermal neutrality temperature is 30 °C with relative humidity of 50 % as shown by Ferreira and Yanagihara [20]. Results presented in Table 10.2 confirm those of Batato [10] making evident that although the energy contribution of radiation, convection, evaporation, and respiration is of the same order of magnitude of the metabolism, their exergy contribution is much smaller.

The environment parameters such as temperature and relative humidity affect the values of exergy rates, as can be seen in Table 10.3. Exergy rates and flow

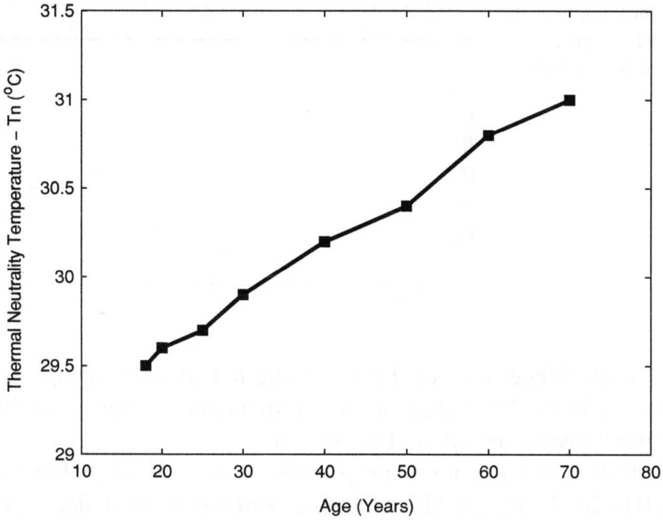

Fig. 10.2 Thermal neutrality temperature as a function of age [29]

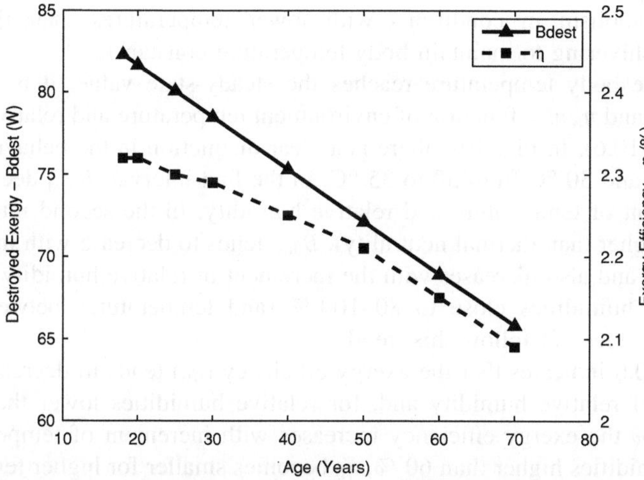

Fig. 10.3 B_{dest} and η_b as a function of life span ($T_o = 32$ °C and $\varphi_o = 70$ %) [29]

rates associated with radiation, convection, and respiration are larger when the difference between skin temperature and environment temperature are higher (lower temperatures) and respiration flow rates are also larger for lower relative humidity. Exergy flow rates due to evaporation are larger for higher temperatures and lower relative humidity.

The exergy rate associated to metabolism and destroyed exergy rate of every member of the model of the human body, shown in Table 10.4, indicates that for thermal neutrality condition compartments such as trunk and head have the highest

Table 10.2 Difference of the contribution of the energy and exergy rates and flow rates [29]

	(W)
Q_r	29.0
Q_c	23.2
E	19.7
R	7.2
B_r	0.35
B_c	0.27
B_e	1.26
B_{res}	0.15

destroyed exergy. Since most of the metabolic reactions occur in these elements (first column of Table 10.4), part of the thermal exergy released is distributed to the rest of the body by the circulatory system.

The time history of B_{dest} and η_b are described in Fig. 10.4a, b. For a temperature step from 30 to 28 °C, B_{dest} tends to increase, while η_b tends to decrease. For a step of temperature from 30 to 32 °C the opposite behavior is observed: B_{dest} decreases while η_b increases with time. From Fig. 10.4a, it is clear that the exergy destroyed becomes larger than the initial metabolism (79.1 W), and this is due to the increase of the metabolism in conditions with lower temperatures than the thermal neutrality (shivering to maintain body temperature constant).

When the body temperature reaches the steady-state value, it is possible to obtain B_{dest} and η_b as a function of environment temperature and relative humidity (Figs. 10.5, 10.6). In Fig. 10.5 there is a clear distinction in the behavior of B_{dest}, between 25 and 30 °C from 30 to 35 °C. In the first interval, B_{dest} decreases with the increment of temperature and relative humidity. In the second interval (temperatures higher than thermal neutrality), B_{dest} tends to decrease with increment of temperature and also increases with the increment of relative humidity. Although, for relative humidities close to 80–100 % (and temperatures between 30 and 35 °C) B_{dest} does not follow this trend.

Figure 10.6 indicates that the exergy efficiency (η_b) tends to decrease with the increment of relative humidity and, for relative humidities lower than approximately 60 % the exergy efficiency increases with increment of temperature. For relative humidities higher than 60 %, η_b becomes smaller for higher temperatures. Figures 10.5 and 10.6 indicate that the body is more efficient and destroys less exergy in lower relative humidities and higher temperatures.

10.2.5 Concluding Remarks

The thermal model of the human body was able to analyze the exergy behavior of the body during life span and under different environment conditions. From the range of analyzed conditions, it is possible to conclude that:

Table 10.3 Effect of environment parameter on exergy terms and on exergy efficiency [29]

T (°C)	φ(%)	B_r(W)	B_c (W)	B_e(W)	B_{res} (W)	Total (W)
27	40	0.56	0.44	1.66	0.32	2.98
27	60	0.60	0.47	0.96	0.18	2.21
30	40	0.32	0.25	1.68	0.22	2.47
30	60	0.35	0.27	1.09	0.11	1.82
33	40	0.03	0.02	3.50	0.17	3.72
33	60	0.03	0.02	2.18	0.07	2.30

Table 10.4 Member contribution to the exergy metabolism and exergy destroyed [29]

	B_M (W)	B_{dest} (W)
Hand	0.26	0.25
Arm	0.52	0.50
Forearm	0.94	0.90
Foot	0.54	0.50
Leg	1.26	1.21
Thigh	2.85	2.74
Head	15.21	14.86
Neck	0.49	0.47
Trunk	50.66	49.52
Overall	79.10	77.07

- The thermal neutrality temperature of a person is a function of age, and the minimum values occur when at the time of life when the person has the higher metabolism (18 years old). For an adult, the temperature of neutrality tends to increase with life.
- As expected, the theory of minimum entropy production is confirmed for an adult. The exergy efficiency also decreases during life span. Both quantities achieve the maximum value at the age of 18 years old.
- The energy contribution of heat rates and mass flow rates are relevant, although the associated exergy flow rates and rates are almost negligible because they are strongly dependent of the environment conditions.
- Both trunk and head have the highest exergy destruction rates, which is an expected result, since most of the metabolic reactions occur in these elements.
- The environment conditions such as temperature and relative humidity influence the exergy behavior of the body. The body tends to destroy more exergy in lower temperatures. The exergy efficiency decreases with the rise of relative humidity (for all cases). And the body has better exergy efficiency in environments with higher temperature and lower relative humidity.

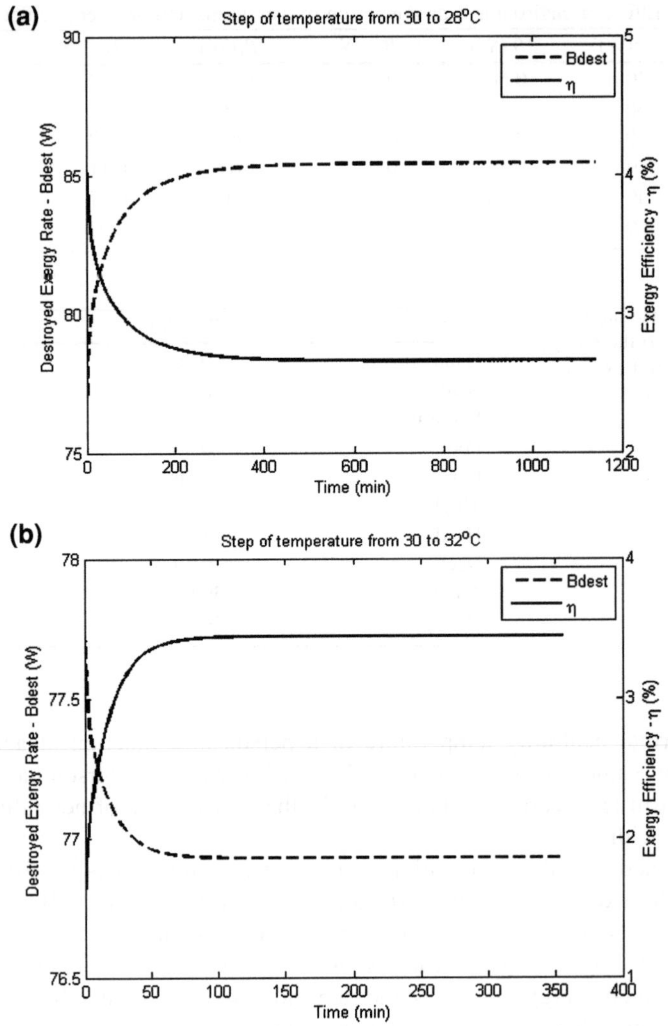

Fig. 10.4 Body response to a step in environment conditions for: lower temperature (**a**) and higher temperature (**b**) [29]

10.3 Exergy Analysis of Human Respiration Under Physical Activity

10.3.1 Introduction

The study described next aims at evaluating the exergy performance of the human respiratory system under physical activity. The analysis involves the rate at which oxygen is supplied to the lungs and transported by the blood to the tissues, and the

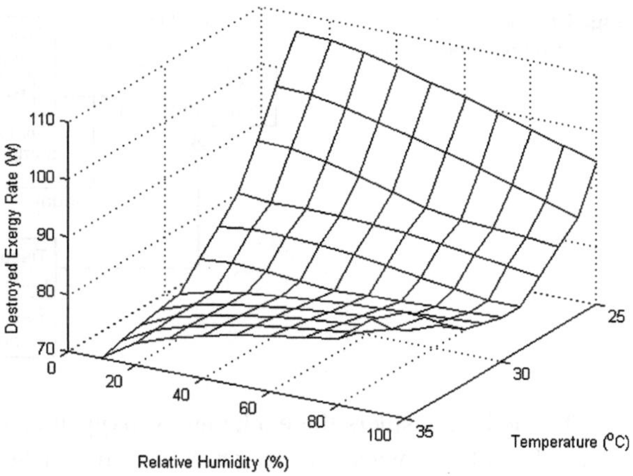

Fig. 10.5 B_{dest} as a function of different environment temperatures and relative humidities conditions

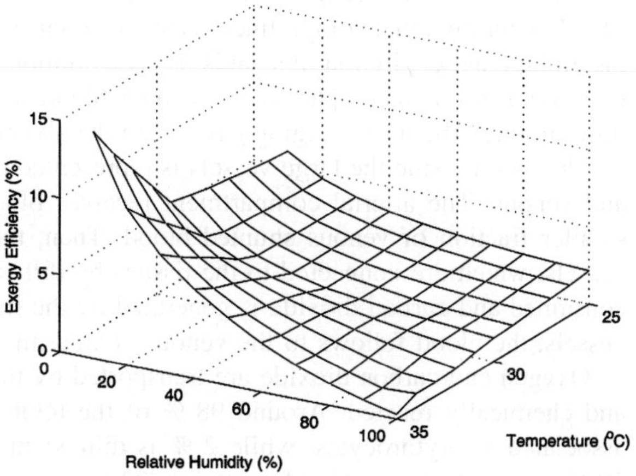

Fig. 10.6 η_b as a function of different environment temperatures and relative humidities conditions

rate of CO_2 elimination. The exergy analysis presented is based on results generated by models of the human respiratory and thermal systems previously developed by Albuquerque-Neto et al. [26], and a simplification of the model developed by Ferreira and Yanagihara [20].

10.3.2 Respiratory System Description

The mathematical model of the human respiratory system used herein was developed by Albuquerque-Neto et al. [26]. In this model, the human body is also represented by compartments. The model does not take into account the same tissue distribution of the thermal model. It is focused on the lung representation.

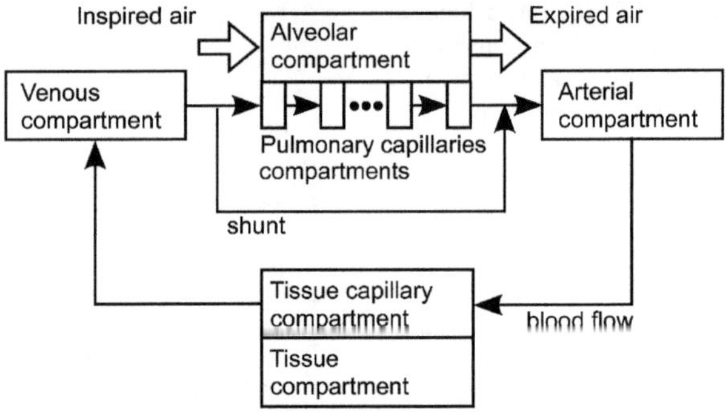

Fig. 10.7 Respiratory system representation [30]

The model considers the exchange of oxygen and carbon dioxide in the human body. The places where those gases are present are divided into compartments. A representation of the model is found in Fig. 10.7.

The air inside the lung alveolus is represented by the alveolar compartment. The flow through that compartment is the alveolar ventilation. It is the result of the inspiration and expiration mechanisms. The pulmonary capillaries are represented by a series of blood compartments. Venous blood enters the first compartment and flows through them exchanging gases with the alveolar compartment by diffusion.

The blood inside the large vessels is represented by two compartments, arterial and venous. The arterial compartment receives blood from the lung, and also a smaller fraction of venous shunted blood. Then, the blood follows to the small vessels, which are connected to the tissues by diffusion. In the tissues, oxygen is consumed and carbon dioxide is generated by the metabolism. Leaving the small vessels, the blood follows to the venous compartment and returns to the lung.

Oxygen and carbon dioxide are transported by the blood and tissues dissolved and chemically reacted. Around 98 % of the total O_2 transported is chemically associated to erythrocytes, while 2 % is diluted in the plasma. As for the CO_2, 70 % is transported as bicarbonate ion, 23 % associated to erythrocytes, and 7 % diluted in the plasma. Several equations are used to relate the gases concentration with their partial pressures.

The ventilation, the cardiac output, and the lung diffusion coefficient depend on the physical activity. They are related to the oxygen consumption rate.

With this model, it is possible to calculate the volumetric concentration and the partial pressure of the oxygen and carbon dioxide in all the compartments. The fraction of the gases in the ambient air, the altitude, and the level of physical activity are model inputs.

10.3.3 Exergy Analysis

Two control volumes were defined for the exergy analysis as shown in Fig. 10.8. The first control volume (CV1) includes the lung, the arterial compartment, and

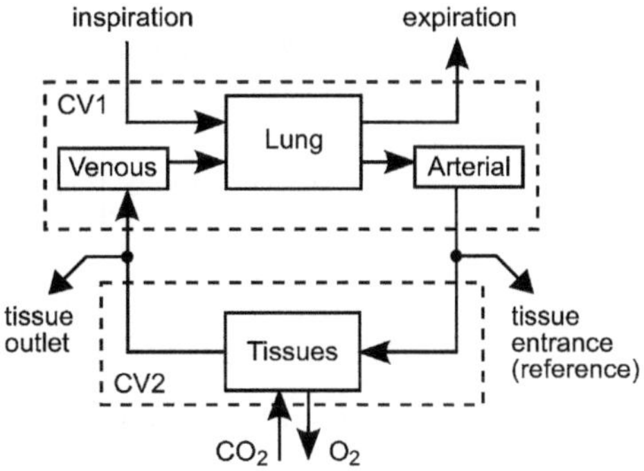

Fig. 10.8 Control volume definition [30]

the venous compartment. Across the control volume boundary, there are flow of blood, which comes from the tissue outlet and goes into the tissue entrance, and air flow, a consequence of inspiration and expiration.

The second control volume (CV2) includes the blood in the small vessels and in the tissues. Blood flows across the control volume boundary. It comes from the arterial compartment and goes to the venous compartment. The tissue metabolism is represented by oxygen and carbon dioxide flow across the boundary. The purpose of this simplification is to disregard the complexity of the metabolic chemical reactions. Hence, the energy conversion due to the oxidation of carbon fuels inside the cells is characterized by the exergy difference between the inlet oxygen and outlet carbon dioxide streams in the tissues.

In order to develop an exergy balance, it is required that data for temperature, pressure, and composition of each stream are available, as well as mass flow rates. The respiratory system model allows the determination of the oxygen and carbon dioxide partial pressure and the volumetric concentration in all the places across the boundaries: inspiration and expiration air, tissue entrance and tissue outlet blood, and tissues. The thermal system model permits the determination of the temperature in all those points. Furthermore, to calculate the exergy of each stream, a reference state must be set. The thermodynamic state of the blood, O_2 and CO_2, at the tissue entrance is defined as the reference state (see Fig. 10.8).

For the exergy analysis, some simplifications were adopted. The blood is considered to be an ideal mixture of liquid and two ideal gases (CO_2 and O_2). Thus, the exergy is associated to different ways in which O_2 and CO_2 chemically combined in the blood were not taken into consideration. These simplifications allow calculating the exergy of O_2 and CO_2 using specific heats and gas constants (R_{O_2} refer to the gas constant—8.314 kJ/kmol K—divided by the molar mass of O_2, the same for CO_2). As for the liquid part, the influence of pressure on the exergy value is considered negligible.

As the reference state was set at the entrance of the tissues, the exergy of the blood at this point is equal to zero, and the exergy of the blood at the lungs entrance is calculated considering the following equations:

$$B_{\text{liq}} = m_{\text{liq}} c_{\text{p,liq}} \left(T_{\text{t}} - T_0 - T_0 \ln \frac{T_{\text{t}}}{T_0} \right) \tag{10.23}$$

$$B_{\text{O}_2,\text{bl}} = m_{\text{O}_2,\text{bl}} \left[c_{\text{p,O}_2} \left(T_{\text{t}} - T_0 - T_0 \ln \frac{T_{\text{t}}}{T_0} \right) + T_0 R_{\text{O}_2} \ln \frac{P_{\text{O}_2,\text{t}}}{P_{\text{O}_2,0}} \right] \tag{10.24}$$

$$B_{\text{CO}_2,\text{bl}} = m_{\text{CO}_2,\text{bl}} \left[c_{\text{p,CO}_2} \left(T_{\text{t}} - T_0 - T_0 \ln \frac{T_{\text{t}}}{T_0} \right) + T_0 R_{\text{CO}_2} \ln \frac{P_{\text{CO}_2,\text{t}}}{P_{\text{CO}_2,0}} \right] \tag{10.25}$$

$$B_{\text{bl}} = B_{\text{liq}} + B_{\text{O}_2,\text{bl}} + B_{\text{CO}_2,\text{bl}} \tag{10.26}$$

Across the boundaries of the lung control volume (CV1), besides the blood flow, there is inspiration and expiration. The exergy of the inspiration has just the O_2 component, and the expiration has both the O_2 and CO_2 components. The exergy of the N_2 and water are not taken into consideration, since these components are not transferred to the blood stream, leaving with the expired air, with almost the same composition of the inlet. The following equations are used to calculate the exergy of the inspiration and expiration components:

$$B_{\text{O}_2,\text{in}} = m_{\text{O}_2,\text{in}} \left[c_{\text{p,O}_2} \left(T_{\text{a}} - T_0 - T_0 \ln \frac{T_{\text{a}}}{T_0} \right) + T_0 R_{\text{O}_2} \ln \frac{P_{\text{O}_2,\text{a}}}{P_{\text{O}_2,0}} \right] \tag{10.27}$$

$$B_{\text{O}_2,\text{ex}} = m_{\text{O}_2,\text{ex}} \left[c_{\text{p,O}_2} \left(T_{\text{ex}} - T_0 - T_0 \ln \frac{T_{\text{ex}}}{T_0} \right) + T_0 R_{\text{O}_2} \ln \frac{P_{\text{O}_2,\text{ex}}}{P_{\text{O}_2,0}} \right] \tag{10.28}$$

$$B_{\text{CO}_2,\text{ex}} = m_{\text{CO}_2,\text{ex}} \left[c_{\text{p,CO}_2} \left(T_{\text{ex}} - T_0 - T_0 \ln \frac{T_{\text{ex}}}{T_0} \right) + T_0 R_{\text{CO}_2} \ln \frac{P_{\text{CO}_2,\text{ex}}}{P_{\text{CO}_2,0}} \right] \tag{10.29}$$

The exergy destroyed in the lung control volume (CV1) is equal to:

$$B_{\text{dest},1} = B_{\text{O}_2,\text{in}} + B_{\text{bl}} - B_{\text{O}_2,\text{ex}} - B_{\text{CO}_2,\text{ex}} \tag{10.30}$$

Across the boundaries of the tissue control volume (CV2), there are the O_2 and CO_2 flow due to the metabolism. The exergy of those flows are calculated by the following expressions:

$$B_{O_2,met} = m_{O_2,met}\left[c_{p,O_2}\left(T_t - T_0 - T_0 \ln\frac{T_t}{T_0}\right) + T_0 R_{O_2} \ln\frac{P_{O_2,t}}{P_{O_2,0}}\right] \quad (10.31)$$

$$B_{CO_2,met} = m_{CO_2,met}\left[c_{p,CO_2}\left(T_t - T_0 - T_0 \ln\frac{T_t}{T_0}\right) + T_0 R_{CO_2} \ln\frac{P_{CO_2,t}}{P_{CO_2,0}}\right] \quad (10.32)$$

The exergy destroyed in the tissue control volume (CV2) is equal to:

$$B_{dest,2} = B_{CO_2,met} - B_{O_2,met} - B_{bl} \quad (10.33)$$

The exergy destroyed rate in the whole respiratory system is the sum of the exergy destroyed rate in each control volume.

10.3.4 Results

An experiment conducted by Nagle et al. [27], and graphically published by Webb [28], was chosen for simulation. The subject was a 34-year-old man, 1.75 m tall, and mass equal to 74 kg. In the beginning of the experiment, the subject was kept still for the measurement of his initial condition. After that, he started to walk on a treadmill simulating a −5 % grade downhill. Then he rested for 15 min and started to walk again, but this time simulating a 5 % grade uphill. His metabolism and the environment heat loss were measured. The exposition to heavier exercise levels is not considered in the present work because the model results do not fit the experimental data.

For simulation, the operative temperature 28 °C and relative humidity 40 % at sea level were considered.

Figures 10.9 and 10.10 show the conditions of the blood in the tissue entrance and outlet. Figure 10.9 also shows the oxygen consumption of the human body during the simulation. It starts to rise when the exercise begins (0.3 h). Those figures contain the blood data used for the exergy analysis. Besides them, the conditions of the inspired and expired air and the flow are also used for the analysis. The state of the inspired air is constant, and the state of the expired air has small variation during simulation. The ventilation and the blood flow follow the O_2 consumption variation.

The exergy destroyed rate was calculated for the lung and tissues during simulation. The result is shown in Fig. 10.11. The sum of both terms is also shown. It represents the exergy destroyed of the whole control volume.

The increase of the overall exergy destroyed is proportional to the internal energy variation due to the metabolism, which is equal to 110 W in the beginning

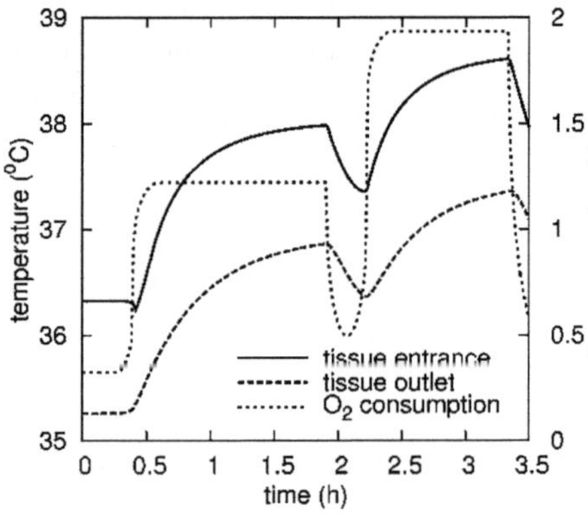

Fig. 10.9 Blood temperature in the tissue entrance and outlet and O_2 consumption during simulation [30]

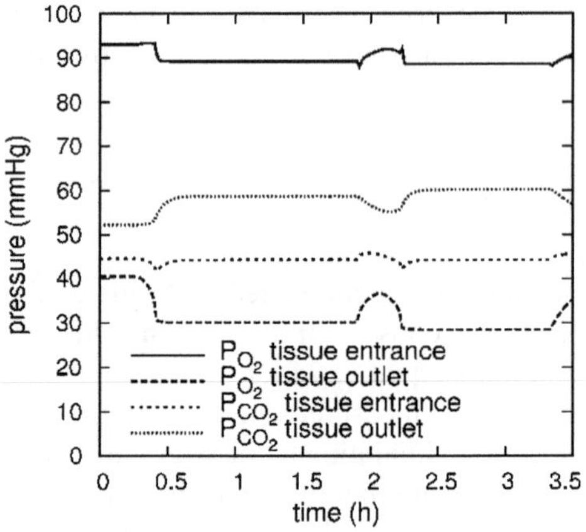

Fig. 10.10 O_2 and CO_2 blood partial pressure in the tissue entrance and outlet during simulation [30]

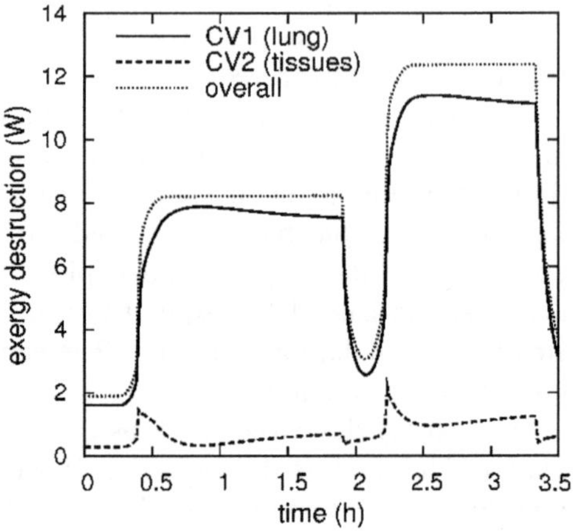

Fig. 10.11 Exergy destroyed in the lung, tissues, and whole body during simulation [30]

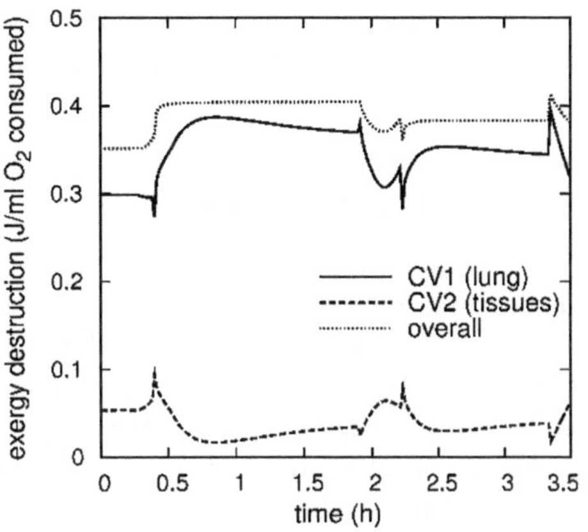

Fig. 10.12 Relation between exergy destroyed and O_2 consumption in the lung, tissues, and whole body during simulation [30]

of the simulation, 417 W in the first level of physical activity, and 661 W in the second level of physical activity. For the first level of exercise, the mean body temperature is around 37 °C and the exergy destroyed is slightly higher than 8 W. For the second level of physical activity, the exergy destroyed is around 12 W. These results indicate that there is an increase of the exergy destroyed in the respiratory system as the level of physical activity increases, which could be associated to higher destroyed exergy in the thermoregulatory system in order to regulate the body temperature.

The relation between the destroyed exergy rate and the O_2 consumption is shown in Fig. 10.12. Despite the increase of the exergy destroyed in the tissues under physical activity (Fig. 10.11), the relation between exergy destroyed and O_2 consumption (Fig. 10.12) shows an increase in the efficiency of the tissue O_2 consumption.

At the beginning of the exercise (Fig. 10.12), the exergy destroyed in the tissues decreases while the exergy destroyed in the lung rises. That is, the tissues become more effective under exercise, and the lung less effective. An explanation for this fact is that the O_2 is more available to the tissues from the blood, than from the environment to the blood. The difference between the blood O_2 partial pressure and the tissue O_2 partial pressure is small. The human blood has some physical–chemical characteristics that make the oxygen to be delivered to the tissues with that small difference. On the other hand, the difference of the O_2 partial pressure between the environment and the blood is larger. The air takes a long way from the mouth to the alveolus, where the gas diffusion with the blood takes place. Furthermore, the alveolar space has a large volume (approximately 2,000 ml at the end of expiration). At each respiratory cycle, just 15 % of the air is renewed. This delay prevents the alveolar air concentration from having abrupt alterations. The CO_2 also has similar characteristics: small difference between blood and tissue CO_2 partial pressure; and a large difference between blood and environment CO_2 partial pressure.

The exergy destroyed of the whole body has a small variation during exercise, a consequence of the lung and tissues opposed variations.

10.3.5 Final Considerations

In this first attempt to analyze the exergy performance of the human respiratory and thermal systems under physical activity, the exergy destroyed was obtained for the lung and the tissues. The results showed that the tissues are more effective under physical activity, while the lung is less effective.

Some simplifications were considered. One is the use of two distinct non-integrated physiological models. Other simplifications are in the exergy analysis: the metabolic chemical reactions are not considered; the rise of the expired air temperature is not related with the metabolism.

The exergy analysis of the human body will be improved as the physiological models become more accurate. For instance, the comparison of the exergy destroyed rates between smokers and nonsmokers can be useful information in order to quantify the reduction in the performance of the respiratory system and its consequences to the human life quality.

References

1. Schördinger E (1944) What is life? The physical aspects of living cell. Cambridge University Press, Cambridge
2. Prigogine I, Wiame J (1946) Biologic et Thermodynamique des phenomenes irreversibles. Experimentia 2:451–453
3. Stoward PJ (1962) Thermodynamics of biological growth. Nature 194:977–978
4. Zotin AI, Zotina RS (1967) Thermodynamics aspects of developmental biology. J Theor Biol 17:57–75
5. Balmer RT (1984) Entropy and aging in biological systems. Chem Eng Commun 17:171–181
6. Aoki I (1987) Entropy balance of white-tailed deer during a winter night. B Math Biol 149:321–327
7. Aoki I (1989) Entropy flow and entropy production in the human body in basal conditions. J Theor Biol 141:11–21
8. Aoki I (1990) Effects of exercise and chills on entropy production in human body. J Theor Biol 145:421–428
9. Aoki I (1991) Entropy principle for human development, growth and aging. J Theor Biol 150:215–223
10. Batato M, Borel L, Deriaz O et al (1990) Analyse exergétique théorique et expérimentale du corps humain. 26:120–130
11. Rahman MA (2007) A novel method for estimating entropy generation rate in human body. Therm Sci 11:75–92
12. Silva C, Annamalai K (2008) Entropy generation and human aging: lifespan entropy and effect of physical activity level. Entropy 10:100–123
13. Silva C, Annamalai K (2009) Entropy generation and human aging: lifespan entropy and effect of diet composition and caloric restriction diets. J Thermodyn 2009:1–10

14. Prek M (2005) Thermodynamic analysis of human heat and mass transfer and their impact on thermal comfort. Int J Heat Mass Tran 48:731–739
15. Prek M (2006) Thermodynamical analysis of human thermal comfort. Energy 31:732–743
16. Prek M, Butala V (2010) Principles of exergy analysis of human heat and mass exchange with the indoor environment. Int J Heat Mass Tran 48:731–739
17. Ferreira MS, Yanagihara JI (2009) A transient three-dimensional heat transfer model of the human body. Int Commun Heat Mass 36:718–724
18. Lems S (2009) Thermodynamic explorations into sustainable energy conversion—learning from living systems. PhD Thesis, Technische Universiteit Delft
19. Luo L (2009) Entropy production in a cell and reversal of entropy flow as an anticancer therapy. Front Phys China 4:122–136
20. Ferreira MS, Yanagihara JI (2009a) Exergy analysis of the human thermal system. In: 2nd Brazilian meeting of biomechanical engineering, Florianópolis (In Portuguese)
21. Pennes HH (1948) Analysis of tissue and arterial blood temperatures in the resting human forearm. J Appl Physiol 1:93–122
22. Wissler EH (1985) Mathematical simulation of human thermal behavior using whole body models. In: Shitzer A, Eberhart RC (eds) Heat transfer in medicine and biology, analysis and applications. Plenum Press, New York
23. Dear JD, Arens E, Hui Z et al (1997) Convective and radiative heat transfer coefficients for individual human body segments. Int J Biometeorol 40:141–156
24. ASHRAE American Society of Heating, Refrigerating and Air-Conditioning Engineers (1993) Physiological principles and thermal comfort. In: Handbook of fundamentals, Atlanta, chapter 8, pp 1–29
25. Harris JA, Benedict FG (1918) A biometric study of human basal metabolism. P Natl Acad Sci U S A 4:370
26. Albuquerque-Neto C, Yanagihara JI, Turri F (2008) A carbon monoxide transport model of the human respiratory system applied to urban atmosphere exposure analysis. J Braz Soc Mech Sci Eng 30:253–260
27. Nagle FJ, Webb P, Wanta DM (1990) Energy exchange in downhill and uphill walking: a calorimetric study. Med Sci Sports Exerc 22:540–544
28. Webb P (1995) The physiology of heat regulation. Am J Physiol Regul Integr Comp Physiol 268:838–850
29. Mady CEK, Ferreira MS, Yanagihara JI et al (2011) Modeling the exergy behaviour of human body. In: Proceedings of the 24th international conference of efficiency, cost, optimization, simulation and environmental impact of energy systems, Novi Sad
30. Albuquerque-Neto C, Pellegrini LF, Ferreira MS et al (2010) Exergy analysis of human respiration under physical activity. Int J Thermodynamics 13:105–109

Author Biography

Silvio de Oliveira Jr. is an Associate Professor at the Polytechnic School of the University of São Paulo, Brazil. He has been developing research activities on heat pumps and refrigeration systems, solar energy, energy conservation in industrial processes, cogeneration systems, and exergy and thermoeconomic analysis of thermal processes. He has supervised seven Doctor Theses (in Mechanical Engineering), 16 Master Dissertations (in Mechanical Engineering, Automotive Engineering and Energy), 9 MBAs in Energy and Cogeneration and Distributed Generation, and 70 Undergraduate Projects in Mechanical Engineering. He is author/co-author of about 190 publications and communications and senior member of the Brazilian Society of Mechanical Sciences and Engineering. He is also member of the Editorial Board of the International Journal of Thermodynamics. He has been involved in the past 10 years with research projects related to energy utilization in biodiesel production plants, sugar and alcohol utilities and production plants, offshore and onshore petroleum platforms, refinery utilities plants, production processes of petroleum derived fuels, co/trigeneration and combined cycle plants, airplane energy systems, and modeling and simulation of twin-screw multiphase pumping systems.

S. de Oliveira Jr., *Exergy*, Green Energy and Technology,
DOI: 10.1007/978-1-4471-4165-5, © Springer-Verlag London 2013

Index

S. de Oliveira Jr., *Exergy*, Green Energy and Technology,
DOI: 10.1007/978-1-4471-4165-5, © Springer-Verlag London 2013